Statistics and Finance
An Interface

Proceedings of the Hong Kong International Workshop on

Statistics and Finance
An Interface

HONGKONG

IWSF

Centre of Financial Time Series, The University of Hong Kong

4–8 July 1999

Editors

Wai-Sum Chan, Wai Keung Li & Howell Tong

Department of Statistics and Actuarial Science

The University of Hong Kong

ICP

Imperial College Press

Published by

Imperial College Press
57 Shelton Street
Covent Garden
London WC2H 9HE

Distributed by

World Scientific Publishing Co. Pte. Ltd.
P O Box 128, Farrer Road, Singapore 912805
USA office: Suite 1B, 1060 Main Street, River Edge, NJ 07661
UK office: 57 Shelton Street, Covent Garden, London WC2H 9HE

British Library Cataloguing-in-Publication Data
A catalogue record for this book is available from the British Library.

STATISTICS AND FINANCE
An Interface

ISBN 1-86094-237-7

This book is printed on acid-free paper.

Printed in Singapore by Regal Press (S) Pte. Ltd.

PREFACE

It was at the height of the Asian financial crisis when we first conceived the idea of holding an international workshop on "Statistics in Finance" in Hong Kong, which was coincidentally called the *Capital of Risk* by Professor Tony Giddens, the Director of the London School of Economics, in his Reith Lecture broadcasted by the B.B.C. World Service live from Hong Kong around the time. The preparations took about a year to complete. It was during that unusually wet week of 4-8 July 1999 that the workshop finally took place. Sixty-two participants, all by invitation, from 14 countries/regions were "locked behind closed doors" in the secluded (by Hong Kong standard) Goldcoast International Hotel in Tuen Mun, Hong Kong, where discussions of important issues, debates on the best methods and exchanges of ideas were only interrupted by sumptuous lunches and dinners provided by the hotel at knock-down prices; we were true believers in seeking opportunities from crises.

The workshop was entirely funded by a generous grant from the University of Hong Kong under their Centre of Distinction scheme, to which we record our most sincere gratitude. The participants, a list of which is given elsewhere in this volume, were drawn from the famous to the upcoming, from the East to the West and from the theoretical to the practical. Chinese strategists believe that there are three principal factors determining the success or otherwise of any human endeavour, namely the timing, the venue and the people. Looking back, it seems that we couldn't have been luckier with our workshop.

This volume represents some of the fruits of this intensely cerebral but extremely enjoyable venture with the group photograph as visual evidence of the latter. For easy reference, we have divided the written papers, all of which have been appropriately updated during and after the workshop and through a fairly rigorous reviewing process, into five groups, namely *Time Series Methodology*, *Long Memory and Value at Risk*, *Volatility*, *Forecasting* and *Applications*. It is our genuine hope that readers will find these papers of relevance to their statistical analyses of financial data as, we believe, they cover topics of current interest in this fast expanding area.

W.S. Chan
W.K. Li
H. Tong

HONGKONG

IWSF

Hong Kong International Workshop on Statistics in Finance
4-8 July 1999

Centre of Financial Time Series
Department of Statistics & Actuarial Science
The University of Hong Kong

ACKNOWLEDGMENTS

List of Members of the Organising Committee of the Workshop

H. Tong (Chairman)
W.K. Li (Deputy Chairman)
W.S. Chan
K.W. Ng
Albert C.S. Wong
H. Yang
David Yeung
Philip L.H. Yu
L. Zhu

List of Participants of the Workshop

Hong-Zhi An
Peter Brockwell
Beda Chan
Kung-Sik Chan
Ngai-Hang Chan
W.S. Chan
Cathy Chen
Bing Cheng
Yin Wong Cheung
Ray Chou
Boris Choy
Casper G. de Vries
Cees Diks
Paul Embrechts
Tom P.W. Fong
Jürgen Franke
Clive W.J. Granger
Marc Hallin
Cars H. Hommes
Wai Cheung Ip
Genshiro Kitagawa
Jens-Peter Kreiss
Kin Lam
Wai Keung Li
Shi-Qing Ling
Shu-Ing Liu
Z.M. Ma
Michael McAleer
Ulrich Muller
K.W. Ng
Ragnar Norberg
Wilfredo Palma
M. Hashem Pesaran
Peter Robinson
Tina Hviid Rydberg
Chou Yiu Sin
Richard L. Smith
Mike K.P. So
George C. Tiao
Howell Tong
Henghsiu Tsai
Ruey S. Tsay
M. Tse
Yiu Kuen Tse
Chi Ming Wong
Chun Shan Wong
Heung Wong
Samuel P.S. Wong
Zhong-Jie Xie
H. Yang
Qiwei Yao
David Yeung
Iris Yeung
Philip L.H. Yu
K.C. Yuen
L. Zhu

List of Referees for the Workshop papers

H.Z. An
P. Brockwell
K.S. Chan
P. Embrechts
J. Gao
C. Hommes
K. Lam
S.Q. Ling
M. McAleer
M.H. Pesaran
T.H. Rydberg
M. So
R.S. Tsay
M. Tse
Y.K. Tse
Albert Wong
Y. Xia
H. Yang
L. Zhu

List of Graduate Students who assisted the Workshop

Billy K.S. Ching
Doris S.Y. Chong
Bill K.S. Chow
Cher M.W. Ng
Pauline W.Y. Tsang
May W.M. Wong

List of Administrative and Technical Staff who assisted the Workshop

Betty S.C. Cheung
Irene M.L. Cheung
Esther Y.W. Cheung
Ada Y.M. Lai
Wilson T.W. Li
Novia W.M. Poon
Moon Y. Ng
Y.K. Wong

CONTENTS

Part I: Time Series Methodology

HEAVY-TAILED AND NON-LINEAR CONTINUOUS-TIME ARMA MODELS FOR FINANCIAL TIME SERIES

P. J. BROCKWELL

Statistics Department, Colorado State University,
CO 80523-1877, USA
E-mail: pjbrock@stat.colostate.edu

Properties of linear continuous-time ARMA (or CARMA) processes driven by second-order Lévy processes are examined. These extend the class of Gaussian CARMA processes to include heavier-tailed series such as those frequently encountered in financial applications. Non-linear Gaussian CAR processes are also considered and illustrated with threshold models fitted to daily returns on the Australian All-ordinaries and Dow-Jones Industrial Indices. AIC comparisons are made with ARCH and GARCH models fitted to the same data.

1 Introduction

A zero-mean Gaussian continuous-time ARMA(p,q) (or CARMA(p,q)) process $\{Y(t)\}$ with $0 \le q < p$ and coefficients $a_1,\ldots,a_p$, $b_0,\ldots,b_q$, is defined (see e.g. Brockwell and Davis [1]) to be a stationary solution of the (suitably interpreted) pth order linear differential equation,

$$a(D)Y(t) = b(D)DW(t), \ t \ge 0, \tag{1.1}$$

where D denotes differentiation with respect to t, $\{W(t)\}$ is standard Brownian motion,

$$a(z) := z^p + a_1 z^{p-1} + \cdots + a_p,$$

$$b(z) := b_0 + b_1 z + \cdots + b_p z^p,$$

and the coefficients b_j satisfy $b_q \neq 0$ and $b_j = 0$ for $q < j \le p$. Since the derivatives $D^j W(t)$ do not exist in the usual sense, we interpret (1.1) as being equivalent to the *observation* and *state* equations,

$$Y(t) = \mathbf{b}'\mathbf{X}(t), \tag{1.2}$$

and

$$d\mathbf{X}(t) - A\mathbf{X}(t)dt = \mathbf{e}\, dW(t), \tag{1.3}$$

where

$$A = \begin{bmatrix} 0 & 1 & 0 & \cdots & 0 \\ 0 & 0 & 1 & \cdots & 0 \\ \vdots & \vdots & \vdots & \ddots & \vdots \\ 0 & 0 & 0 & \cdots & 1 \\ -a_p & -a_{p-1} & -a_{p-2} & \cdots & -a_1 \end{bmatrix}, \quad \mathbf{e} = \begin{bmatrix} 0 \\ 0 \\ \vdots \\ 0 \\ 1 \end{bmatrix}, \quad \mathbf{b} = \begin{bmatrix} b_0 \\ b_1 \\ \vdots \\ b_{p-2} \\ b_{p-1} \end{bmatrix}$$

and we assume that $\mathbf{X}(0)$ is a Gaussian random vector such that

$$\mathbf{X}(0) \text{ is uncorrelated with } \{W(t), t \geq 0\}. \tag{1.4}$$

The state equation (1.3) is an Ito differential equation for $\mathbf{X}(t)$. In the case $p = 1$, A is defined to be $-a_1$. Because of the linearity of (1.3), its solution has the simple form,

$$\mathbf{X}(t) = e^{At}\mathbf{X}(0) + \int_0^t e^{A(t-u)}\mathbf{e}\, dW(u), \tag{1.5}$$

where the integral is defined as the L^2 limit of approximating Riemann-Stieltjes sums. The process $\{\mathbf{X}(u), u \geq 0\}$ also satisfies the relations,

$$\mathbf{X}(t) = e^{A(t-s)}\mathbf{X}(s) + \int_s^t e^{A(t-u)}\mathbf{e}\, dW(u), \quad \text{for all } t > s \geq 0, \tag{1.6}$$

which clearly show (by the independence of increments of $\{W(t)\}$) that $\{\mathbf{X}(u)\}$ is Markov.

It is well-known (see e.g. Brockwell [2]) that the equations (1.4) and (1.6) have a weakly stationary solution if and only if the eigenvalues $\lambda_1, \ldots, \lambda_p$ of A (which are the same as the zeroes of the autoregressive polynomial $z^p + a_1 z^{p-1} + \cdots + a_p$) all have negative real parts, i.e if and only if

$$\Re(\lambda_i) < 0, \; i = 1, \ldots, p. \tag{1.7}$$

If $\{\mathbf{X}(t)\}$ is such a solution then it is easy to show that

$$E(\mathbf{X}(0)) = \mathbf{0} \tag{1.8}$$

and

$$E(\mathbf{X}(0)\mathbf{X}'(0)) = \Sigma := \int_0^\infty e^{Ay}\mathbf{e}\,\mathbf{e}'e^{A'y}dy. \tag{1.9}$$

Conversely if $\mathbf{X}(0)$ satisfies the conditions (1.4), (1.8) and (1.9), then the process $\{\mathbf{X}(t)\}$ defined by (1.5) is weakly stationary and satisfies the relations,

$$E[\mathbf{X}(t)] = \mathbf{0}, \; t \geq 0, \tag{1.10}$$

and

$$E[\mathbf{X}(t+h)\mathbf{X}(t)'] = e^{Ah}\Sigma,\ h \geq 0. \tag{1.11}$$

From (1.2) the mean and autocovariance function of the CARMA(p,q) process $\{Y(t)\}$ are then given by

$$E[Y(t)] = 0,\ t \geq 0 \tag{1.12}$$

and

$$\gamma_Y(h) = E[Y(t+h)Y(t)] = \mathbf{b}'\ e^{A|h|}\Sigma\ \mathbf{b}. \tag{1.13}$$

If the zeroes of the autoregressive polynomial are all distinct then the autocovariance function of $\{Y(t)\}$ has the simple form (see Brockwell [3]),

$$\gamma_Y(h) = \sum_{\lambda:a(\lambda)=0} \frac{e^{\lambda|h|}b(\lambda)b(-\lambda)}{a'(\lambda)a(-\lambda)}. \tag{1.14}$$

Since the results of the preceding paragraph depend only on the first and second moments of the process $\{W(t)\}$, we can define a general (not necessarily Gaussian) CARMA(p,q) process as follows.

Definition 1.1. If $\{W(t)\}$ is a zero-mean process with uncorrelated increments such that $E[(W(t) - W(s))^2] = t - s$, for all $t \geq s \geq 0$, and p and q are integers such that $0 \leq q < p$, then $\{Y(t), t \geq 0\}$ is a CARMA(p,q) process with parameters $a_1, \ldots, a_p, b_0, \ldots, b_q$, if and only if $\{Y(t)\}$ satisfies (1.2) with $\{\mathbf{X}(t)\}$ a weakly stationary solution of the equations (1.4) and (1.6).

The necessary and sufficient condition for existence of such a process is that the eigenvalues $\lambda_1, \ldots, \lambda_p$ of the matrix A satisfy the condition (1.7). The mean and autocovariance function of $\{Y(t)\}$ are given by (1.12) and (1.13). (If we replace the condition $EW(t) = 0$ in Definition 1.1 by $EW(t) = ct$, the only effect is a corresponding change in $E\mathbf{X}(t)$ and $EY(t)$.) In the case when $\{W(t)\}$ is standard Brownian motion, the process $\{Y(t)\}$ is clearly also strictly stationary.

The process $\{Y(t)\}$ is said to be a CARMA process with mean μ if and only if $\{Y(t) - \mu\}$ is a CARMA process.

Since our concern in this paper is with the distributional properties of non-Gaussian CARMA processes we must impose stronger conditions on $\{W(t)\}$ than those in Definition 1.1. We therefore suppose that the increments of $\{W(t)\}$ are independent rather than uncorrelated and strengthen the assumption (1.4) to the condition,

$$\mathbf{X}(0) \text{ is independent of } \{W(t), t \geq 0\}. \tag{1.15}$$

This leads to the class of Lévy-driven CARMA(p, q) process, defined as follows.

Definition 1.2. If $\{W(t)\}$ is a second order Lévy process, and p and q are integers such that $0 \leq q < p$, then $\{Y(t), t \geq 0\}$ is a Lévy-driven CARMA(p, q) process with parameters $a_1, \ldots, a_p, b_0, \ldots, b_q$, if and only if $\{Y(t)\}$ satisfies (1.2) with $\{\mathbf{X}(t)\}$ a strictly stationary second order solution of the equations (1.6) and (1.15).

It is clear that the condition (1.7) is necessary for the existence of a Lévy driven CARMA(p, q) process $\{Y(t)\}$ since it is necessary for the existence of a weakly stationary solution of (1.6) and (1.15) In the following section we show that condition (1.7) is also sufficient for the existence of $\{Y(t)\}$, determine the finite-dimensional joint characteristic functions of the process and give some illustrative examples. Maximum likelihood inference for Lévy driven CARMA processes is discussed, in conjunction with that for Gaussian non-linear CAR processes, in Section 4.

An excellent account of Lévy processes can be found in the lecture notes of Ito [4] and in the more recent book of Bertoin [5]. The basic properties which we need are given in Section 2.

First order stochastic differential equations with non-negative Lévy input process have been widely used in storage theory (Cinlar and Pinsky [6], Harrison and Resnick [7], Brockwell et al.[8]) and more recently as a basis for non-Gaussian stochastic volatility models by Barndorff-Nielsen and Shephard [9], who consider a wide variety of such models and their financial applications. Lévy-driven CARMA processes are of particular interest because they have the same auto-covariance functions as corresponding Gaussian processes but exhibit a wide range of non-Gaussian marginal distributions such as the more heavy tailed distributions frequently encountered in financial data.

Gaussian CARMA processes have been extensively used by Jones [10] and others for modelling irregularly spaced data, since the Kalman recursions allow relatively simple calculation of the likelihood of such data. Lévy-driven CARMA processes can be used in the same way, but likelihood (as distinct from Gaussian likelihood) calculation is rather more complicated.

2 Lévy-driven CARMA Processes

Suppose that $\{W(t), t \geq 0\}$ is a Lévy process. Then the process $\{W(t)\}$ has homogeneous independent increments and the characteristic function of $W(t)$, $\phi_t(\theta) := E(\exp(i\theta W(t))$, has the form (see e.g. Ito [4]),

$$\phi_t(\theta) = \exp(t\xi(\theta)), \quad \theta \in \mathbb{R}, \tag{2.1}$$

where

$$\xi(\theta) = i\theta m - \frac{1}{2}\theta^2\sigma^2 + \int_{\mathbf{R}_0} (e^{i\theta x} - 1 - \frac{ix\theta}{1+x^2})\nu(dx), \tag{2.2}$$

for some $m \in \mathbb{R}$, $\sigma \geq 0$, and measure ν on the Borel subsets of $\mathbb{R}_0 = \mathbf{R}\backslash\{0\}$. The measure ν is called the Lévy measure of the process W and has the property,

$$\int_{\mathbb{R}_0} \frac{u^2}{1+u^2}\nu(du) < \infty.$$

If ν is the zero measure then $\{W(t)\}$ is Brownian motion with $E(W(t)) = mt$ and $\text{Var}(W(t)) = \sigma^2 t$. If $m = \sigma^2 = 0$ and $\int_{\mathbb{R}_0} \frac{|u|}{1+u^2}\nu(du) < \infty$, then $W(t) = at + P(t)$, where $\{P(t)\}$ is a compound Poisson process with jump-rate $\nu(\mathbf{R}_0)$, jump-size distribution $\nu/\nu(\mathbf{R}_0)$, and $a = -\int_{\mathbf{R}_0} \frac{u}{1+u^2}\nu(du)$. Another important example is the gamma process $\{W(t)\}$, for which

$$\xi(\theta) = \int_{\mathbf{R}_0} (e^{i\theta x} - 1)\nu(dx),$$

$\nu(du) = \alpha u^{-1}e^{-\beta u}du, u > 0$, and $W(t)$ has probability density function given by $\beta^{\alpha t}x^{\alpha t-1}e^{-\beta x}/\Gamma(\alpha t), x > 0$. This is an example of a Lévy process whose sample-paths have infinitely many jumps in every interval of positive length. If $\{W_1(t)\}$ and $\{W_2(t)\}$ are two independent and identically distributed gamma processes then $W_1 - W_2$ is a symmetrized gamma process with Lévy measure, $\nu(du) = \frac{1}{2}\alpha|u|^{-1}e^{-\beta|u|}du$.

Our goal is to study the properties of a CARMA(p,q) process driven by a second-order Lévy process $\{W(t)\}$. The cumulant generating function (cgf) of $W(1)$ then satisfies the condition,

$$|\xi''(0)| = \text{Var}(W(1)) < \infty.$$

The corresponding Lévy-driven CARMA process $\{Y(t), t \geq 0\}$ is defined as in Section 1, where we observed that a necessary condition for its existence is the condition (1.7) on the eigenvalues of the matrix A. The following theorem establishes the sufficiency of this condition and determines the finite dimensional characteristic functions of the process.

Theorem 2.1 *If $\{W(t)\}$ is a second order Lévy process with cgf $\xi(\theta)$ as in (2.1), then the Lévy-driven CARMA process specified by Definition 1.2 exists if and only if condition (1.7) is satisfied, in which case the cumulant generating function of $Y(t_1), Y(t_2), \ldots, Y(t_n)$, $(0 \leq t_1 < t_2 < \cdots < t_n)$ is*

$$\ln E[\exp(i\theta_1 Y(t_1) + \cdots + i\theta_n Y(t_n))] =$$

$$\int_0^\infty \xi\left(\sum_{i=1}^n \theta_i \mathbf{b}' e^{A(t_i+u)}\right) \mathbf{e} du + \int_0^{t_1} \xi\left(\sum_{i=1}^n \theta_i \mathbf{b}' e^{A(t_i-u)}\right) \mathbf{e} du +$$

$$\int_{t_1}^{t_2} \xi\left(\sum_{i=2}^n \theta_i \mathbf{b}' e^{A(t_i-u)}\right) \mathbf{e} du + \cdots + \int_{t_{n-1}}^{t_n} \xi\left(\sum_{i=2}^n \theta_i \mathbf{b}' e^{A(t_i-u)}\right) \mathbf{e} du. \tag{2.3}$$

In particular, the marginal distribution of $Y(t)$ *has cgf,*

$$\ln E[\exp(i\theta Y(t))] = \int_0^\infty \xi(\theta \mathbf{b}' e^{Au} \mathbf{e}) du. \tag{2.4}$$

Proof It suffices to prove the sufficiency of condition (1.7). If it is satisfied, then the first term on the right of (1.5) converges in L^2 to zero as $t \to \infty$ and the integral term, which we shall denote by $\mathbf{V}(t)$, converges in L^2 to a random vector $\mathbf{V}$. Consequently $\mathbf{X}(t)$ converges in L^2 and hence in distribution to $\mathbf{V}$. By the homogeneity of the Lévy process $\{W(t)\}$, $\mathbf{V}(t)$ has the same distribution as

$$\mathbf{U}(t) = \int_0^t e^{Au} \mathbf{e} dW(u),$$

so that $\mathbf{U}(t)$ also converges in distribution as $t \to \infty$ to $\mathbf{V}$. The cgf of $\mathbf{U}(t)$ is easily calculated since

$$\mathbf{U}(t) = \lim(\text{m.s.})_{\Delta\to 0} \sum_{i=1}^n \exp(Au_i)\mathbf{e}(W(u_i) - W(u_{i-1})),$$

where $0 = u_0 < u_1 < \cdots < u_n = t$ and $\Delta = \max(u_i - u_{i-1})$. Hence, for all $\boldsymbol{\theta} \in \mathbf{R}^p$,

$$\ln E[\exp(i\boldsymbol{\theta}' \mathbf{U}(t)] = \lim_{\Delta\to 0} \sum_{i=1}^n \xi(\boldsymbol{\theta}' e^{Au} \mathbf{e})(u_i - u_{i-1})$$

$$= \int_0^t \xi(\boldsymbol{\theta}' e^{Au} \mathbf{e}) du.$$

and so $\mathbf{X}(t)$ converges in distribution as $t \to \infty$ to a random vector with cgf,

$$\kappa(\boldsymbol{\theta}) = \int_0^\infty \xi(\boldsymbol{\theta}' e^{Au} \mathbf{e}) du. \tag{2.5}$$

This implies that the distribution defined by (2.5) is the unique stationary distribution for the Markov process $\{\mathbf{X}(t)\}$ defined by (1.6) and (1.15). Hence if $\mathbf{X}(0)$ has cgf (2.5), the process $\{\mathbf{X}(t)\}$ is strictly stationary. Since $Y(t) = \mathbf{b}'\mathbf{X}(t)$, the process $\{Y(t)\}$ is then strictly stationary with cgf (2.4).

To determine the joint stationary distribution of $\mathbf{X}(t_1), \mathbf{X}(t_2), \ldots, \mathbf{X}(t_n)$, $0 \le t_1 < t_2 < \cdots < t_n$, we write

$$\sum_{k=1}^{n} i\boldsymbol{\theta}_k'\mathbf{X}(t_k) = i\sum_{k=1}^{n} \boldsymbol{\theta}_k'\mathbf{X}(0) + \int_0^{t_1} \sum_{k=1}^{n} \boldsymbol{\theta}_k' e^{A(t_k-u)}\mathbf{e}dW(u) +$$

$$\int_{t_1}^{t_2} \sum_{k=2}^{n} \boldsymbol{\theta}_k' e^{A(t_k-u)}\mathbf{e}dW(u) + \cdots + \int_{t_{n-1}}^{t_n} \boldsymbol{\theta}_n' e^{A(t_n-u)}\mathbf{e}dW(u). \tag{2.6}$$

Since the terms on the right are independent, we find, by the same calculations which led to (2.5), that the joint cgf of $\mathbf{X}(t_1), \ldots, \mathbf{X}(t_n)$ is

$$\ln E[\exp(i\boldsymbol{\theta}_1'\mathbf{X}(t_1) + \cdots + i\boldsymbol{\theta}_n'\mathbf{X}(t_n)] =$$

$$\int_0^{\infty} \xi\left(\sum_{i=1}^{n} \boldsymbol{\theta}_i' e^{A(t_i+u)}\right)\mathbf{e}du + \int_0^{t_1} \xi\left(\sum_{i=1}^{n} \boldsymbol{\theta}_i e^{A(t_i-u)}\right)\mathbf{e}du +$$

$$\int_{t_1}^{t_2} \xi\left(\sum_{i=2}^{n} \boldsymbol{\theta}_i' e^{A(t_i-u)}\right)\mathbf{e}du + \cdots + \int_{t_{n-1}}^{t_n} \xi\left(\sum_{i=2}^{n} \boldsymbol{\theta}_i' e^{A(t_i-u)}\right)\mathbf{e}du. \tag{2.7}$$

Since $Y(t_i) = \mathbf{b}'X(t_i)$, $i = 1, 2, \ldots, n$, we find at once from (2.7) that the joint cgf of $Y(t_1), \ldots, Y(t_n)$ is as specified in (2.4). □

Example 1 In the special case when ν is the zero measure, $\{Y(t)\}$ reduces to the Gaussian CARMA(p, q) process with mean $b_0 m/a_p$ as defined in Section 1.

Example 2 If $\{W(t)\}$ is a compound Poisson process with finite jump-rate λ and bilateral exponential jump-size distribution with probability density, $f(x) = \frac{1}{2}\beta e^{-\beta|x|}$, then by Theorem 2.1, the corresponding CAR(1) process has marginal cgf,

$$\kappa(\theta) = \int_0^{\infty} \xi(\theta e^{-a_1 u})du,$$

where $\xi(\theta) = \lambda\theta^2/(\beta^2 + \theta^2)$. Straightforward evaluation of the integral gives

$$\kappa(\theta) = -\frac{\lambda}{2a_1}\ln\left(1 + \frac{\theta^2}{\beta^2}\right),$$

showing that $Y(t)$ is distributed as the difference between two independent gamma distributed random variables with exponent $\lambda/(2a_1)$ and scale parameter β. In particular, if $\lambda = 2a_1$, the marginal distribution is bilateral exponential.

Example 3 Numerous examples of marginal distributions for CAR(1) processes driven by non-negative Lévy processes can be found in the paper of Barndorff-Nielsen and Shephard [9] who use them in conjunction with stochastic volatility models. Some other examples with applications in storage theory can be found in the papers of Cinlar and Pinsky [6], Harrison and Resnick [7] and Brockwell et al. [8].

Example 4 For general Lévy-driven CARMA processes, explicit determination of the marginal and joint distributions will not be possible. However Theorem 2.1 still allows calculation of moments and the marginal density can be simulated using the approximating sums.

Remark 1. Lévy-driven CARMA processes have exactly the same autocorrelation structure as a Gaussian CARMA process with the same coefficients $a_1, \ldots, a_p, b_0, \ldots, b_q$ (see (1.13) and (1.14)). This implies in particular that a Lévy-driven CARMA(p,q) process sampled at times $0, 1, 2, \ldots$, will be a discrete-time ARMA(p,r) process for some $r < p$. Lévy-driven CARMA processes however exhibit a much larger range of marginal and joint distributions than their Gaussian counterparts.

Example 5 Figures 1 and 2 show the histogram and sample autocorrelation function of the absolute daily returns $(100 \ln(P(t)/P(t-1))$ on the Hang Seng Index for the period July 1st, 1997 - April 9th, 1999. It has been observed by Granger et al. [11] that, as in this example, such absolute daily returns frequently follow an approximately exponential distribution with a slowly decaying positive autocorrelation function. The sample autocorrelation function (Figure 2) can be well approximated by that of a CARMA(2,1) model with coefficients $a_1 = 2.66$, $a_2 = .30$, $b_0 = 1.00$ and $b_1 = 2.80$, estimated by maximization of the Gaussian likelihood. In an attempt to approximate the empirical marginal distribution, the two parameters of a gamma process $\{W(t)\}$ were adjusted so that the simulated marginal distribution of the corresponding gamma-driven CARMA(2,1) process had approximately the appropriate shape. A good match (see Figure 3) was obtained by choosing the distribution of $W(1)$ to have exponent .060 and scale parameter 10. A more systematic approach to maximum likelihood estimation for such models is described in Section 4. The lack of evidence for long memory in the sample autocorrelation function of this data is very likely due to its relatively short length and is consistent with the suggestion of Granger et al. [11] that the long memory observed in longer realizations of these series may be due to shifting levels.

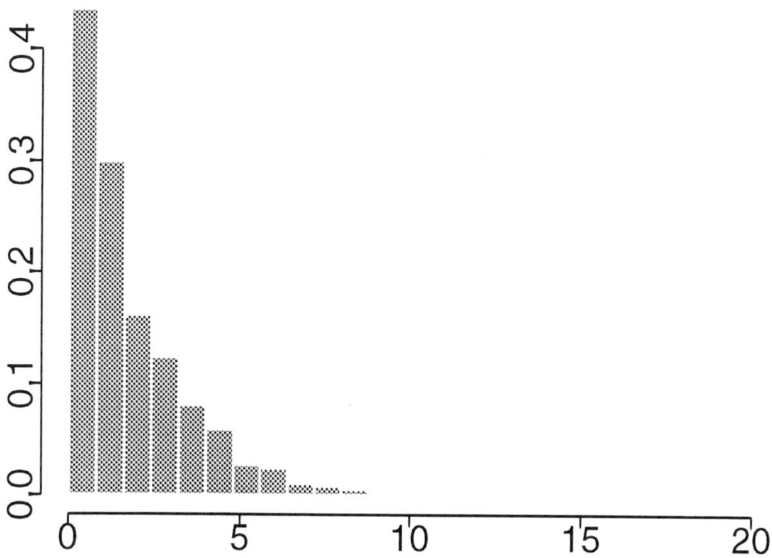

Figure 1: *Histogram of the daily absolute returns on the Hang-Seng Index, July 1st, 1997 - April 9th, 1999.*

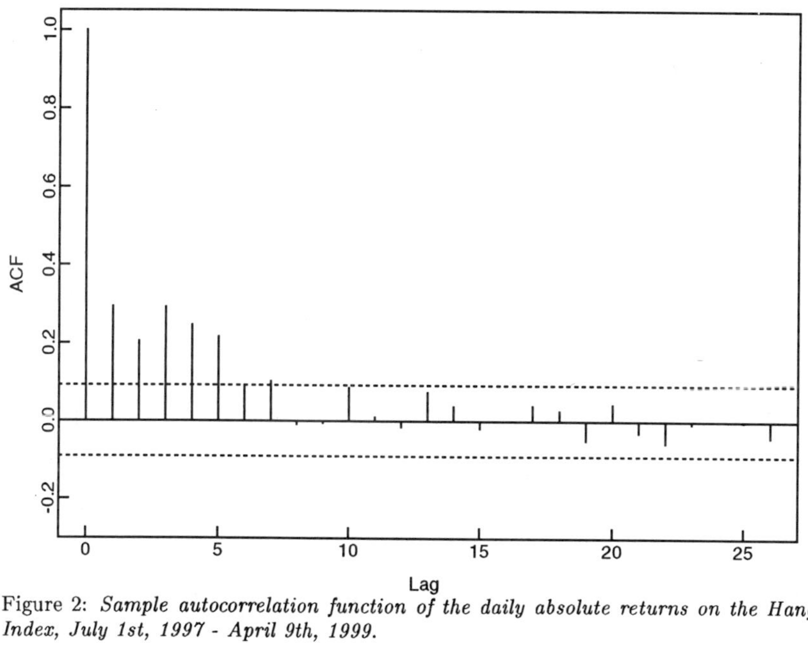

Figure 2: *Sample autocorrelation function of the daily absolute returns on the Hang Seng Index, July 1st, 1997 - April 9th, 1999.*

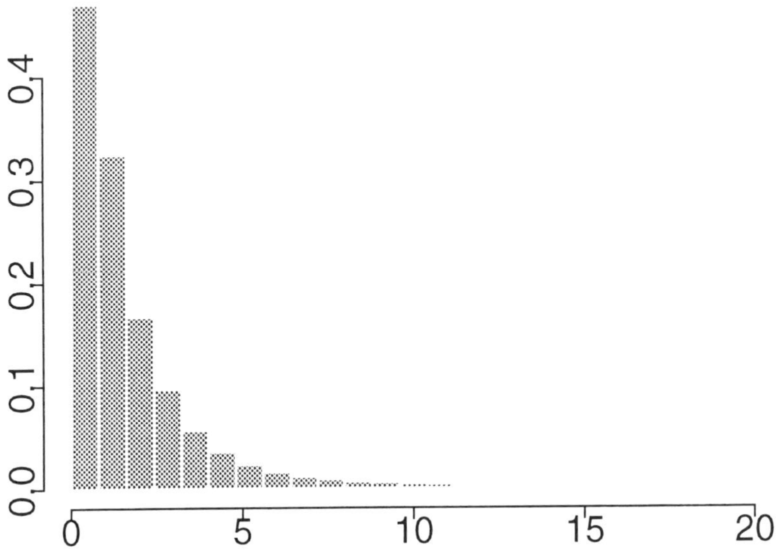

Figure 3: *Histogram of 10000 simulated daily absolute returns on the Hang-Seng Index based on the Lévy-driven CARMA(2,1) model of Example 5.*

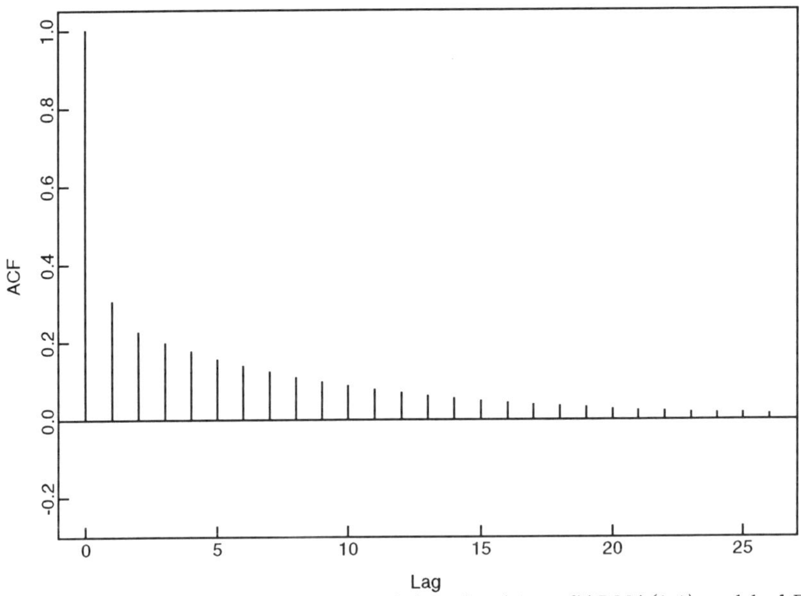

Figure 4: *The autocorrelation function of the Lévy-driven CARMA(2,1) model of Example 5.*

3 Nonlinear Gaussian CAR Models

The Gaussian CAR(p) process with coefficients $a_1, \ldots, a_p$ and $b(= b_0)$ and mean μ was defined in Section 1. It is a stationary solution of the stochastic differential equation,

$$(D^p + a_1 D^{p-1} + \cdots + a_p)Y(t) = bDW(t) + a_p\mu, \tag{3.1}$$

and has the state space representation,

$$Y(t) = [1\ 0\ \cdots\ 0]\mathbf{Y}(t),\ t \geq 0, \tag{3.2}$$

where $\{\mathbf{Y}(t)\}$ is the unique stationary solution of

$$d\mathbf{Y}(t) = (A\mathbf{Y}(t) + a_p\mu\mathbf{e})dt + b\mathbf{e}\ dW(t), \tag{3.3}$$

A and $\mathbf{e}$ are defined as in Section 1 and all the eigenvalues of A are assumed to have negative real parts.

A natural family of non-linear CAR processes is obtained by allowing the parameters $a_1, \ldots, a_p$ and μ in (3.1) to depend on $Y(t)$. In particular if we partition the real line into subintervals, $(-\infty, y_1]$, $(y_1, y_2]$, $\ldots$, (y_m, ∞), on each of which the parameter values are constant, then we obtain a continuous-time analogue of the threshold models of Tong [12] which we shall refer to as a CTAR(p) process. The equation (3.1) with parameters depending on $Y(t)$ has the same state-space representation (3.2) and (3.3) as in the linear case, except for the dependence of $a_1, \ldots, a_p$ and μ on $Y(t)$.

We shall assume that b is non-zero and independent of $Y(t)$, although, as we shall observe later, there are two important cases, namely the threshold AR(1) and AR(2) processes, for which the latter assumption can be relaxed. We shall also assume that the functions $a_1, \ldots, a_p$ and μ are bounded and measurable on $\mathbf{R}$.

We can drop the stationarity assumption which we imposed in the linear case, although conditions under which there is a stationary solution of (3.3) are of course still of interest and have been considered, for CTAR(1) processes by Stramer et al.[13] and for a more general class of continuous-time threshold ARMA processes by Stramer et al.[14].

The definition of a non-linear CAR process in terms of (3.2) and (3.3) immediately raises the question of whether or not a solution exists and if so whether or not it is unique. By the fundamental theorem on existence of strong solutions of stochastic differential equations (see e.g. Oksendal[15], Theorem 5.2.1), equation (3.3) has a unique strong solution, for given $\mathbf{Y}(0)$, if the coefficients of dt and $dW(t)$ satisfy standard Lipschitz and growth conditions.

The Lipschitz conditions will not however be satisfied for threshold models because of discontinuities in at least one of the functions $a_1, \ldots, a_p$ and μ. In this case we must look for weak solutions of the state equation (3.3). With the aid of the Cameron-Martin-Girsanov formula, we can construct a weak solution of (3.3) which is valid whether or not there are discontinuities in $a_1, \ldots, a_p$ and μ and which gives a characterization of the transition function of the Markov process $\mathbf{Y}(t)$. In particular, the first and second moments of the components of $\mathbf{Y}(t)$ conditional on $\mathbf{Y}(0)$ can be expressed explicitly as expected values of functionals of standard Brownian motion. For details see Brockwell [3]. This representation suggests computing moments of the transition function of $\mathbf{Y}(t)$ by simulating realizations of standard Brownian motion, evaluating the appropriate functionals and averaging over the realizations.

Remark 2. It can be shown, using random time transformations, that the restriction to constant functions $b(Y(t))$ can be relaxed for the CTAR(1) model (Stramer et al., [13]) and the CTAR(2) model (Brockwell and Williams [19]) with a single threshold at level r . It seems likely that the condition can also be relaxed more generally but no proof has yet been given.

An alternative approximate method for computing conditional moments of the state-vector of a CTAR(p) process (Brockwell and Hyndman [20]) uses the sequence of Euler approximations, defined by the observation and state equations (cf. (3.2) and (3.3)),

$$Y_n(t) = \begin{bmatrix} 1 & 0 & \cdots & 0 \end{bmatrix} \mathbf{Y}_n(t), \; t = 0, \frac{1}{n}, \frac{2}{n}, \ldots, \tag{3.4}$$

and

$$\mathbf{Y}_n(t + \frac{1}{n}) = [I + \frac{1}{n} A(Y_n(t))]\mathbf{Y}_n(t) + [\frac{1}{\sqrt{n}} Z(t) b(Y_n(t)) + \frac{1}{n} c(Y_n(t))] \; \mathbf{e}, \tag{3.5}$$

where $c = a_p \mu$ and $\{Z(t)\}$ is an iid sequence of random variables with $P(Z = 1) = P(Z = -1) = 0.5$.

The process $\{\mathbf{Y}_n(t)\}$ in (3.5) is clearly Markovian and the conditional expectations,

$$\mathbf{m}_n(\mathbf{y}, t) = E(\mathbf{Y}_n(t) | \mathbf{Y}_n(0) = \mathbf{y}),$$

satisfy the backward Kolmogorov equations,

$$\begin{aligned} \mathbf{m}_n(\mathbf{y}, t + n^{-1}) = & \frac{1}{2} \mathbf{m}_n(\mathbf{y} + n^{-1}(A(y)\mathbf{y} + c(y))\mathbf{e}) + n^{-1/2} b(y)\mathbf{e}, t) \\ & + \frac{1}{2} \mathbf{m}_n(\mathbf{y} + n^{-1}(A(y)\mathbf{y} + c(y))\mathbf{e}) - n^{-1/2} b(y)\mathbf{e}, t), \end{aligned} \tag{3.6}$$

with the initial condition,

$$\mathbf{m}_n(\mathbf{y}, 0) = \mathbf{y}. \tag{3.7}$$

These equations clearly determine the moments $\mathbf{m}_n(\mathbf{y}, t)$ uniquely. Higher order moments satisfy the same equation (3.6), and a modification of the initial condition (3.7) in which the right-hand side is appropriately replaced. For example

$$S_n(\mathbf{y}) := E[\mathbf{Y}_n(t)\mathbf{Y}_n(t)'|\mathbf{Y}_n(0) = \mathbf{y}],$$

satisfies (3.6) with the initial condition,

$$S_n(\mathbf{y}, 0) = \mathbf{y}\mathbf{y}'.$$

Brockwell and Williams[19] showed for the CTAR(2) process that if we extend the definition of the process $\{\mathbf{Y}^{(n)}\}$ so that it is defined for all real $t \geq 0$ by $\mathbf{Y}^{(n)}(t) = \mathbf{Y}^{(n)}([nt]/n)$, where $[nt]$ is the integer part of t, then $\{\mathbf{Y}^{(n)}\}$ converges in distribution as $n \to \infty$ to a solution of (3.3). (Convergence in distribution here means weak convergence of the associated probability measures on $D_{\mathbb{R}^2}[0, \infty)$.)

For particular cases of CTAR(1) and CTAR(2) processes with a single threshold Brockwell and Stramer[21] found close agreement between the calculation of moments based on simulation of Brownian motion and the calculation of moments based on (3.4) and (3.5) with $n = 10$. The adequacy of approximation can be checked by increasing the value of n but for n larger than 10, the exact solution of (3.6) becomes prohibitive and instead we rely on simulation of the process $\mathbf{Y}^{(n)}$ itself. This method seems preferable to the one based on simulation of Brownian motion as the variance of the functional whose expectation is to be computed is frequently large.

Sufficient conditions for geometric ergodicity of the CTAR(1) process with a single threshold and for the CTAR(p) process with single threshold and constant b are given by Stramer et al. [13,14] respectively. For the CTAR(1) process defined by

$$dY(t) + a(Y(t))dt = b(Y(t))dW(t) + c(Y(t))dt, \tag{3.8}$$

where

$$(a(y), b(y), c(y)) = \begin{cases} (a^{(1)}, b^{(1)}, c^{(1)}) & \text{if } y < r, \\ (a^{(2)}, b^{(2)}, c^{(2)}) & \text{if } y \geq r, \end{cases}$$

these conditions reduce to

$$\lim_{|x| \to \infty} [a(x)x^2 - 2c(x)x] > 0, \tag{3.9}$$

and the stationary distribution then has the density,

$$\pi(x) = kb^{-2}(x)\exp\{-b^{-2}(x)[a(x)x^2 - 2c(x)x]\}, \tag{3.10}$$

where k is the uniquely determined constant such that $\int_{-\infty}^{\infty}\pi(x)dx = 1$. For the CTAR(2) process it suffices for all the eigenvalues of the two matrices $A^{(1)}$ and $A^{(2)}$ to have negative real parts, where $A^{(1)}$ and $A^{(2)}$ are respectively the values of the coefficient matrix A in the defining equation (3.3) when $Y(t)$ is below and above the threshold .

Remark 3. Many extensions of the CTAR(p) process are of potential interest. One can for example define a continuous-time threshold ARMA(p, q) process by allowing dependence on $Y(t)$ of the matrices B and A in the state-space representation,

$$Y(t) = \begin{bmatrix} 1 & 0 & \cdots & 0 \end{bmatrix} \mathbf{Y}(t),\ t \geq 0, \tag{3.11}$$

where $\mathbf{Y}(t)$ is a stationary solution of the vector AR(1) equation,

$$d\mathbf{Y}(t) = BAB^{-1}\mathbf{Y}(t)dt + B\mathbf{e}\ dW(t),\ t \geq 0 \tag{3.12}$$

and

$$B = \begin{bmatrix} b_0 & b_1 & b_2 & \cdots & b_{p-1} \\ 0 & 1 & 0 & \cdots & 0 \\ 0 & 0 & 1 & \cdots & 0 \\ \vdots & \vdots & \vdots & & 0 \\ 0 & 0 & 0 & \cdots & 1 \end{bmatrix} \quad \text{if } b_0 \neq 0.$$

(If $b_0 = 0$ and i is the smallest integer such that $b_i \neq 0$, then we replace the first component of the $(i+1)^{\text{th}}$ row in the definition of B by 1.) Properties of such processes are however less understood and they do not have the same direct relationship to a univariate stochastic differential equation with coefficients depending on $Y(t)$ as does the CTAR(p) process. Other extensions would allow the thresholds to depend in a more general way on the state vector and allow the observations to be vector-valued.

4 Inference for Continuous-time ARMA Processes

For linear Gaussian CARMA processes, maximum likelihood estimation based on observations $Y(t_1), \ldots, Y(t_N)$ can be carried out very conveniently using the Kalman recursions and the innovations form of the likelihood (see Jones[10]). For both the linear heavy-tailed and non-linear Gaussian models considered in Sections 2 and 3 respectively, maximum likelihood estimation can be carried out as described below, provided the state process $\mathbf{Y}(t)$ in the representation (3.11) and (3.12) has a transition density which can be computed. One approach to this computation is via simulation. Another is to approximate the transition function by a Gaussian transition function with moments computed as described in Section 3. In the applications described in this section the latter approach was used, with first and second moments of the transition function of the state vector computed from an Euler approximation with $n = 20$.

In each case the process $Y(t)$ can be expressed as in (3.11) and (3.12), where the state process,

$$\mathbf{Y}(t) = \begin{bmatrix} Y(t) \\ \mathbf{V}(t) \end{bmatrix},$$

is Markov. If we assume that $\{\mathbf{Y}(t)\}$ has a transition density and write $p(y_{r+1}, \mathbf{v}_{r+1}, t_{r+1} - t_r | y_r, \mathbf{v}_r)$ for the density of $(Y(t_{r+1}), \mathbf{V}(t_{r+1})')'$, given $\mathbf{Y}(t_r) = (y_r, \mathbf{v}_r')'$, then the joint probability density f_r of the random variables, $Y(t_r), \mathbf{V}(t_r), Y(t_{r-1}), Y(t_{r-2}), \ldots, Y(t_1)$ satisfies the recursions,

$$f_{r+1}(y_{r+1}, \mathbf{v}_{r+1}, y_r, y_{r-1}, \ldots, y_1)$$
$$= \int p(y_{r+1}, \mathbf{v}_{r+1}, t_{r+1} - t_r | y_r, \mathbf{v}_r) f_r(y_r, \mathbf{v}_r, y_{r-1}, \ldots, y_1) d\mathbf{v}_r, \qquad (5.1)$$

For a given set of observed values $y_1, \ldots, y_N$, at times $t_1, \ldots, t_N$, the functions $f_2, \ldots, f_N$ are functions of $\mathbf{v}_2, \ldots, \mathbf{v}_N$ respectively. These functions can easily be computed recursively from (5.1) in terms of f_1 and the functions $p(y_{r+1}, \cdot, t_{r+1} - t_r | y_r, \cdot)$. The likelihood of the observations $y_1, \ldots, y_N$, is then clearly

$$L(\theta; y_1, \ldots, y_N) = \int_{\mathbf{v}_N} f_N(\mathbf{v}_n) d\mathbf{v}_n. \qquad (5.2)$$

The filtered value of the unobserved vector $\mathbf{V}(t_r)$, $r = 1, \ldots, N$, (i.e. the conditional expectation of $\mathbf{V}(t_r)$ given $Y(t_i) = y_i$, $i = 1, \ldots, r$) is readily obtained from the function f_r as

$$\tilde{\mathbf{v}}_r = \frac{\int \mathbf{v} f_r(\mathbf{v}) d\mathbf{v}}{\int f_r(\mathbf{v}) d\mathbf{v}}. \qquad (5.3)$$

The Markov property of $\{\mathbf{Y}(t)\}$ then gives the predictor of $Y(t_{r+1})$,

$$\tilde{y}_{r+1} = m((y_r, \tilde{\mathbf{v}}_r')', t_{r+1} - t_r). \qquad (5.4)$$

The preceding calculations are all dependent on the determination of the transition densities, $p(y_{r+1}, \mathbf{v}_{r+1}, t_{r+1} - t_r | y_r, \mathbf{v}_r)$ with y_{r+1} and y_r equal to the observations at times t_{r+1} and t_r, $r = 1, \ldots, N-1$ respectively, and the evaluation of the successive $(p-1)$-dimensional integrals appearing in the recursions (5.1). For $p = 2$, these are one-dimensional integrals and, by discretizing v, can be evaluated as a sequence of matrix multiplications.

In order to compute the Gaussian likelihood, it suffices to determine the first two moments of p and this can be done for non-linear Gaussian CAR processes by either of the methods discussed in Section 4 or by simulation. If we take $f_1(y, \mathbf{v})$ to be the Dirac delta function assigning mass one to $(y_1, \mathbf{0}')'$, then the likelihood in (5.2) is the density of $Y(t_2), \ldots, Y(t_n)$, *conditional* on $Y(t_1) = y_1$ and $\mathbf{V}(t_1) = \mathbf{0}$.

Since the recursions (5.1) apply to the *exact* joint densities f_r (provided the densities p exist), we can in principle carry out the same computations (5.2) and (5.3) with the required values of the density functions estimated by simulation in order to compute the likelihood and conditional expectations for a large class of non-Gaussian and non-linear CARMA processes. The following models were all fitted by maximization of the conditional Gaussian likelihood and numerical approximation of the first and second order moments of the transition density using Euler approximations.

Example 7 The Australian All Ordinaries Index is an average of share prices on the Australian Stock Exchange. Its closing values, $\{P(t), t = 0, \ldots, 520\}$, recorded daily for 521 trading days ending July 18th, 1994, were studied by Brockwell and Williams [19]. They found that a CTAR(2) model for the return series gave a substantial improvement in terms of AIC value over the best linear ARMA model. The performance of the model in forecasting out-of-sample values was not noticably better in terms of mean squared error than the linear model. On the other hand the empirical conditional one-step mean squared errors were better given positive values of the current return than those given negative values. In this example we examine the series

$$U(t) = 100 \ln(P(t)/P(t-1))$$

for the daily closing prices of both the All Ordinaries and Dow-Jones Industrial Indices for the period July 1st, 1997, through April 9th, 1999. The improvement shown (in terms of AIC) by the threshold Gaussian CAR(2) model over the best linear Gaussian ARMA model (white noise) is more spectacular in this case and is summarized in the following table comparing AIC values for the white noise, CTAR(2), ARCH(1), GARCH(1,1) and GARCH$_t$(1,1) models.

	WN	CTAR(2)	ARCH(1)	GARCH(1,1)	GARCH$_t$(1,1)
AO	1319.2	1233.6	1231.9	1228.8	1229.5
DJ	1516.6	1449.7	1500.2	1470.9	1439.5

It is seen from the table that the improvements obtained by the CTAR(2) models are comparable to those achieved by GARCH. However the good performance of the GARCH model with heavy tails in representing the Dow-Jones returns suggests the potential for development of non-linear Lévy-driven models also. This is more evident still in modelling more volatile series such as the Nikkei and Hang Seng indices over the same time period. The CTAR(2) models fitted to the All Ordinaries and Dow-Jones returns are shown below.

All Ordinaries Returns

$$D^2U(t) + 0.04DU(t) + 2.00U(t) = 9.40DW(t) + 1.55, \; U(t) < -2.4, \quad (5.5)$$

$$D^2U(t) + 2.94DU(t) + 5.60U(t) = 4.70DW(t) + 0.15, \; U(t) > -2.4. \quad (5.6)$$

Dow-Jones Returns

$$D^2U(t) + .04DU(t) + 3.90U(t) = 11.15DW(t) + 2.30, \; U(t) < -2.3, \quad (5.7)$$

$$D^2U(t) + 2.49DU(t) + 5.30U(t) = 5.20DW(t) + 0.45, \; U(t) > -2.3. \quad (5.8)$$

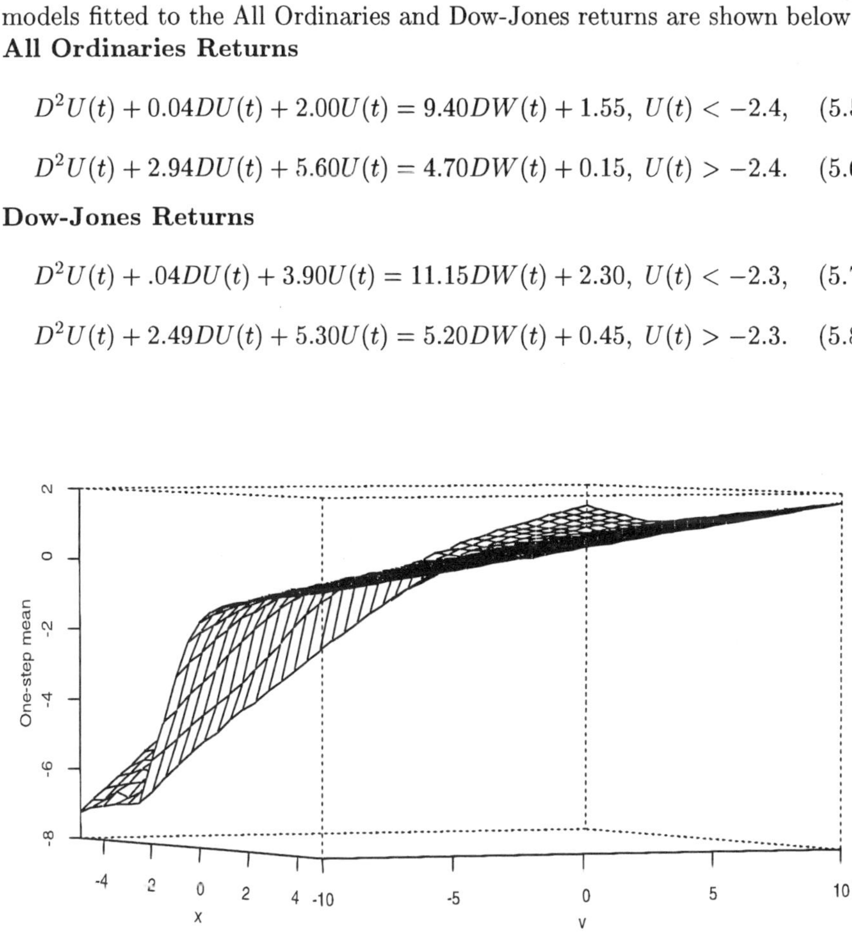

Figure 5: *Conditional mean of $U(t+1)$ given $U(t) = x$, $U'(t) = v$, All Ordinaries returns*

Figures 5 and 6 show respectively the conditional mean and standard deviation of $U(t+1)$ given the values x and v of $U(t)$ and $U'(t)$ for the first

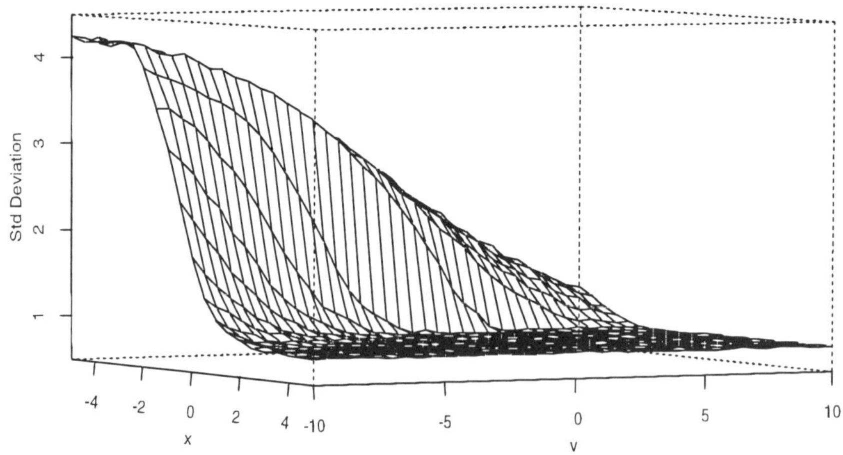

Figure 6: *Conditional standard deviation of* $U(t+1)$ *given* $U(t) = x$, $U'(t) = v$, *All Ordinaries returns*

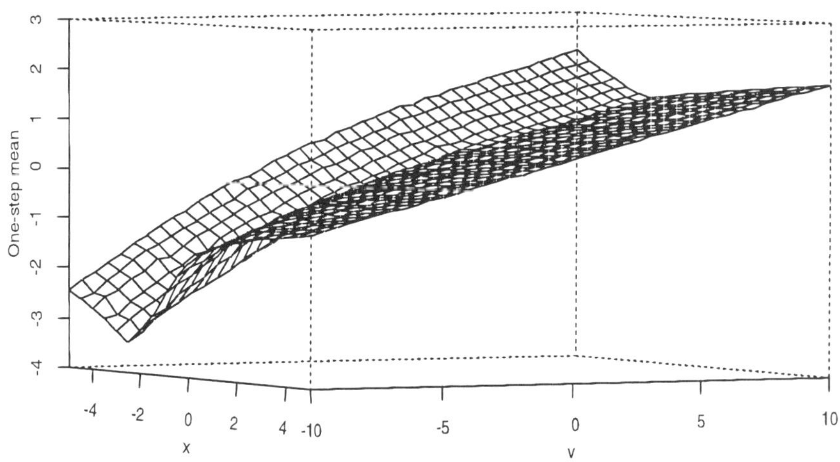

Figure 7: *Conditional mean of* $U(t+1)$ *given* $U(t) = x$, $U'(t) = v$, *Dow-Jones returns*

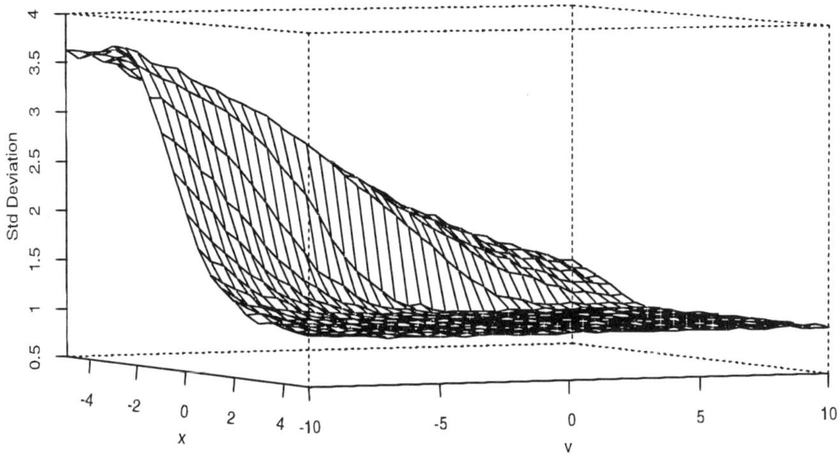

Figure 8: *Conditional standard deviation of* $U(t+1)$ *given* $U(t) = x$, $U'(t) = v$, *Dow-Jones returns*

of these models and Figures 7 and 8 do the same for the second one. Both models illustrate the increase in standard deviation as both $U(t)$ and $U'(t)$ become more negative. Since we can estimate the current unobserved value of $U'(t)$ on the basis of $U(t)$ and previous observations of U, these Figures allow us to estimate the standard error of one-step prediction based on the data up to the current time.

Acknowledgments

This work was partially supported by NSF Grant DMS 9972015. I am also indebted to Tina Rydberg for drawing my attention to Barndorff-Nielsen and Shephard[9].

References

1. P.J. Brockwell and R.A. Davis *Introduction to Time Series and Forecasting*, (Springer-Verlag, New York, 1996).
2. P.J. Brockwell, *J. Statist. Planning and Inference* **8**, 157 (1994).
3. P.J. Brockwell in *Handbook of Statistics: Stochastic Processes, Theory and Methods*, ed. C.R. Rao and D.N. Shanbhag (Elsevier, Amsterdam, 1999).

4. K. Ito *Stochastic Processes*, (Aarhus Universitet, Matematisk Institute, Lecture Notes 16, 1969).
5. Jean Bertoin *Lévy Processes*, (Cambridge University Press, Cambridge, 1996).
6. E. Cinlar and M. Pinsky, *J. Appl. Prob* **9**, 422 (1972).
7. J.M. Harrison and S.I. Resnick, *Math. Operat. Res.* **1**, 347 (1976).
8. P.J. Brockwell, S.I. Resnick and R.L. Tweedie, *Adv. Appl. Prob.* **14**, 392 (1982).
9. O.E. Barndorff-Nielsen and N. Shephard, Working Paper in Economics, Nuffield College, Oxford, 1999.
10. R.H. Jones in *Applied Time Series Analysis II*, ed. D.F. Findley (Academic Press, New York, 1981).
11. C.W. Granger, Z. Ding and S. Spear, Paper presented at Hong Kong International Workshop on Statistics in Finance, 1999.
12. H. Tong *Threshold Models in Non-linear Time Series Analysis*, (Springer Lecture Notes in Statistics **21**, Springer-Verlag, New York, 1983).
13. O. Stramer, P.J. Brockwell and R.L. Tweedie, *Adv. Appl. Prob.* **28**, 728 (1996).
14. O. Stramer, R.L. Tweedie and P.J. Brockwell, *Statistica Sinica* **6**, 715 (1996).
15. Bernt Oksendal *Stochastic Differential Equations*, 5th ed., (Springer-Verlag, Berlin, 1998).
16. Ioannis Karatzis and Steven E. Shreve *Brownian Motion and Stochastic Calculus*, (Springer-Verlag, New York, 1991).
17. D.W. Btroock and S.R.S. Varadhan *Multidimensional Diffusion Processes*, (Springer-Verlag, Berlin, 1979).
18. H. Tong *Non-linear Time Series:A Dynamical System Approach*, (Clarendon Press, Oxford, 1990).
19. P.J. Brockwell and R.J. Williams, *Adv. Appl. Prob.* **29**, 205 (1997).
20. P.J. Brockwell and R.J. Hyndman, *International J. Forecasting* **8**, 157 (1992).
21. P.J. Brockwell and O. Stramer, *Ann. Inst. Stat. Math.* **47**, 1 (1995).

NONLINEAR STATE SPACE MODEL APPROACH TO FINANCIAL TIME SERIES WITH TIME-VARYING VARIANCE

GENSHIRO KITAGAWA and SEISHO SATO
The Institute of Statistical Mathematics
4-6-7 Minami-Azabu, Minato-ku, Tokyo 106-3569 Japan
E-mail: kitagawa,sato@ism.ac.jp

By expressing a time series model with time-varying variance in nonlinear state space model form, it is possible to decompose the financial time series into trend, stationary component and the noise with varying volatility. The models with changing volatility are significantly better than the ordinary trend models in terms of the AIC, indicating that it is possible to obtain a good model by an explicit modeling of volatility changes. In the analysis of Nikkei 225 Japanese stock price index data, a clear relation between the difference of the trend and the local variance was found. Therefore we further developed models that take into account the relation between the trend and the variance. For the Nikkei 225 data, the AIC value was further decreased suggesting the presence of causal relation. On the other hand, for exchange rate data, such relation was not found.

1 Introduction

In most of the analysis of financial time series such as the stock prices, models are usually considered after applying nonlinear transformation, $r_n = \Delta \log y_n = \log y_n - \log y_{n-1}$. Well known examples are the ARCH model (Engle (1982))

$$\begin{aligned} r_n &= \sigma_n w_n \\ \log \sigma_n^2 &= \alpha + \beta w_{n-1}^2 \end{aligned} \tag{1}$$

and the stochastic volatility model (Harvey and Shepherd (1996))

$$\begin{aligned} r_n &= \sigma_n w_n \\ \log \sigma_n^2 &= \alpha + \beta \log \sigma_{n-1}^2 + v_n. \end{aligned} \tag{2}$$

Here, it is usually assumed that w_n is distributed as normal distribution with the mean 0 and the variance 1.

These models implicitly assume that the financial time series is a random walk and that the trend and the volatility are mutually independent. However, in actual financial time series, these assumptions are not always adequate and by modeling the relation between the trend and the volatility, it is very likely to obtain a better model.

Recently, based on the use of non-Gaussian state space model and a related nonlinear filter, the maximum likelihood method is exploited for the estimation of the unknown parameters. A popular method of estimating the stochastic volatility model is to apply the nonlinear transformation to r_n as

$$\log r_n^2 = \log \sigma_n^2 + \log w_n^2 \tag{3}$$

where log w_n^2 is distributed as a double exponential distribution

$$\log w_n^2 \sim \frac{1}{\sqrt{2\pi}} \exp\left\{ \frac{w}{2} - \frac{e^w}{2} \right\}. \tag{4}$$

Usually this non-Gaussian distribution is approximated by a Gaussian distribution with the mean $\eta = 1.2704$ and the variance $\pi^2/2$. Then it can be approximated by a linear-Gaussian state space model and the Kalman filter can be applied for state estimation. However, by using the nonlinear state space model, it is possible to express in state space model without any approximation. The stochastic volatility model (2) can be considered as special form of nonlinear state space model with nonlinear observation model if we consider $log\sigma_n^2$ as the state. Then the estimate of the parameter of the stochastic volatility model and the estimate of the volatility can be achieved by applying nonlinear filter/smoother to the nonlinear state space model.

2 The Models

2.1 Model with Trend and Stochastic Volatility

In this section, we consider a model to decompose time series with time-varying variance as

$$y_n = T_n + \sigma_n \varepsilon_n, \qquad \varepsilon_n \sim N(0,1), \tag{5}$$

where T_n is the trend at time n, ε_n is the Gaussian white noise with mean 0 and variance 1, and σ_n is changing with time. This model can be applied to either the original series y_n or the logarithm of the series $r_n = \log y_n$. However, for simplicity, hereafter we shall consider models for y_n. In this model, the change of the volatility of the time series y_n is expressed by the time change of the factor σ_n. Therefore, in this article, we consider σ_n as a kind of volatility and call σ_n^2 the time-varying variance.

Following the seasonal adjustment model (Kitagawa and Gersch (1984)), it is assumed that the trend component T_n follows the stochastic difference equation with order k,

$$\Delta^k T_n = e_{1n}, \qquad e_{1n} \sim N(0, \tau_1^2). \tag{6}$$

Further, concerning the time-varying variance σ_n^2, $\log \sigma_n^2$ is assumed to follow the stochastic difference equation with order ℓ,

$$\Delta^\ell \log \sigma_n^2 = e_{2n}, \qquad e_{2n} \sim N(0, \tau_2^2), \tag{7}$$

where e_{1n} and e_{2n} are the white noise sequence with variance τ_1^2 and τ_2^2, respectively. The order k and the variance τ_1^2 control the smoothness of the trend, and ℓ and τ_2^2 control the smoothness of the change of the variance. As the orders of the trend model and the time-varying variance, k and ℓ, we usually use 1 or 2.

2.2 Model with Trend, Stationary Component and Stochastic Volatility

In general, economic time series is consisted of many components with various frequencies. Therefore, the model with only trend component, which is consisted of mainly with low frequency waves, and a white noise component, which contains all frequency components equally, cannot fit very well. Therefore, we shall extend the model in Eq. (5) and consider a model with trend, stationary component and white noise as follows;

$$y_n = T_n + p_n + \sigma_n \varepsilon_n, \qquad \varepsilon_n \sim N(0, 1), \tag{8}$$

where T_n, σ_n and ε_n are the same as the ones in Eq. (5) and p_n is a stationary component and follows an autoregressive model,

$$p_n = \sum_{j=1}^{m} a_j p_{n-j} + e_{3n}, \qquad e_{3n} \sim N(0, \tau_3^2). \tag{9}$$

Here e_{3n} is assume to be a white noise sequence with the mean 0 and the variance τ_3^2. It is expected that a broad class of characteristics can be expressed by this model.

3 State Space Models

3.1 State Space Representation of Time-varying Variance Model

The stochastic volatility model introduced in section 2.2 can be expressed in a state space model form. Define a $(k + m + \ell\ )$-dimensional state vector by

$$x_n = (T_n, \ldots, T_{n-k+1} | p_n, \ldots, p_{n-m+1} | \log \sigma_n^2, \ldots, \log \sigma_{n-\ell+1}^2)^T. \tag{10}$$

Further, assume that T_n, p_n and $\log \sigma_n^2$ follow Eqs. (6), (9) and (7), respectively, and $(k + m + \ell) \times (k + m + \ell)$ matrix F and $(k + m + \ell) \times 3$ matrix G are defined by

$$F = \left[\begin{array}{ccc|ccc|ccc} c_1^k & c_2^k \ \ldots & c_k^k & & & & & & \\ 1 & & & & & & & & \\ & \ddots & & & 0 & & & 0 & \\ & & 1 & & & & & & \\ \hline & & & a_1 & a_2 \ \ldots & a_m & & & \\ & & & 1 & & & & & \\ & 0 & & & \ddots & & & 0 & \\ & & & & & 1 & & & \\ \hline & & & & & & c_1^\ell & c_2^\ell \ \ldots & c_\ell^\ell \\ & & & & & & 1 & & \\ & 0 & & & 0 & & & \ddots & \\ & & & & & & & & 1 \end{array}\right], \quad G = \left[\begin{array}{ccc} 1 & 0 & 0 \\ 0 & 0 & 0 \\ \vdots & \vdots & \vdots \\ 0 & 0 & 0 \\ \hline 0 & 1 & 0 \\ 0 & 0 & 0 \\ \vdots & \vdots & \vdots \\ 0 & 0 & 0 \\ \hline 0 & 0 & 1 \\ 0 & 0 & 0 \\ \vdots & \vdots & \vdots \\ 0 & 0 & 0 \end{array}\right], \qquad (11)$$

where c_i^j is the binomial coefficient defined by $(x-1)^j = -\sum_{i=0}^{j} c_i^j x^{j-i}$, then the time-changes of these components can be expressed by

$$x_n = F x_{n-1} + G v_n, \qquad (12)$$

where v_n is a 3-dimensional white noise defined by

$$v_n = (e_{1n}, e_{2n}, e_{3n})^T, \qquad (13)$$

and its variance covariance matrix is given by $Q = \mathrm{diag}\{\tau_1^2, \tau_2^2, \tau_3^2\}$.

On the other hand, since T_n, p_n and $\log \sigma_n^2$ are components of the state vector x_n as defined by Eq. (10), from the model in Eq. (8), the time series y_n can be expressed by a nonlinear observation model as

$$y_n = T_n + p_n + \sigma_n \varepsilon_n = h(x_n, \varepsilon_n). \qquad (14)$$

The Eqs. (12) and (14) constitute a state space model with a state vector x_n. Here, Eq. (12) is a system model with the system noise v_n. On the other hand, Eq. (14) is the observation equation with ε_n being the observation noise. σ_n is obtained by a nonlinear transformation of $\log \sigma_n^2$, which is a component of a state vector x_n, and the observation y_n cntaines a nonlinear term which is obtained as the product of σ_n and the noise component ε_n. Therefore, our models shown in Sec. 2.1 and 2.2 can be expressed in a nonlinear state space model.

4 Estimation of the State and Decomposition into Components

4.1 Estimation of the State

In general, the problem of estimating the state x_n based on the observations up to time point j, $Y_j = \{y_1, \ldots, y_j\}$ is called the state estimation. Especially, for particular cases, $j < n$, $j = n-1$, $j = n$ and $j > n$, the state estimation problem is called prediction, one-step-ahead-prediction, filter and smoothing.

In the state space modeling, once the state x_n is estimated from the observations Y_j, then from the definition of the state in Eq. (10), the trend component T_n, stationary component p_n and time-varying variance σ_n^2 or "volatility" σ_n can be immediately obtained. Namely, the first component $x_n(1)$, the $(k+1)$-st component $x_n(k+1)$ and the $(k+m+1)$-st component $x_n(k+m+1)$ are the estimates of the trend, stationary component and the logarithm of the time-varying variance, respectively. In other words, by this modeling the decomposition of time series and estimation of the volatility can be realized simultaneously.

However, in contrast to the case of linear Gaussian state space model for which a computationally efficient Kalman filter can be applied for state estimation (Anderson and Moore (1979)), use of nonlinear filter is necessary for nonlinear state space model in Eq. (14).

4.2 Estimation of the Components by a Monte Carlo Filter

As mentioned in section 3, for the estimation with our stochastic volatility models with trend component, it is necessary to use a nonlinear state space model. If the state dimension is less than or equals to 3, by the use of non-Gaussian filter based on numerical integration, it is possible to obtain precise estimates of the state even for general state space model (Kitagawa (1987), Kitagawa, Sato and Nagahara (1998)). However, for higher dimensional nonlinear state space models, it is not practical to apply direct numerical integration method. Monte Carlo filter was developed for such a situation (Kitagawa (1996)). See also Gordon, Salmond and Smith (1993) and Hürzeler and Künsch (1998).

In the Monte Carlo filter, each of the one-step-ahead predictor $p(x_n|Y_{n-1})$, the filter $p(x_n|Y_n)$ and the system noise distribution $p(v_n)$ is approximated by many "particles." These "particles" are points in high-dimensional space corresponding to the state x_n and the noise v_n. For $p(v_n)$, since its density function is specified from the given model, we can generate m random numbers from $p(v_n)$, $V_n^{(1)}, \ldots, V_n^{(m)}$. Then the empirical distribution function defined from $V_n^{(1)}, \ldots, V_n^{(m)}$,

$$P_n(x) = \frac{1}{m} \sum_{j=1}^{m} I(x; V_n^{(j)}) \tag{15}$$

can be considered as an approximation to the true distribution function

$$P(x) = \int_{-\infty}^{x} p(t)dt, \tag{16}$$

where $I(x; a)$ is the indicator function defined by

$$I(x; a) = \begin{cases} 0 & x < a \\ 1 & x \geq a \end{cases}. \tag{17}$$

On the other hand, it is not so straightforward as the case of V_n to generate the particles $P_n^{(j)}$ and $F_n^{(j)}$ approximating the predictive distribution $p(x_n|Y_{n-1})$ and the filter distribution $p(x_n|Y_n)$. But, by the procedure shown below, it is possible to generate particles which can be considered as independent realizations from these distributions (Kitagawa 1996).

1. Approximation of initial distribution: Generate $F_0^{(j)} \sim p(x_0|Y_0)$.
2. Approximation of system noise: Generate $V_n^{(j)} \sim p(v)$.
3. Approximation of predictive distribution: Compute $P_n^{(j)} = f(F_{n-1}^{(j)}, V_n^{(j)})$.
4. Computation of Bayes factor: Compute $\alpha_n^{(j)} = p(y_n|x_n = P_n^{(j)})$.
5. Re-approximation of the filter distribution by resampling: Generate $\{F_n^{(j)}\}$ from $\{F_n^{(j)}\}$.

In step 1, m particles are generated by using random number distributed as the initial distribution $p(x_0|Y_0)$. Step 2–5 is repeated for N times depending on the sample size. In step 2, generate realizations of system noise by using random number. Then in step 3, substitute the particles generated in step 2 and step 5 (step 1 for n=1) into the right hand side of the equation and obtain $P_n^{(j)}$. In step 4, compute the Bayes factors of the particle obtained in step 3. These Bayes factors can be interpreted as relative importance of these particles after observing y_n. In step 5, generate m particles by sampling with replacement by using the Bayes factors $\alpha_n^{(j)}$ as probability of each particle.

Then these particles can be considered as independent realizations from filter distribution. In actual state estimation, this Monte Carlo filter algorithm can be extended to smoothing estimates, $S_n^{(j)}$, (Kitagawa 1996).

As mentioned in the previous section, the trend component and the stochastic volatility can be obtained from the estimated state vector. Namely, the particles obtained in the above mentioned algorithm, $P_n^{(j)}$, $F_n^{(j)}$ and $S_n^{(j)}$ are $k+m+\ell$ dimensional vector, and its first, the $(k+1)$-st and the $(k+m+1)$-st components are $T_{n|L}^{(j)}$, $p_{n|L}^{(j)}$, $\log \sigma^2{}_{n|L}^{(j)}$, respectively.

Then m one-dimensional particles $\{T_{n|L}^{(1)}, \ldots, T_{n|L}^{(m)}\}$, $\{p_{n|L}^{(1)}, \ldots, p_{n|L}^{(m)}\}$ and $\{\log \sigma^2{}_{n|L}^{(1)}, \ldots, \log \sigma^2{}_{n|L}^{(m)}\}$ are the ones expressing the trend, stationary components and the volatility. Here, for example, $T_{n|L}^{(j)}$ denotes the first component of the jth particle, and corresponding to $L = n-1$, n and N, they approximate the one-step-ahead predictor, the filter and the smoother, respectively.

Then

$$\frac{1}{m}\sum_{j=1}^{m} I(x; T_{n|L}^{(j)}), \quad \frac{1}{m}\sum_{j=1}^{m} I(x; p_{n|L}^{(j)}), \quad \frac{1}{m}\sum_{j=1}^{m} I(x; \sigma_{n|L}^{(j)}) \tag{18}$$

are approximations to the marginal distribution functions of $T_{n|L}^{(j)}$, $p_{n|L}^{(j)}$ and $\log \sigma^2{}_{n|L}^{(j)}$. In particular, its mean values are obtained by

$$\frac{1}{m}\sum_{j=1}^{m} T_{n|L}^{(j)}, \quad \frac{1}{m}\sum_{j=1}^{m} p_{n|L}^{(j)}, \quad \frac{1}{m}\sum_{j=1}^{m} \sigma_{n|L}^{(j)}. \tag{19}$$

4.3 Estimation of the Parameters of the Model

Once the one-step-ahead predictor of the state x_n, $p(x_n|Y_{n-1})$, is obtained, the one-step-ahead predictor of the time series y_n is obtained by

$$p(y_n|Y_{n-1}) = \int p(y_n|x_n) p(x_n|Y_{n-1}) dx_n. \tag{20}$$

In the Monte Carlo filter, the one-step-ahead predictor $p(x_n|Y_{n-1})$ is approximated by m particles, $P_n^{(1)}, \ldots, P_n^{(m)}$. Therefore, from Eq. (15), we have an approximation

$$p(y_n|Y_{n-1}) \approx \int p(y_n|x_n) \frac{1}{m}\sum_{j=1}^{m} I(x_n; P_n^{(j)}) dx_n$$

$$= \frac{1}{m} \sum_{j=1}^{m} p(P_n^{(j)} | Y_{n-1}). \tag{21}$$

On the other hand, if N observations $Y_N = \{y_1, \dots, y_N\}$ are given, the likelihood of the model is expressed as

$$L(\theta) = p(Y_N) = p(Y_{N-1})p(y_N | Y_{N-1}) = \cdots = \prod_{n=1}^{N} p(y_n | Y_{n-1}). \tag{22}$$

Therefore, the log-likelihood of the stochastics volatility model is defined by using the one-step-ahead predictive distribution, $p(y_n | Y_{n-1})$,

$$\ell(\theta) = \sum_{n=1}^{N} \log p(y_n | Y_{n-1}) \approx \frac{1}{m} \sum_{n=1}^{N} \sum_{j=1}^{m} \log p(y_n | P_n^{(j)}). \tag{23}$$

Then by using a numerical optimization method such as the quasi-Newton method, the maximum likelihood estimate of the parameter θ can be obtained by maximizing this log-likelihood function. However, in actuality, it is difficult to obtain a precise approximation to the maximum likelihood estimate since the log-likelihood function obtained by Monte Carlo filter, Eq. (21), contains the sampling error. However, because the log-likelihood is not so sensitive to the change of the values of the variance parameters τ_1^2, τ_2^2, τ_3^2, it is usually possible to obtain reasonable estimates of the parameters by discrete search.

Another way to mitigate this problem is to use the self-organizing state space model (Kitagawa (1998)). In this method, the original state vector is augmented with the unknown parameter vector as

$$z_n = \begin{bmatrix} x_n \\ \theta \end{bmatrix}. \tag{24}$$

Then the state space model for this augmented state vector can be easily obtained from the original model. Monte Carlo filter can provide state estimates for this augmented state space model. Since the augmented state space model contains both the original state and the unknown parameter, it means that the state estimation and parameter estimation can be realized simultaneously.

5 Data Analysis

5.1 Nikkei 225 Japanese Stock Index Data

Figure 1 shows the Nikkei 225 Index data (January 1987–August 1990) and the estimated trend and noise components obtained by the seasonal adjustment program DECOMP (Akaike et al. 1985). The second order trend model

$$\begin{aligned} y_n &= t_n + w_n \\ t_n &= 2t_{n-1} - t_{n-2} + v_n \end{aligned} \tag{25}$$

is used, and it is assumed that the noises w_n and v_n are distributed with Gaussian white noise with constant variances, σ^2 and τ^2, respectively. The maximum likelihood estimators of the variances are $\hat{\sigma}^2 = 4.70 \times 10^4$ and $\hat{\tau}^2 = 1.93 \times 10^4$ and the AIC was 14190.

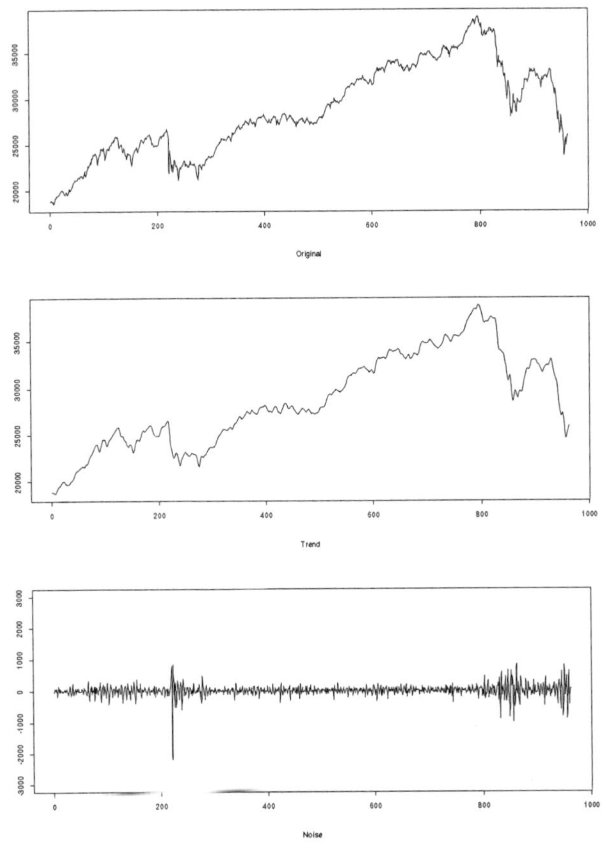

Figure 1: Trend model: original data, trend and noise

Figure 1 (a)–(c) respectively show the original time series, y_n, trend component t_n and the noise w_n. It can be seen that after significant drop of stock

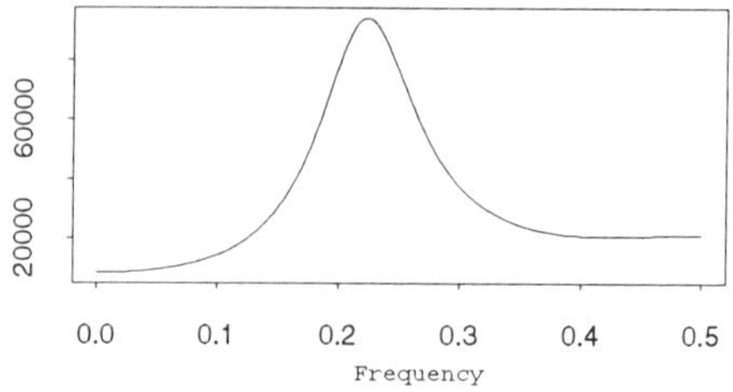

Figure 2: Spectrum of the noise

Table 1. Various Stochastic Volatility Models and AIC

Model	AIC
Constant Variance Model	
Trend+Noise	14190
Trend+AR+Noise	13882
Stochastic Volatility Model	
Trend+Noise	
Gaussian Model	13580
Cauchy Distribution	13648
Model in Eq. (31) + Mixture Distribution	13553
Trend+AR+Noise	
Mixture Distribution	13412
Model in Eq. (36)	13352
Model in Eq. (37)	13339

prices after the Black Monday and the crush of bubble, the amplitude of the noise increased by several times and the volatility was increased.

Figure 2 shows the power spectrum of the noise sequence w_n obtained by using the autoregressive model with order 3. A significant peak is seen around at $f = 0.22$ (4.5 day period), and these two figures reveal both the assumptions of whiteness and constancy of the variance do not seem to apply for this series.

Figure 3 shows the results obtained by using DECOMP with $M_2 = 2$, namely by using a model with stationary AR component

$$y_n = t_n + p_n + w_n. \tag{26}$$

Here the variance is assume to be a constant over time. As shown in Table 1, the AIC of the model was 13882, compared with the case of Eq. (25) AIC

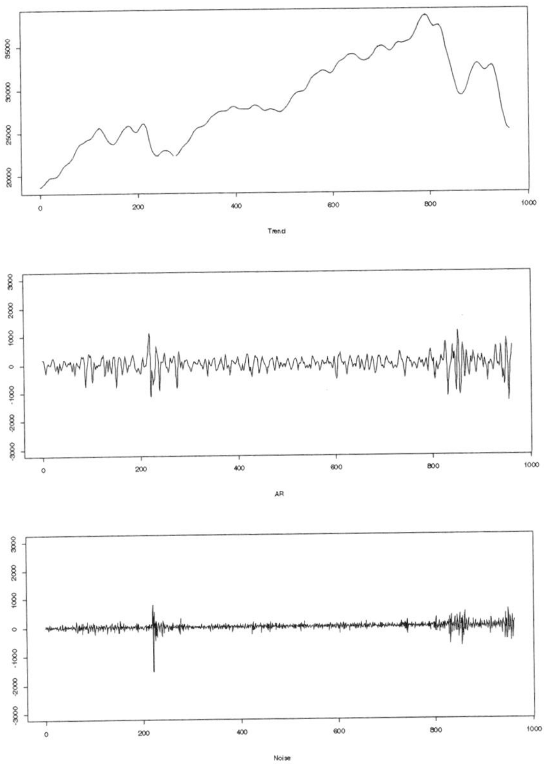

Figure 3: Estimated trend, AR component and the noise by the model (26)

decreased by 308, indicating that the fit of the model is significantly improved. By this model, the cyclic movement seen in the original data was expressed as a stationary AR component p_n. As a result, the trend estimate becomes much smoother than the one in Figure 1 and the noise component w_n. However, even with this decomposition, corresponding to the change of the volatility, the amplitudes of p_n and w_n change significantly.

We then estimated models with time-varying variance. In the trend plus stochastic volatility model given in Eq. (5) with $k = \ell = 2$, an approximate maximum likelihood estimates of the parameters are $\hat{\tau}_1^2$=9000, $\hat{\tau}_2^2$=0.0026 and AIC=13580. Compared with AIC of the ordinary model with constant variance, AIC=14190, decreases by 610, indicating significant improvement of the fit of the model.

Figure 4 shows the estimated trend, noise and the "volatility," σ_n, obtained by this model. The increase of the volatility after the Black Monday and after the crush of babble is clearly shown. Even with this simplest version, the stochastic volatility model introduced in this paper can automatically estimate changing volatility.

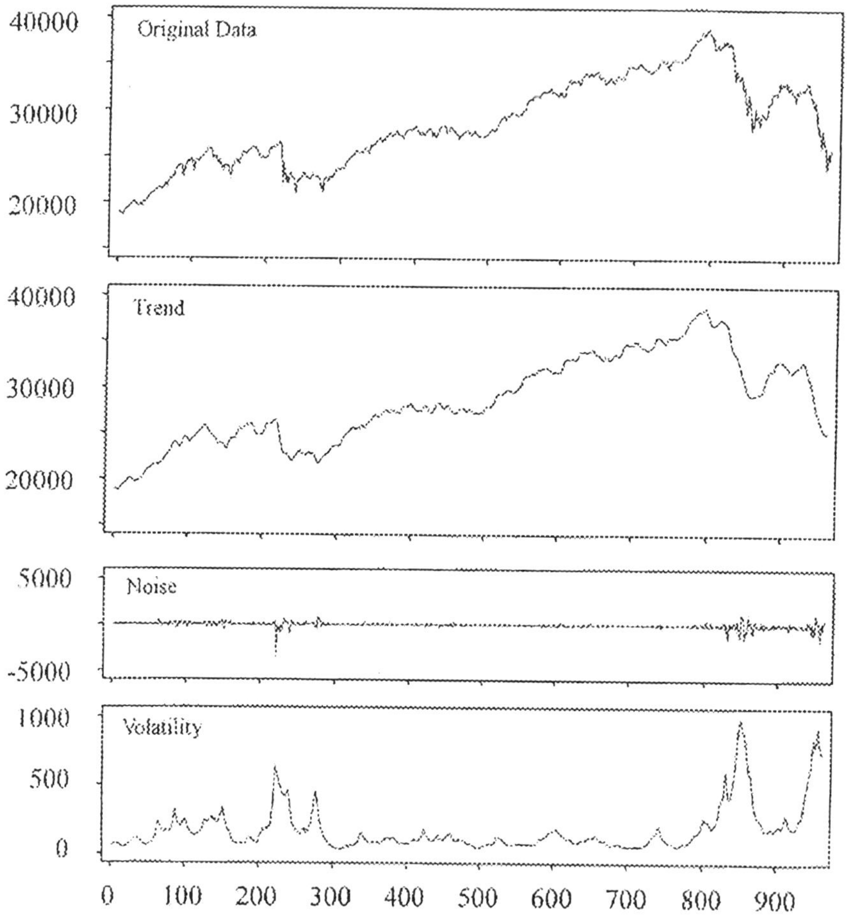

Figure 4: Trend+Stochastic Volatility model

6 Extension of the Trend Model

In this section, to treat abrupt changes of the slope of the trend and the level shift seen in Figure 4, we shall extend the trend model and introduce non-Gaussian distribution for system noise.

6.1 *Modeling for Level Shift*

Consider the second order trend model

$$T_n = 2T_{n-1} - T_{n-2} + e_{n1} \tag{27}$$

and define the first order difference of T_n by $\Delta T_n = T_n - T_{n-1}$ then we have

$$\begin{aligned} T_n &= T_{n-1} + (T_{n-1} - T_{n-2}) + e_{n1} = T_{n-1} + \Delta T_{n-1} + e_{n1} \\ \Delta T_n &= T_n - T_{n-1} = (T_{n-1} + \Delta T_{n-1} + e_{n1}) - T_{n-1} = \Delta T_{n-1} + e_{n1}. \end{aligned} \tag{28}$$

Then the model in Eq. (27) is equivalent to (Harvey (1989))

$$\begin{aligned} T_n &= T_{n-1} + \Delta T_{n-1} + e_{n1} \\ \Delta T_n &= \Delta T_{n-1} + e_{n1}. \end{aligned} \tag{29}$$

The state space representation for this model is given by

$$x_n = \begin{bmatrix} T_n \\ \Delta T_n \end{bmatrix}, \quad F = \begin{bmatrix} 1 & 1 \\ 0 & 1 \end{bmatrix}, \quad G = \begin{bmatrix} 1 \\ 1 \end{bmatrix}. \tag{30}$$

In the ordinary trend model, the change of the trend is caused by only the change of the "slope" of the trend.

Therefore, we introduce another noise term e_{n4} and consider an extended model (Harvey (1989)),

$$\begin{aligned} T_n &= T_{n-1} + \delta T_{n-1} + e_{n1} + e_{n2} \\ \delta T_n &= \delta T_{n-1} + e_{n1}. \end{aligned} \tag{31}$$

Here, e_{n1} is the change of the slope, e_{n2} is the noise causing level shift. In this model, δT_n corresponds to the change of the trend T_n at time t, but different from ΔT_n, it is not exactly the difference of the trend T_n. With this model, we can freely model the trend by distinguishing the level shift and change of the slope.

6.2 *Modeling Sudden Change of Trend and Level Shift by Non-Gaussian Distribution*

By extending the trend model in Eq. (29), we obtain a model which can treat the changes of slope and level separately. However, this model still does not take into account the sudden changes like the one seen after the Black Monday, and it is very difficult to treat it properly. This is because, by the Gaussian distributions used for system noise e_{n1} and e_{n2}, the rate of the occurrence of large noise are completely specified by its variance (τ_1^2 or τ_2^2), and thus cannot express exceptionally occurred sudden fall of stock prices. In this section, we shall introduce non-Gaussian distribution for system noise. As a non-Gaussian distribution for that purpose, Cauchy distribution

$$p(v|\tau^2) = \frac{\tau}{\pi} \frac{1}{v^2 + \tau^2} \tag{32}$$

is frequently used. Compared with the Gaussian distribution, the density function of the Cauchy distribution has higher density near origin and heavier tails and converges to 0 more slowly as the absolute value of v becomes large. Therefore this distribution naturally express the situation that the for most cases the noise is small but it may take very large or small noise with small probability (Kitagawa (1999), Kitagawa and Matsumoto (1996)). In the context of financial time series, it means that the non-Gaussian distribution model can naturally take into account of the occurrence of sudden fall or increase with very low probabilities.

As a further generalization of the distribution, we can consider the type VII Pearson family of distributions with the density function

$$p(v|\tau^2, b) = \frac{\tau^{2b-1}\Gamma(b)}{\Gamma(b-1/2)\Gamma(1/2)} \frac{1}{(v^2+\tau^2)^b}. \tag{33}$$

In this distribution, we can change the shape of the distribution by specifying the parameter b (Kitagawa 1987).

Further, by using the following mixture distribution, it is possible to express a situation where the probabilities of the sudden fall and sudden raise are different

$$p(v) = (1-\alpha)\phi_1(v) + \alpha\phi_2(v), \tag{34}$$

where α is the probability of occurring unusual changes, ϕ_1 is the distribution for normal situation, ϕ_2 is the distribution for the abnormal situation. As ϕ_1, we usually use a normal distribution with unknown variance. On the other hand, as ϕ_2 we may use a normal distribution with larger variance or the one with non zero mean value. We may also use a uniform distribution expressing the situation where in case of sudden fall or raise, noise may take any value. In the analysis below, we used a uniform distribution with probability only in negative region assuming that the abnormality occur only in the negative side,

$$\phi_2(v) \sim U(-d, 0). \tag{35}$$

Of course, it is also possible to use a mixture of more than two distributions. Also, we may model by using type IV Pearson family of distributions which has unsymmetric density (Nagahara 1996).

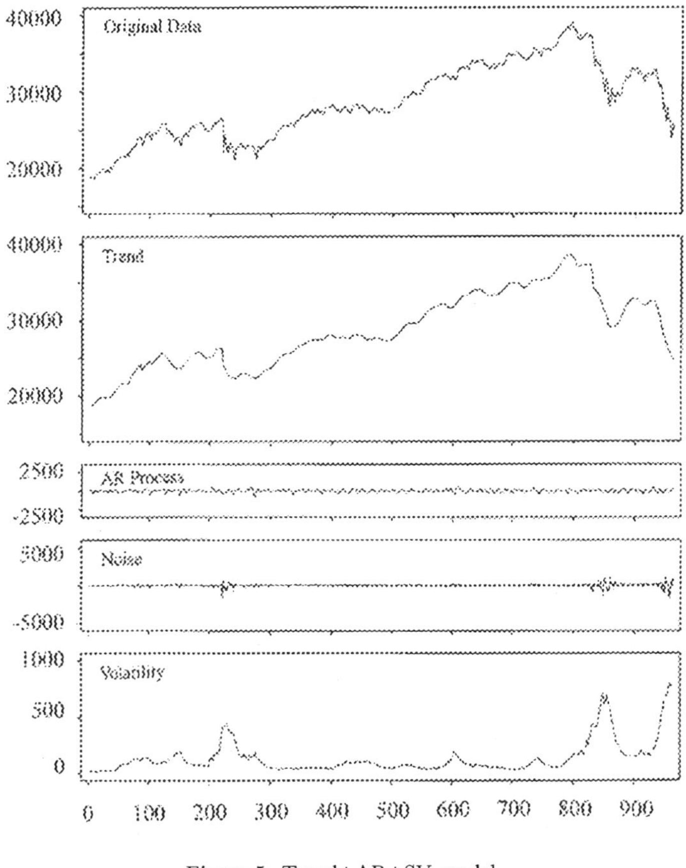

Figure 5: Trend+AR+SV model

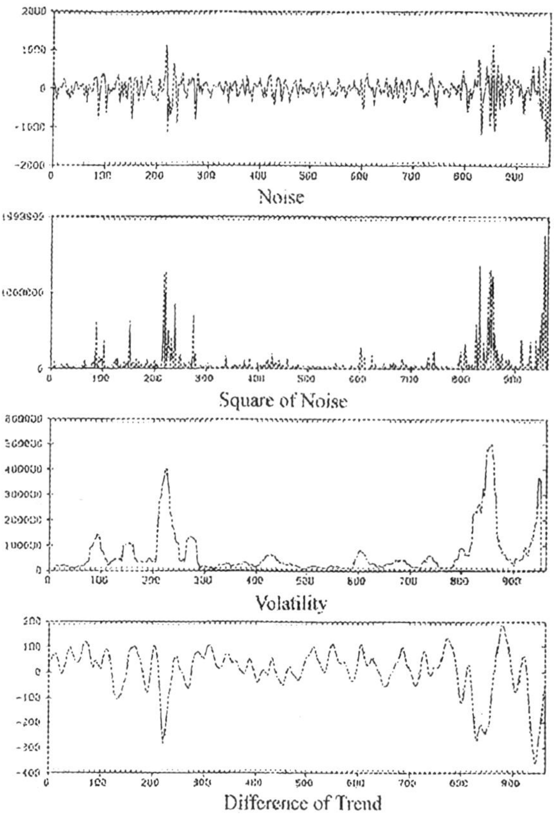

Figure 6: Volatility vs Difference of Trend

6.3 Examples

As shown in Table 1, similarly to the case of Figure 4, when we used the Cauchy distribution instead of the Gaussian distribution for system noise of the trend model with stochastic volatility, we have AIC=13648, indicating that this model is worse than the standard model. This indicates that the distribution of large deviations are nonsymmetric. On the other hand, by changing the model to Eq. (31) which has a level shift and using a mixture of Gaussian and uniform distributions for system noise, AIC value decreases by 50 and we have AIC=13553. Further, by applying the model with trend, stationary component and the noise distributed as mixture distribution, AIC is further reduced by 140 (AIC=13412) indicating the significant improvement of the fit. Figure 5 shows the trend, stationary AR component, noise and the volatility estimated by this model. It can be seen that the time change of the volatility is very large and it increases as much as 10 times of the normal time after the Black Monday and the crush of bubble.

On the other hand, Figure 6 shows the first difference of the trend Δt_n, squares of the noise series w_n^2, and an estimate of the local variance obtained by smoothing it. Comparison between the local variance and the difference of the trend reveals a negative correlation. Figure 7 shows the scatter plot of log-variance versus the difference of the trend. This figure also indicates the negative correlation between the variance and the difference of the trend.

The above argument suggests the possibility of obtaining better model by introducing a model which takes into account that the variance of the series fluctuating around the trend is time varying and that its variance is affected by the change of the trend.

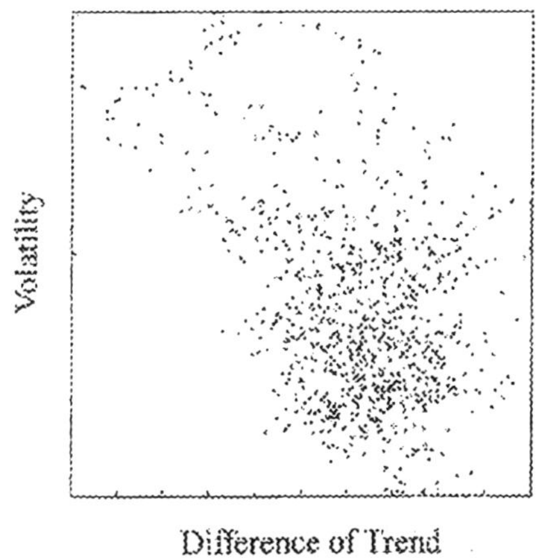

Figure 7: Relationship between Trend and Volatility

7 A Model Taking Account of the Relation Between the Changes of Trend and Volatility

The state space model is unified tool to express various kind of time series models, and actually by using the state space model, it is possible to generalize the stochastic volatility model in various ways. In this section, we generalize

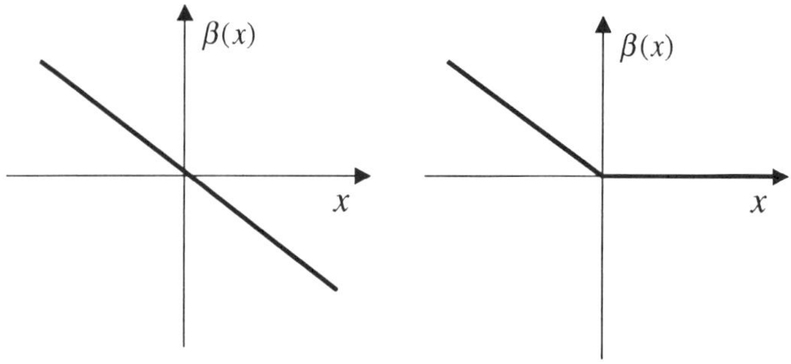

Figure 8: Various function forms: (a) Linear, (b)nonlinear

the model to take into account the relation between the difference of the trend and the volatility.

7.1 Effect of the Change of Trend to the Variance

Taking into account of the effect of the change of trend to that of volatility shown in Figures 6 and 7, we consider the following model;

$$\log \sigma_n^2 = \log \sigma_{n-1}^2 + \beta(\Delta_{n-1}) + u_n \tag{36}$$

where, $\beta(x)$ is, in general, a nonlinear function of x. As shown in Figure 8 (a), if $\beta(x) = cx$, it means that the effect to the change of log-variance is proportional to the change of the trend. In particular, if c=0, then it is assumed that the changes of trend and the variance are independent. On the other hand, if we use a nonlinear function such as (b), the logarithm of the variance increases only when the trend decreases and that its increase is proportionally to the decrease of the trend.

Then the basic model and the component models can be expressed as

$$\begin{aligned} y_n &= T_n + p_n + \sigma_n w_n \\ T_n &= T_{n-1} + \delta T_{n-1} + v_{sn} + v_{\ell n} \\ \delta T_n &= \delta T_{n-1} + v_{sn} \\ \log \sigma_n^2 &= \log \sigma_{n-1}^2 + \beta(\delta_{n-1}) + e_{2n} \\ p_n &= \sum_{j=1}^{m} a_j p_{n-j} + e_n. \end{aligned} \tag{37}$$

For Nikkei 225 data, the AIC of the model is 13352, and is decreased by 60 compared with the model which assumes that the two components are independent. Figure 9 shows the estimated trend, noise and the volatility obtained with this model.

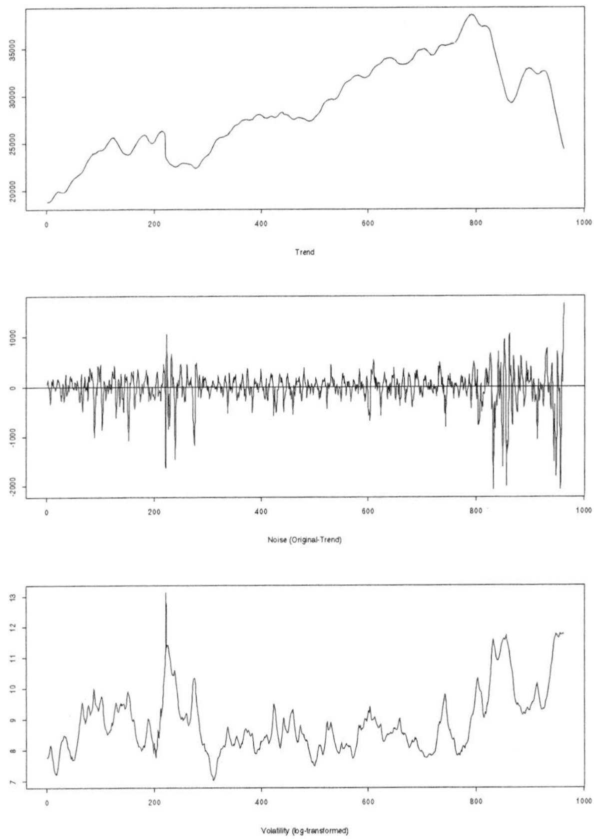

Figure 9: Nikkei225 - Correlation model (Trend and Volatility)

7.2 *Effect of Variance to System Noise*

In the previous subsection, we modeled the series by assuming that the change of the trend affects to the mean value of the change of the log-variance. This type of modeling will be useful when it is actually seen that the sudden decrease of the trend results in the increase of the variance. However, the model can be also generalized by introducing a model in which the change of the trend directly affect the distribution of the system noise, *i.e.*,

$$\begin{aligned}\log \sigma_n^2 &= \log \sigma_{n-1}^2 + u_n \\ u_n &\sim (1 - \beta(\delta T_{n-1}))\phi_0(u) + \beta(\delta T_{n-1})\phi_1(u).\end{aligned} \tag{38}$$

In this model, the mixture weight of two distributions $\beta(x)$ changes with the value of δT_n. Therefore, by defining $\beta(x)$ with $0 \leq \beta(x) \leq 1$, it is possible to change the distribution of $\log \sigma_n^2$ depending on the value of δT_n. The AIC of this model was 13339, further reduction of the AIC value was achived compared with the models shown in the previous subsection.

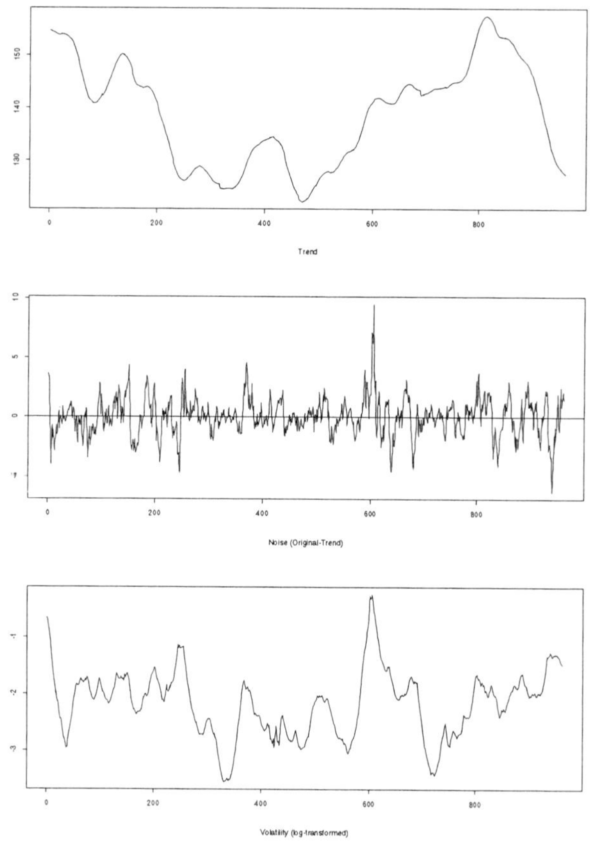

Figure 10: Exchange rate - Correlation model (Trend and Volatility)

8 Analysis of Exchange Rate Data

So far we analyzed the Nikkei 225 stock index data. We applied the same method for the analysis of Japan-U.S. exchange rate (January 1, 1987 – August 31, 1991). Figure 10 shows the results. Concerning the analyzed interval, the change of the volatility is small compared with that of the stock price data. Figure 11 shows, similarly to Figure 7, the plot of the volatility vs difference of the trend. Negative correlation between the volatility and the difference of the trend is not seen in this case, suggesting that the presence of negative correlation seen in Figure 7 does not always occur.

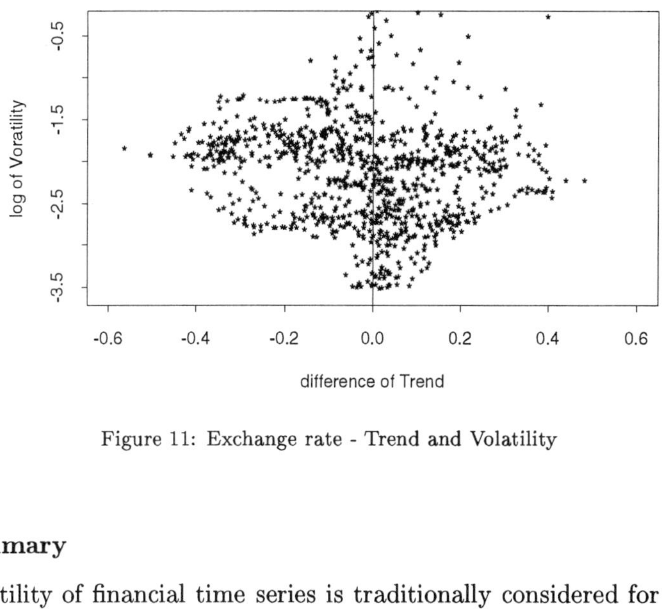

Figure 11: Exchange rate - Trend and Volatility

9 Summary

The volatility of financial time series is traditionally considered for stock returns defined as the difference of the series after taking logarithm of the original series. In this paper, we present models that decompose the original series directly into trend, stationary component and white noise with time-varying variance. These models can be expressed by nonlinear state space model and its state vector can be estimated by Monte Carlo filter and smoothing algorithms. Since the state vector of these models contain the trend, stationary component and volatility, by the Monte Carlo smoothing, it is possible to estimate the volatility of the series automatically.

The analysis of Nikkei 225 data in latter half of 1980's shows negative correlation between the change of the volatility and the change of the trend. Therefore, we modeled the series by assuming that the change of the trend affects the change of the volatility. By these modifications, the values of AIC gradually decreases and eventually obtained significantly better model compared with the standard trend model.

We did the same analysis on the exchange rate data. In this case, however, we could not find any significant relation between the change of the trend and that of the volatility.

References

1. H. Akaike et al., *Computer Science Monographs*, No. 11. (1985).
2. B.D.O. Anderson and J.B. Moore, *Optimal Filtering*, (Prentice-Hall, 1979).
3. R.F. Engle, Econometrica **50**, 987 (1982)
4. N.J. Gordon, D.J. Salmond and A.F.M. Smith, IEEE Proc.-F **140**, 107 (1993).
5. A.C. Harvey, *Forecasting, Structural Time Series Models and the Kalman Filter*, (Cambridge University Press, 1989).
6. A. Harvey and N. Shepherd,*Jour. Buss. Econ. Statist.* **14**, 429 (1996).
7. M. Hürzeler and H.R. Künsch, *Jour. Comp. Garph. Statist.* **7**, 175 (1998).
8. E. Jacquier, N. Polson and P.E. Rossi, *Jour. Buss. Econ. Statist.* **12**, 371 (1994).
9. G. Kitagawa, *Jour. Amer. Statist. Assoc.* **82**, 1032 (1987).
10. G. Kitagawa, JCGS **5**, 1 (1996).
11. G. Kitagawa, JASA **93**, 1203 (1998).
12. G. Kitagawa, *The Practice of Time Series Analysis*, eds., H. Akaike and G. Kitagawa, 181, (Springer-Varlag, New York, 1999).
13. G. Kitagawa, S. Sato and Y. Nagahara, *Kinyu Kenkyu* (*Monetary and Economic Studies*), *Bank of Japan(in Japanese)* **18**, 45 (1998).
14. G. Kitagawa and W. Gersch, JASA **79**, 378 (1984).
15. G. Kitagawa and N. Matsumoto, JASA **91**, 521 (1996).
16. Y. Nagahara., *Financial Eng. and the Japanese Markets* **3**, 121 (1996).
17. Y. Nagahara and G. Kitagawa, *Jour. of Comp. Finance* **2**, 33 (1999).

NONPARAMETRIC ESTIMATION AND BOOTSTRAP FOR FINANCIAL TIME SERIES

J.-P. KREIß
Institut für Mathematische Stochastik
Technische Universität Braunschweig
Pockelsstrasse 14
D-38106 Braunschweig
Germany
E-mail: j.kreiss@tu-bs.de

In this paper nonparametric statistical methods for financial time series are investigated. In the beginning it is argued that conditional mean and conditional variance (volatility) of the underlying process are quantities of interest. After presenting briefly the idea of local polynomial estimation and giving an asymptotic result of the behaviour of such kernel estimates for quite general mixing observation we deal in the main part of the paper with the construction of confidence intervals for the conditional mean and conditional variance or deviation function. Since usual pointwise confidence intervals are not completely satisfactory it is suggested to use simultaneous confidence bands. For both pointwise and simultaneous confidence bands a bootstrap procedure (wild bootstrap) is proposed, which can be used to calculate critical values. Besides theoretical results we apply the proposals to real financial data.

1 Introduction

There exists a huge literature on financial time series. Following the seminal work of Black and Scholes (1973) and Merton (1973) an adequate model for the price S_t at time t of a financial asset, e.g. stock price, stock index or exchange rate, is given through a so-called geometric Brownian motion with volatility σ and drift μ, i.e.

$$S_t = S_0 e^{\sigma W_t + (\mu - \sigma^2/2)\cdot t} \ , \ t \in [0,T] \ . \tag{1}$$

(W_t) denotes a standard Brownian motion. This model immediately leads to the following discretized version for so-called log-returns R_t

$$R_t = \log S_t - \log S_{t-1} = \mu - \sigma^2/2 + \sigma \cdot e_t \ , \ t = 1, 2, \dots \ , \tag{2}$$

where the real-valued random variables (e_t) are independent and normally distributed with zero mean and unit variance. One feature of this model is that the conditional mean of S_t given the past is equal to $S_{t-1} \cdot e^{\mu}$. Thus μ can be interpreted as the average yield rate for the period $t-1$ to t. The fundamental time series model (2) for the log-returns has been extended into several directions. One main extension was to model the volatility σ no longer by a constant but by a function depending on the past. Knowledge of this volatility function is extremely crucial in the theory of option pricing. Another point was to drop the assumption of normality for (e_t). In this respect the introduction of ARCH-processes (autoregressive with conditional heteroskedasticity) by Engle (1982) and generalizations of it, like GARCH (generalized ARCH) (cf. Bollerslev (1986)), EGARCH, IGARCH, FIGARCH and so on, have gained increasing prominence for modelling financial time series in recent years. Generally speaking, these models are defined as follows

$$R_t = \mu + \varepsilon_t \ , \ t = 1, 2, ..., \tag{3}$$

where $\varepsilon_t = \sigma_t e_t$ for an independent and identically distributed (i.i.d.) sequence (e_t), sometimes assumed to be normally distributed, with zero mean and unit variance. In the GARCH-model σ_t fulfills the following recursion

$$\sigma_t = \sqrt{\alpha_0 + \sum_{i=1}^{p} \alpha_i \varepsilon_{t-i}^2 + \sum_{j=1}^{q} \beta_j \sigma_{t-j}^2} \ . \tag{4}$$

Of course some additional assumptions on the non-negative coefficients α_i and β_j are needed in order to ensure for example stationarity of the model (3) with volatility according to (4). If $q = 0$ we refer to the ARCH(p)-model. In applications it seems that quite often an ARCH(1)- or a GARCH(1,1)-process is used, e.g. to model log-returns of exchange rates. The GARCH(1,1)-process is stationary if $\alpha_0 > 0$ and $\alpha_1 + \beta_1 < 1$ (of course $\alpha_1, \beta_1 \geq 0$). Engle, Lilien and Robins (1987) introduced the further generalization to so-called ARCH-in-mean models, which has been extended to GARCH-in-mean and investigated by Duan (1995). He also presented a link of this econometric model with the literature of option pricing.

If we go back to the basic model (2) and replace σ by a volatility process σ_t, then it seems quite natural to lay down the following discrete financial time series model for the log-returns of a financial asset

$$R_t = \mu - \sigma_t^2/2 + \sigma_t e_t \ , \ t = 1, 2, ... \ . \tag{5}$$

Here σ_t denotes a random variable which is measurable with respect to the past up to time $t-1$, represented through the σ-field $\mathcal{F}_{t-1}$. The special structure of the conditional mean of the log-returns as a simple function of the conditional variance (i.e. the volatility) ensures (as above), that the conditional mean of the underlying asset price S_t given the past up to time $t-1$ equals $S_{t-1} \cdot e^{\mu}$. In some situations it may make sense to assume that the conditional gain $E[S_t|\mathcal{F}_{t-1}]$ also depends on the volatility, e.g. in such a way that a more risky

asset with high volatility may have a higher conditional mean. Duan's (1995) proposal in introducing an additional unit risk premium in the conditional mean, while using the GARCH-modelling, is exactly in this direction.
Thus, as a more general discrete time series model for financial time series may serve the following

$$R_t = \mu_t + \sigma_t \cdot e_t \ , \ t = 1, 2, \ldots \ . \tag{6}$$

Here μ_t and σ_t are assumed to be random variables measurable with respect to the past $\mathcal{F}_{t-1}$ and (e_t) denotes a sequence of i.i.d. random variables with zero mean and unit variance.
One relatively simple case is where σ_t as well as μ_t are nothing else but non-negative (resp. real-valued) functions of lagged log-returns. This situation is just a nonparametric autoregression with conditional heteroscedasticity, i.e.

$$R_t = m(R_{t-1}) + s(R_{t-1}) \cdot e_t \ , \ t = 1, 2, \ldots \ . \tag{7}$$

Another possibility is that the conditional variance of R_t is a smooth function of the last innovation $\varepsilon_{t-1} = \sigma_{t-1} e_{t-1}$, i.e.

$$R_t = \mu_t + \varepsilon_t = \mu_t + \sigma_t e_t \ , \ t = 1, 2, \ldots \text{ and } \sigma_t = s(\varepsilon_{t-1}) \ . \tag{8}$$

Here the innovations (ε_t) itself follow an ARCH(1)-model. μ_t may be specified to be a parametric or a nonparametric function of R_{t-1}, or a higher number of lagged values of the log-returns, or (more complicated) a parametric or nonparametric function of σ_{t-1}. The assumption that μ_t is a constant or even equal to zero seems not to be reasonable as is argued above. Of course the model (8) is much more involved than the nonparametric autoregressive setup (7), because we end up with parametric or even worse nonparametric regression with errors in variables, since ε_t as well as σ_t are not observable.

Since the prices of derivative securities depend crucially on the form of the instantaneous volatility of the underlying process, we think that in modeling one should leave this functions as unrestricted as possible. With this respect we completely agree with papers of Ait-Sahalia (1996, 1998). At least it is suggested to run a completely nonparametric model parallel to the parametric analysis and to compare parametric and nonparametric drift and volatility estimates in order to validate parametric models, which are with no doubt extremely helpful in practice. In other words it is suggested to use nonparametric techniques in the area of model selection and model testing, either visually or quantitatively.
We are going to present the main ideas on the basis of the most simplest nonparametric model (7).

The outline of the paper is as follows. In Section 2 we discuss nonparametric estimates for the conditional mean and conditional variance of a stochastic process.

In order to validate parametric models visually, we need to have confidence intervals or even better simultaneous confidence bands for our nonparametric estimators. Since asymptotic theory for simultaneous confidence bands is not available yet and far from to be easy, we suggest quite a flexible bootstrap approach (wild bootstrap) in order to calculate corresponding critical values. This is the contents of Section 3. It also contains some simulation results and applications to real data.

2 Nonparametric Estimation for Time Series

In this part of the paper we briefly recall nonparametric estimation for dependent observations. To this end let us assume that we have a realization $R_0, R_1, ..., R_T$ of a stationary time series at hand. We focus our attention to local polynomial estimators. Such estimators have been introduced in a paper by Stone (1977). Tsybakov (1986), Korostelev and Tsybakov (1993), Fan (1992, 1993) and Fan and Gijbels (1992,1995) discussed the behaviour of local polynomial estimation for nonparametric regression in full detail. Masry (1996) and Härdle and Tsybakov (1997) applied the idea to nonparametric autoregression. A local polynomial estimator $\hat{m}_h(x)$ of $m(x)$ is given as $\hat{a}_0$, where $(\hat{a}_0, ..., \hat{a}_p)$ minimzes

$$\sum_{t=1}^{T} K\left(\frac{x - R_{t-1}}{h}\right)\left(R_t - \sum_{q=0}^{p} a_q \left(\frac{x - R_{t-1}}{h}\right)^q\right)^2 . \tag{9}$$

Here $K : \mathbf{R} \to [0, \infty)$ denotes a probability density, usually compactly supported and symmetric, as we will assume throughout the whole paper, (e.g. the quartic kernel $K(x) = 15/16(1 - x^2)^2 1_{[-1,1]}(x)$) and $h > 0$ denotes the smoothing parameter, the so-called bandwidth. It is an easy task to obtain that

$$\hat{m}_h(x) = \sum_{t=1}^{T} w_h(x, R_{t-1}, \{R_0, \ldots, R_{T-1}\}) R_t = \left[(D_x' K_x D_x)^{-1} D_x' K_x \underline{R}\right]_1 ,$$

where $\underline{R} = (R_1, \ldots, R_T)'$,

$$D_x = \begin{pmatrix} 1 & \frac{x-R_0}{h} & \cdots & (\frac{x-R_0}{h})^p \\ \vdots & \vdots & \ddots & \vdots \\ 1 & \frac{x-R_{T-1}}{h} & \cdots & (\frac{x-R_{T-1}}{h})^p \end{pmatrix},$$

$$K_x = \text{diag}\left[K(\frac{x-R_0}{h}),\dots,K(\frac{x-R_{T-1}}{h})\right]$$

and $[b_0, ..., b_p]_1 = b_0$.
If we choose $p = 0$ then we obtain the usual Nadaraya-Watson kernel smoother for the conditional expectation $E[R_t|R_{t-1} = x]$, namely

$$\hat{m}_h^0(x) = \frac{\sum_{t=1}^T K(\frac{x-R_{t-1}}{h})R_t}{\sum_{t=1}^T K(\frac{x-R_{t-1}}{h})} . \tag{10}$$

The so-called local linear estimator is obtained if we choose $p = 1$. This estimator can be written in the following form

$$\hat{m}_h^1(x) = \hat{m}_h^0(x) + \hat{b}_h(x)(x - \hat{\xi}_h(x)) , \tag{11}$$

where $\hat{b}_h(x)$ turns out to be an estimator of the first derivative of m and $\hat{\xi}_h(x) = \frac{\sum_{t=1}^T K(\frac{x-R_{t-1}}{h})R_{t-1}}{\sum_{t=1}^T K(\frac{x-R_{t-1}}{h})}$. $\hat{\xi}_h(x)$ can be interpreted as a center of the design points around x and the second summand in formula (11) is a correction of the local polynomial estimator for nonsymmetric design, which indicates that local linear estimates overcome the difficulties of the Nadaraya-Watson kernel estimates at the boundary of the design.

An estimator of the conditional deviation $s(x) = \sqrt{\text{Var}[R_t|R_{t-1} = x]}$ can be obtained as follows

$$\hat{s}_h(x) = \left(\sum_{t=1}^T w_h(x, R_{t-1}, \{R_0, ..., R_{T-1}\})(R_t - \hat{m}_h(R_{t-1}))^2\right)^{1/2} . \tag{12}$$

See also Fan and Yao (1998) for an efficient way to estimate the conditional variance.

In order to obtain the asymptotic behaviour of these local polynomial smoothers we need some assumptions. Denote $m(x) = E[R_t|R_{t-1} = x]$.

(A1) $\{R_t : t \geq 0\}$ is a (strictly) stationary time series. We denote by F the common cumulative distribution function of the R_t with density p. Furthermore, we assume absolute regularity (i. e. β-mixing) for $\{R_t\}$ and that the β-mixing coefficients decay at a geometric rate.

(A2) $E[\varepsilon_t^2|R_{t-1} = x] = s^2(x) \in (0, \infty)$ for all $x \in \mathbf{R}$, $E[\varepsilon_t^4|R_{t-1} = x]$ is bounded for all x.

(A3) The stationary density p of R_t is continuously differentiable, bounded and satisfies $p(x) \geq C > 0$ for all $x \in [a, b]$.

(A4) m is $p+1$-times differentiable with $\sup_{x \in [a-\delta, b+\delta]}\{|m^{(p+1)}(x)|\} < \infty$, for some $\delta > 0$.

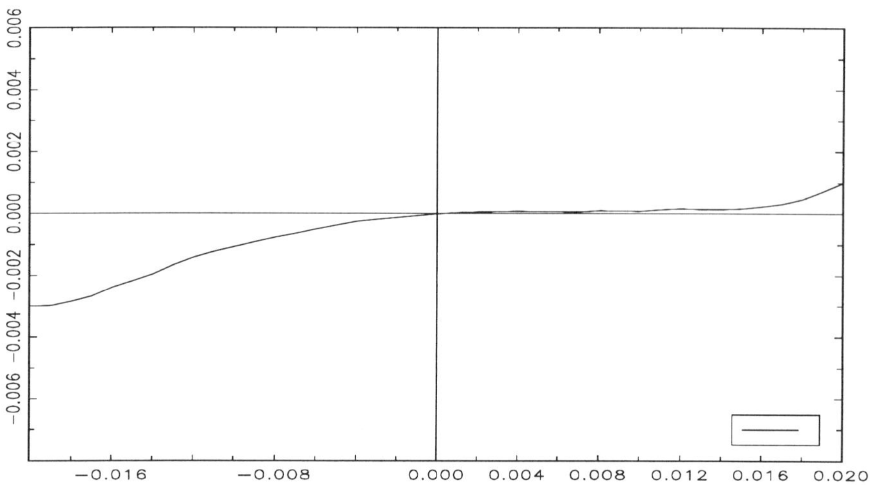

Figure 1
Local constant estimate $\hat{m}_h(x), h = 0.012$, of the conditional mean of daily log-returns for the German stock index DAX (1990-92)

Theorem 1: Assume (A1)-(A4). We have the following asymptotic distributions for the local polynomial estimator. For all $x \in [a,b]$:

$$\sqrt{Th}(\hat{m}_h(x) - m(x)) \Rightarrow \mathcal{N}(b_p(x), \tau_p^2(x)) \tag{13}$$

where for p odd and $h = h(T) \sim T^{-1/(2p+3)}$:

$$b_p(x) = \frac{m^{(p+1)}(x)}{(p+1)!} \cdot C_p^1 \text{ and } \tau_p^2(x) = \frac{s^2(x)}{p(x)} \cdot C_p\,, \tag{14}$$

while for p even and $h = h(T) \sim T^{-1/(2p+5)}$:

$$b_p(x) = \left\{ \frac{m^{(p+1)}(x)p'(x)}{(p+1)!p(x)} + \frac{m^{(p+2)}(x)}{(p+2)!} \right\} \cdot C_p^1 \tag{15}$$

and $\tau_p^2(x)$ as above. The constants C_p^1 and C_p only depend on the kernel K and $C_{2r} = C_{2r+1}$, $C_{2r}^1 = C_{2r+1}^1$ for all integers r.

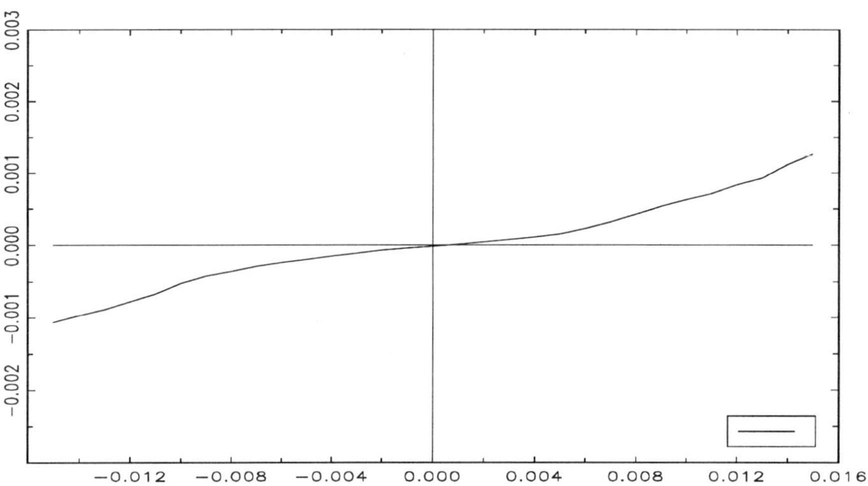

Figure 2
Local constant estimate $\hat{m}_h(x), h = 0.008$, of the conditional mean of daily log-returns for the US \$-DM FX rate (1989-98)

Remark: (i) A similar result for nonparametric regression is well-known in the literature (cf. Fan and Gijbels (1995)).
(ii) It is noteworthy, that the asymptotic distribution for the nonparametric estimators does not reflect the dependence structure and is exactly the same as in the independent regression model with stochastic design, where the design distribution is equal to the stationary density of the process. This fact has been shown for the first time in Robinson (1983).
(iii) Under additional smoothness assumptions upon s, similar results for kernel smoothers for s and s^2 are available.

Sketch of the proof of Theorem 1: Make use of the following decomposition into variance and bias part

$$\begin{aligned}&\sqrt{Th}\,(\hat{m}_h(x) - m(x))\\ =\;& \sqrt{Th}\sum_t w_h(x, R_{t-1}, \{R_0, ..., R_{T-1}\})(R_t - m(R_{t-1}))\\ &+\sqrt{Th}\sum_t w_h(x, R_{t-1}, \{R_0, ..., R_{T-1}\})(m(R_{t-1}) - m(x))\end{aligned}$$

$$= \sqrt{Th}\left(\sum_t \bar{w}_h(x,R_{t-1})(R_t - m(R_{t-1})) + \sum_t \bar{w}_h(x,R_{t-1})(m(R_{t-1}) - m(x))\right)$$
$$o_P(1)\ ,$$

where we abbreviate $\bar{w}_h(x,R_{t-1}) = \sum_{q=0}^{p} d_q^{\infty}(x) K\left(\frac{x-R_{t-1}}{h}\right)\left(\frac{x-R_{t-1}}{h}\right)^q$ and $d_q^{\infty}(x) = \left(\left(ED_x^{'}K_xD_x\right)^{-1}\right)_{1,q+1}$, cf. Neumann and Kreiss (1998). A CLT for mixing random variables (cf. Politis et al. (1997)) and a Taylor expansion of m together with the symmetry of K now lead to the desired result. □

To illustrate the kernel smoothing we choose two examples, namely the daily log-returns of the german stock index DAX from 1990 until 1992 and the daily log-returns of the US-Dollar/DM exchange rate for a ten years period (1989-1998). The estimated conditional expectations $E[R_t|R_{t-1} = x]$ are shown in Figure 1, Figure 2, respectively.

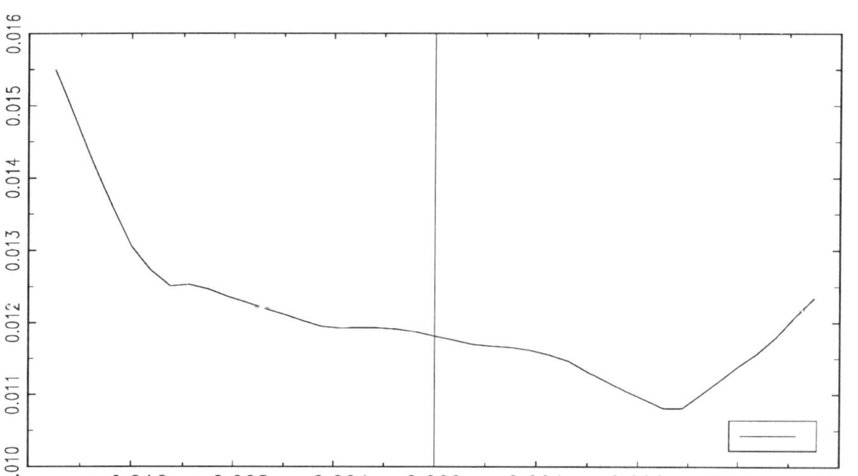

Figure 3
Local constant estimate $\hat{s}_h(x), h = 0.010$, of the conditional deviation of daily log-returns for the DAX (1990-92)

The bandwidth in both Figures 1 and 2 is chosen from a cross-validation criterion. Of course one may obtain similar plots for the estimates of the conditional deviation (i.e. the square root of the conditional variance) (cf. Figures 3 and 4). We would like to judge on these figures, which structure of the functions $\hat{m}_h$

and/or $\hat{s}_h$ is really significant. To answer such questions we need confidence bands. One usual way is to plot the pointwise standard deviation or a $(1-\alpha)$ pointwise confidence interval around the estimates. Such pointwise confidence bands, or better more precisely pointwise confidence intervals, can be derived from asymptotic theory or from a bootstrap procedure as will be shown later.

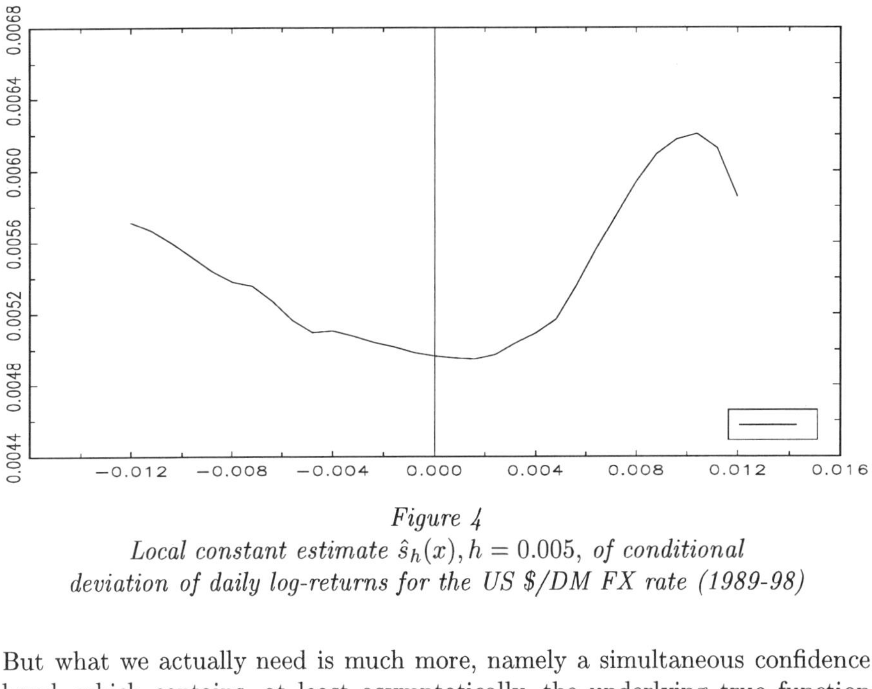

Figure 4
Local constant estimate $\hat{s}_h(x), h = 0.005$, *of conditional deviation of daily log-returns for the US \$/DM FX rate (1989-98)*

But what we actually need is much more, namely a simultaneous confidence band, which contains, at least asymptotically, the underlying true function m or s with probability not less than $1-\alpha$. This means that we have to approximate not only the pointwise distribution of the estimators, but the distribution of quantities like

$$\sup_{x\in[a,b]} |\hat{m}_h(x) - m(x)| \text{ or } \sup_{x\in[a,b]} \frac{|\hat{m}_h(x) - m(x)|}{\sqrt{\hat{V}_h(x)}}, \tag{16}$$

where $\hat{V}_h(x)$ denotes an estimator of the asymptotic variance of $\hat{m}_h(x)$.
From the proof of Theorem 1 we can see that a reasonable estimator of the variance of the local polynomial estimator $\hat{m}_h$ of the conditional mean function

m is given through

$$\sum_{t=1}^{T} w_h^2(x, R_{t-1}, \{R_0, ..., R_T\}) \left(R_t - \hat{m}_h(R_{t-1})\right)^2 \; . \tag{17}$$

Since asymptotic theory for kernel smoothers is not that accurate, even for moderate sample sizes, and not available for the distribution of the quantities (16) we propose in the next section a simple bootstrap procedure in order to approximate these distributions asymptotically. Since we want to be as general as possible, we are looking for a bootstrap procedure which even works if an underlying model of the form

$$R_t = m(R_{t-1}) + s(R_{t-1}) \cdot e_t \; ,$$

(e_t) i.i.d with zero mean and unit variance, does not hold.
In the case that we are interested in the conditional variance function instead of the conditional mean function then we have to apply the same methodology to quantities like

$$\sup_{x \in [a,b]} |\hat{s}_h^2(x) - s^2(x)| \text{ or } \sup_{x \in [a,b]} \frac{|\hat{s}_h^2(x) - s^2(x)|}{\sqrt{\hat{W}_h(x)}} \; ,$$

where an asymptotically consistent estimator of the variance is now given through

$$W_h(x) = \sum_{t=1}^{T} w_h^2(x, R_{t-1}, \{R_0, ..., R_T\}) \left((R_t - \hat{m}_h(R_{t-1})^2 - \hat{s}_h^2(R_{t-1})\right)^2 \; .$$

3 A Bootstrap Procedure for Nonparametric Estimators

Since we are dealing with nonparametric estimates, we are in the nice situation that up to first order the asymptotic behaviour of these estimates does not reflect the dependence structure at all, cf. Theorem 1. This means that we may use in the bootstrap world a regression setup, which is much easier to deal with. We have only to care about the correct design distribution, which has to mimic the stationary distribution of the underlying process. All these arguments strongly suggest to make use of the so-called wild bootstrap procedure.

To motivate this particular resampling scheme first note the different nature of the stochastic and the bias term. It is possible to consistently mimic the distribution of the stochastic term by the bootstrap as we will see below. In contrast, the bias can only be dealt with if some degrees of smoothness of

m are not used by $\hat{m}_h$. From nonparametric regression and density estimation it is known, that there are two main approaches to handle the bias problem, namely undersmoothing and explicit correction.

Let $R_0, R_1, ..., R_T$ denote a realization of the process. The wild bootstrap procedure starts from independently generated random variables $\eta_1, ..., \eta_T$ with $E\eta_1 = 0$ and $E\eta_1^2 = 1$, e.g. $\eta_t \sim \mathcal{N}(0,1)$. Of course it is far from to be necessary to use the standard normal distribution for the random variables η_t. Any other centered and standardized distribution and even a discrete distribution taking only two values can be chosen. [Often, for a higher order performance, the distribution of η_t is chosen such that additionally $E\ \eta_t^3 = 1$; for a discussion of this point and for choices of the distribution of η_t, compare Mammen (1992).] Using these random values we define independent bootstrap innovations according to $\varepsilon_t^* = (R_t - \hat{m}_h(R_{t-1})) \cdot \eta_t$, $t = 1, ..., T$. By definition we then have

$$E^*\varepsilon_t^* = 0 \quad , \quad E^*(\varepsilon_t^*)^2 = \hat{\varepsilon}_t^2 = (R_t - \hat{m}_h(R_{t-1}))^2 \ , \ t = 1, ..., T \ .$$

Here and in the following the notation E^* is used to underline the conditional character of the distribution $\mathcal{L}(\varepsilon_1^*, ..., \varepsilon_T^* | R_0, ..., R_T)$. An appropriate counterpart to model (7) in the bootstrap world is given through

$$R_t^* = \hat{m}_g(R_{t-1}) + \varepsilon_t^* \ , \ t = 1, ..., T \ .$$

Some remarks are now in order. Observe that the bootstrap observations R_t^* are (conditionally) independent, i.e. the dependence structure of the underlying process is not preserved. Moreover, in the bootstrap world, the model is a usual nonparametric regression model with (conditionally) fixed design. Here we introduce a second bandwidth g, which is quite essential if we intend to catch the bias term of the asymptotic distribution correctly. Generally speaking we have to choose $g \gg h$.

The bootstrap resample can be used to calculate

$$\hat{m}_h^*(x) = \sum_{t=1}^{T} w_h(x, R_{t-1}, \{R_0, ..., R_T\}) \cdot R_t^* \ .$$

From Franke, Kreiß and Mammen (1997) we have the following result.

Theorem 2: Under assumptions (A1) to (A4) we have for all $x \in [a, b]$ if $h \sim T^{-1/(2p+3)}$ for p odd and $h \sim T^{-1/(2p+5)}$ for p even and if $h/g \to 0$:

$$\sqrt{Th}\,(\hat{m}_h^*(x) - \hat{m}_g(x)) \Rightarrow \mathcal{N}(b_p(x), \tau_p^2(x))$$

in probability. The bias part $b_p(x)$ and the variance part $\tau_p^2(x)$ do exactly coincide with the quantities (14), (15) in Theorem 1.

To illustrate the result we display plots of pointwise 90% bootstrap confidence intervals around the estimates $\hat{m}_h$ for the DAX and the US \$-DM FX-rate (cf. Figures 5 and 6). From both Figures 5 and 6 one may be tempted to conclude that (with a confidence level of 90%) the assumption of a constant conditional mean function for both data sets has to be rejected. But such a decision should be based on a simultaneous confidence band and not on pointwise confidence intervals.

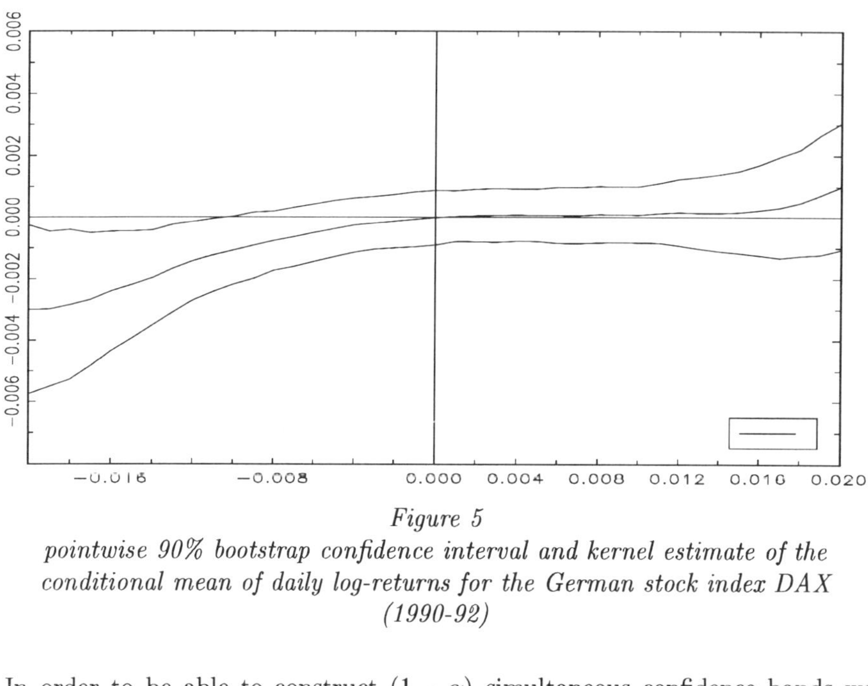

Figure 5
pointwise 90% bootstrap confidence interval and kernel estimate of the conditional mean of daily log-returns for the German stock index DAX (1990-92)

In order to be able to construct $(1 - \alpha)$ simultaneous confidence bands we establish a strong approximation of the nonparametric local polynomial estimators in the real and in the bootstrap world regarded as processes on some compact interval $[a, b]$. To do so we again split the centered kernel smoother into the variance and bias part. The most complicated part is to construct a strong approximation of the variance parts in the real and in the bootstrap world. Let $(r_0, ..., r_T)$ be the realization of $(R_0, ..., R_T)$ at hand. The exact strong approximation result can be cited from Neumann and Kreiss (1998).

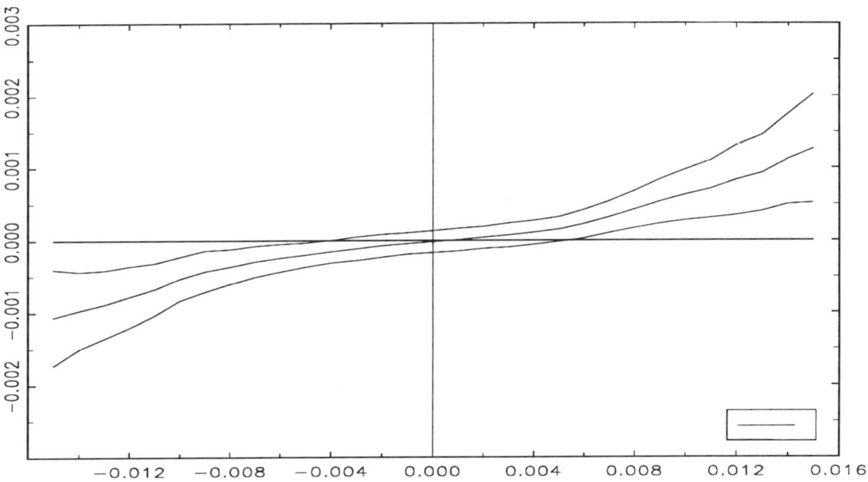

Figure 6
pointwise 90% bootstrap confidence interval and local constant estimate of the conditional mean of daily log-returns for the US \$-DM FX rate (1989-98)

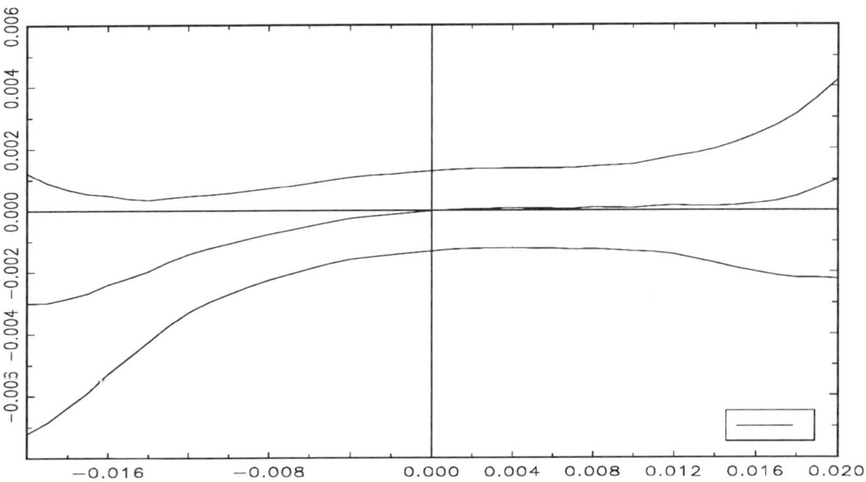

Figure 7
simultaneous 90% bootstrap confidence band and local constant estimate of the conditional mean for daily log-returns of the German stock index DAX (1990-92)

Theorem 3 (Neumann and Kreiss (1998)): Suppose that (A3), (A4) and

(A1') $\{R_t : t \geq 0\}$ is a (strictly) stationary time-homogeneous Markov chain. We denote by F the common cumulative distribution function of the R_t with density p. Furthermore, we assume absolute regularity (i. e. β-mixing) for $\{R_t\}$ and that the β-mixing coefficients decay at a geometric rate.

(A2') For all $M < \infty$ there exist finite constants C_M such that for $\varepsilon_t = R_t - m(R_{t-1})$: $\sup_{x \in \mathbf{R}} \left\{ E\left(|\varepsilon_t|^M \mid R_{t-1} = x \right) \right\} \leq C_M$

are fulfilled. Then we have (on a sufficiently rich probability space) that there exists a pairing of $R_0, \varepsilon_1, ..., \varepsilon_T$ with $\varepsilon_1^*, ..., \varepsilon_T^*$, such that

$$\sup_{x \in [a,b]} \left| \sum_t w_h(x, R_{t-1}, \{R_0, ..., R_T\}) \varepsilon_t - \sum_t w_h(x, r_{t-1}, \{r_0, ..., r_T\}) \varepsilon_t^* \right|$$
$$= \mathcal{O}_P\left((Th)^{-3/4} \log T \right)$$

holds uniformly in $(r_0, ..., r_T) \in \Omega_T$ where $P\{(R_0, ..., R_T) \in \Omega_T\} = \mathcal{O}(T^{-\lambda})$ for all $\lambda > 0$.

Remark: Actually, it can be seen from the proofs that a certain finite number M of uniformly bounded moments in (A2') would suffice, but for the sake of simplicity we decided to state the result in the given form.

With this result now we are able to tackle the construction of nonparametric simultaneous confidence bands for m. It is known, cf. Hall (1991), that first order asymptotic theory for the supremum of an approximating Gaussian process leads to an error in coverage probability of order $(\log T)^{-1}$. As can be seen from Neumann and Kreiss (1998), we obtain an algebraic rate of convergence by using the proposed wild bootstrap procedure.

In order to catch the bias term correctly we use an oversmoothed estimator $\hat{m}_g, g \gg h$, as in Theorem 2 for the underlying regression function in the bootstrap world. This has to be done in order to achieve that the (p+1)th order derivative $\hat{m}_g^{(p+1)}$ estimates $m^{(p+1)}$ consistently. Furthermore we denote by $\hat{V}(x)$ an estimator of the variance $V(x)$ of $\hat{m}_h(x)$, for example

$$\hat{V}(x) = \sum_t w_h^2(x, r_{t-1}, \{r_0, ..., r_T\})(r_t - \hat{m}_h(r_{t-1}))^2 .$$

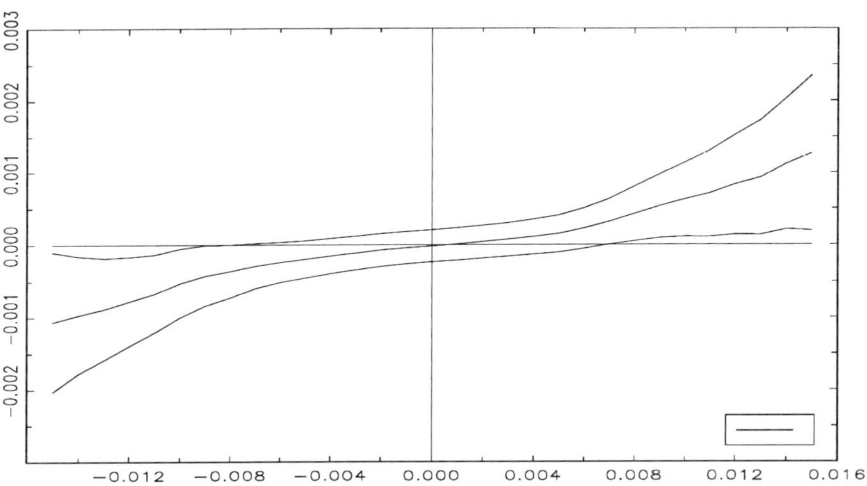

Figure 8
simultaneous 90% bootstrap confidence band and local constant estimate of the conditional mean of daily log-returns for the US \$-DM FX rate (1989-98)

Finally, if we denote by t_α^* the $(1-\alpha)$-quantile of the distribution of

$$\sup_{x\in[a,b]} |\hat{m}_h^*(x) - \hat{m}_g(x)| \,/ \sqrt{\hat{V}(x)}$$

then we obtain a bias-corrected simultaneous bootstrap confidence band for m with an asymptotic coverage probability of $1-\alpha$. This simultaneous band reads as follows

$$\left[\hat{m}_h(x) - \sqrt{\hat{V}(x)} \cdot t_\alpha^* \ , \ \hat{m}_h(x) + \sqrt{\hat{V}(x)} \cdot t_\alpha^*\right] \ . \tag{18}$$

The size of this band is proportional to the estimated standard deviation of $\hat{m}_h$, which seems to be quite reasonable. This band should follow in its size the local variability of $\hat{m}_h$. Thus the simultaneous confidence band can serve as a visual diagnostic tool to detect regions where it is difficult to estimate the underlying function either because of large variance of the $\varepsilon_t' s$ or because of too sparse a design.

Simultaneous bootstrap confidence bands for the conditional mean of log-returns for the DAX and the US \$-DM exchange rate are shown in Figures 7 and 8. It is clearly seen, that the hypothesis of a constant conditional mean function for the log-returns of the FX-rate can be rejected (confidence level 90%), while the decision for the DAX is not in the same direction. The method allows to choose the bandwidth h by any of the popular criteria like cross-validation. Usually data-driven bandwidths $\hat{h}$ approximate a certain non-random bandwidth h_T. If $(\hat{h} - h_T)/h_T$ converges with an appropriate rate, then the estimators $\hat{m}_{\hat{h}}$ and $\hat{m}_{h_T}$ are sufficiently close to each other, such that the above results remain valid.

In financial applications the interest probably is much more directed towards the conditional variance or deviation function (volatility) then towards the conditional mean function. Thus we are going to describe the construction of a reasonable bootstrap approximation of the distribution of the local polynomial estimator $\hat{s}_h$ defined in (12). Since the conditional mean cannot be assumed to be constant, we stay with the following model

$$R_t = m(R_{t-1}) + s(R_{t-1}) \cdot e_t \ .$$

Since we have the equality

$$(R_t - m(R_{t-1}))^2 \ = \ s^2(R_{t-1}) + s^2(R_{t-1}) \cdot (e_t^2 - 1) \ ,$$

where $(e_t^2 - 1)$ has zero mean and is assumed to have finite second moment $E(e_t^2 - 1)^2 = Ee_t^4 - 1$, which is usually larger than zero.

It is reasonable to apply the above methodology to the bivariate time series $(R_t, (R_t - \hat{m}_h(R_{t-1}))^2)$ and to regress nonparametrically the second component on R_{t-1}. The resulting nonparametric estimates for s, i.e.

$$\left(\sum_t w_h(x, R_{t-1}, \{R_0, ..., R_T\}) \, (R_t - \hat{m}_h(R_{t-1}))^2 \right)^{1/2}$$

together with a 90% pointwise bootstrap confidence interval are displayed in Figure 9 and 10, respectively. The wild bootstrap approximation of these pointwise distributions can be obtained quite similar as above. Again let $\eta_1, ..., \eta_T$ denote independent random variables with zero mean and unit variance. Define bootstrap innovations according to

$$\varepsilon_t^* = \left[(R_t - \hat{m}_{h_1}(R_{t-1}))^2 - \hat{s}_h^2(R_{t-1}) \right] \cdot \eta_t \ , \ t = 1, ..., T$$

and define bootstrap realizations as

$$Y_t^* \equiv \left[(R_t - \hat{m}_{h_1}(R_{t-1}))^2 \right]^* = \hat{s}_g^2(R_{t-1}) + \varepsilon_t^* \ , \ t = 1, ..., T \ .$$

Here h_1 denotes the bandwidth used for the auxiliary estimator $\hat{m}_{h_1}$, which may or may not be chosen independently of h, while the third bandwidth g

necessarily has to be chosen such that $h/g \to 0$, because of the same reason as above, namely that the bias part of the asymptotic distribution is caught correctly. Finally define

$$\hat{s}_h^* = \left| \sum_t w_h(x, R_{t-1}, \{R_0, ..., R_T\}) \cdot Y_t^* \right|^{1/2} .$$

As above, we have from Franke, Kreiß and Mammen (1997)

Theorem 4: Assume (A1)-(A4) and

(A5) s is $p+1$-times differentiable with $\sup_{x \in [a-\delta, b+\delta]} \{|s^{(p+1)}(x)|\} < \infty$, for some $\delta > 0$.

Then, if $h/g \to 0$ and $h \sim T^{-1/(2p+3)}$ (p odd), $h \sim T^{-1/(2p+5)}$ (p even), respectively,

$$d_K \left(\sqrt{Th}(\hat{s}_h^*(x) - \hat{s}_g(x)), \sqrt{Th}(\hat{s}_h(x) - s(x)) \right) \to 0 ,$$

as $T \to \infty$, in probability. Here d_K denotes the Kolmogorov distance, i.e. for two distributions P and Q the distance $d_K(P, Q)$ is defined as $\sup_{x \in R} |P(-\infty, x] - Q(-\infty, x]|$.

In analogy to the conditional mean function we display plots of 90% pointwise confidence intervals and simultaneous confidence bands for the conditional deviation function $s(x) = (\text{Var}(R_t | R_{t-1} = x))^{1/2}$ for daily log-returns of the DAX and the US \$ -DM exchange rate (cf. Figures 9 and 10 for pointwise confidence intervals and Figures 11 and 12 for simultaneous confidence bands).

Remark. It has been shown, that the conclusions of Theorems 1,2 and 4 does not depend on whether the underlying time series model is exactly of the form

$$R_t = m(R_{t-1}) + s(R_{t-1}) \cdot e_t \tag{10}$$

for a sequence of i.i.d. random variables (e_t) with zero mean and unit variance. Instead, some smoothness of the conditional mean function (the conditional variance function, respectively) and geometrically β-mixing suffice. This immediately implies that the wild bootstrap proposal of this paper is robust against model misspecification as far as pointwise distributions, i.e. pointwise confidence intervals, are addressed.

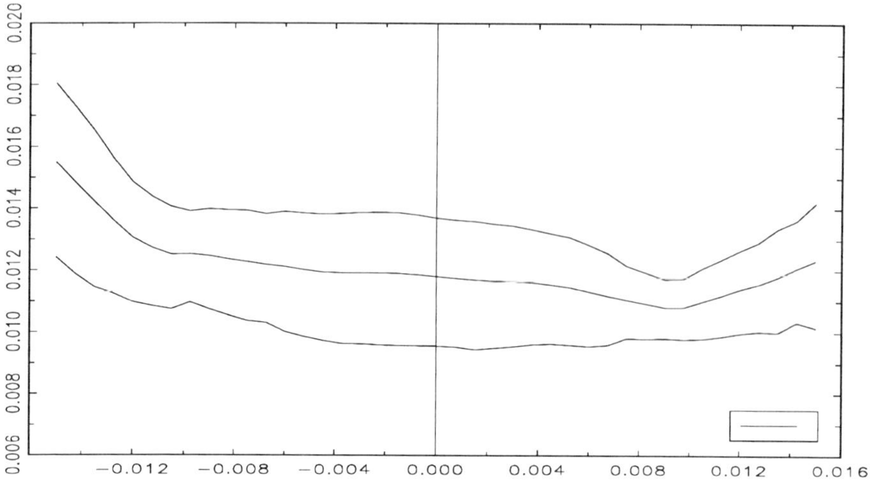

Figure 9
pointwise 90% bootstrap confidence band and local constant estimate of the conditional deviation for the German stock index DAX (1990-92)

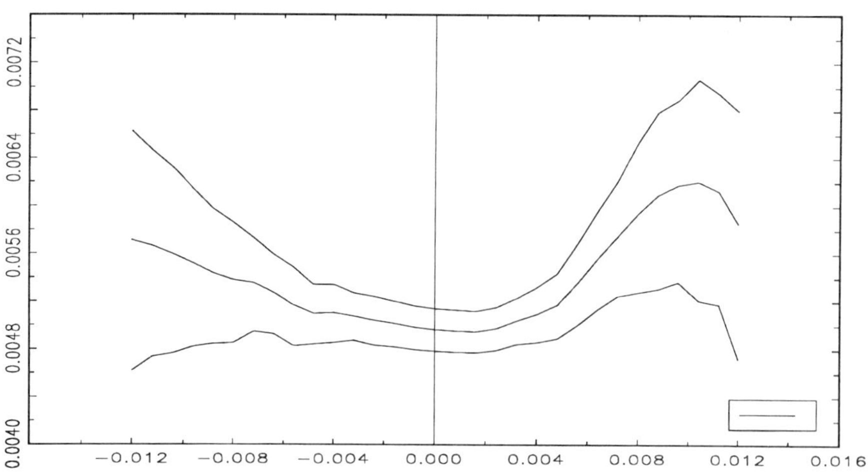

Figure 10
pointwise 90% bootstrap confidence band and local constant estimate of the conditional deviation for the US \$-DM FX rate (1989-98)

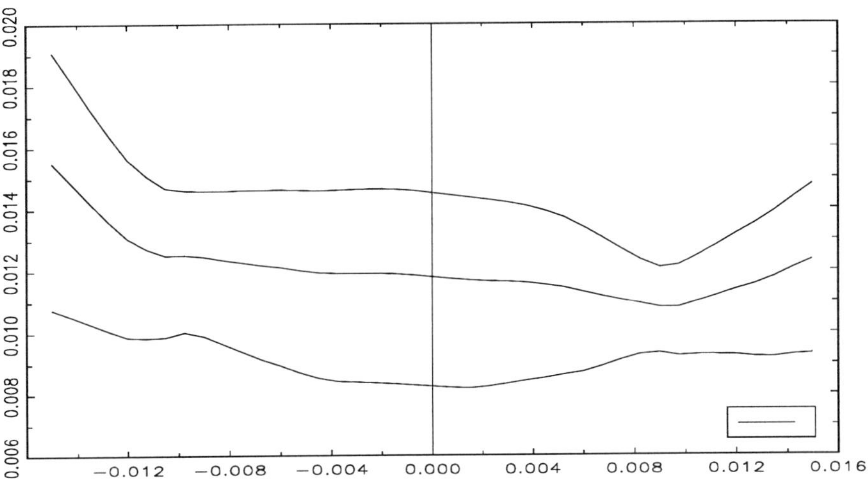

Figure 11
simultaneous 90% bootstrap confidence band and local constant estimate of the conditional deviation for the German stock index DAX (1990-92)

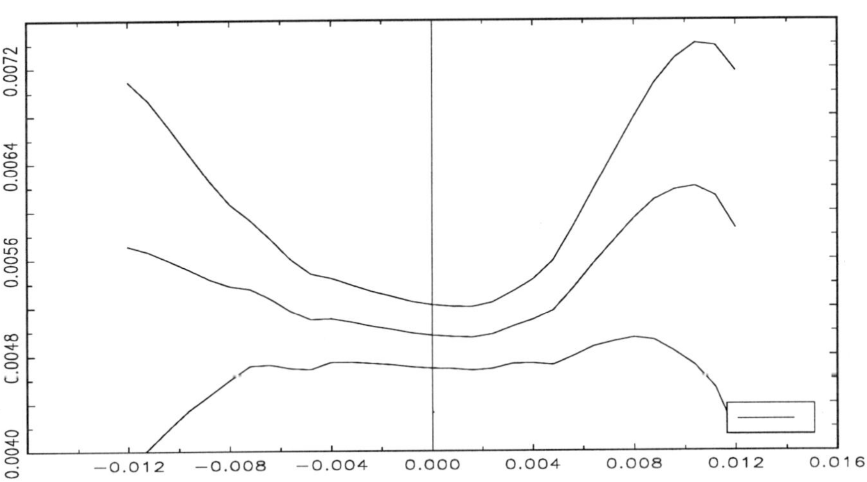

Figure 12
simultaneous 90% bootstrap confidence band and local constant estimate of the conditional deviation for the US \$-DM FX rate (1989-98)

Several other bootstrap proposals for this specific time series situation are possible. At first glance one may think that a bootstrap procedure that preserves the dependence structure of the underlying process would yield better approximation results or at least may be more intuitive. Indeed, such a construction is possible, cf. Franke, Kreiß and Mammen (1997) and Franke, Kreiß, Mammen and Neumann (1998). In both papers the following approach is investigated. Based on properly chosen estimators $\tilde{m}$ and $\tilde{s}$ and bootstrap innovations (e_t^*) i.i.d. with cumulative distribution function $\tilde{F}$ (a consistent estimator of the underlying distribution function F) one can successfully define an alternative bootstrap *process* as follows

$$R_t^* = \tilde{m}(R_{t-1}^*) + \tilde{s}(R_{t-1}^*)e_t^* \; . \tag{20}$$

It is shown in the cited papers, that if the underlying process (R_t) is of the form (19) then this bootstrap proposal works for pointwise and sup-distance distributions, too. This means that a similar construction of pointwise bootstrap confidence intervals and simultaneous bootstrap confidence bands is possible. Nevertheless, this bootstrap proposal fails if the underlying structure is not of the form (19). The reason for this is that the stationary distribution of an underlying mixing process (R_t) and the bootstrap stationary distribution (which actually is the stationary distribution of a Markov chain, cf. (20)) are in generally not close to each other, even not asymptotically. But such a result is really necessary in order to prove consistency of a bootstrap procedure for nonparametric estimators.

Eventually one should mention that there also is a drawback of the wild bootstrap proposal. Namely, if we are interested in global estimators for the conditional mean or variance function, like parametric functions $m_{\hat{\vartheta}}$ or $s_{\hat{\vartheta}}$ then we carefully have to separate between the case where the true underlying function m (or s) belongs to the parametric class or not. As far as the functions belong to the parametric class the proposed wild bootstrap procedure asymptotically works. In the case of model misspecification, i.e. the function of interest does not belong to the parametric class and the parametric class is only viewed as a fit to the true underlying situation, then even the asymptotic distribution usually depends on the whole dependence structure of the underlying process. This immediately implies that the wild bootstrap cannot work, because the dependence structure of the underlying process is not mimicked at all. In this situation it does not help if the underlying structure is of the form (19) (of course with m or s not belonging to the parametric class!). But exactly for this case, i.e. model (19) with or without m and/or s belonging to the parametric class, the nonparametric process bootstrap proposal (20) asymptotically works, cf. Franke, Kreiß, Mammen, Neumann (1998).

The presented results concerning the approximation of the distribution of the sup-distance, which are necessary for the construction of simultaneous bootstrap confidence bands, are clearly based on the Markov property of the time series, since the application of the Skorokhod embedding requires that $E[\varepsilon_t|R_0, ..., R_{t-1}] \equiv 0$. On the other hand, even if the data-generating process does not obey a structural model like (19), it makes sense to fit a nonparametric model. It would be interesting to know whether in the context of sup-type statistics the wild bootstrap remains valid in such a case of an inadequate model. Recall that under mixing and some extra condition, Robinson (1983) showed that the effect of weak dependence vanishes asymptotically for nonparametric estimators. Hart (1995) called this effect whitening by windowing. This means that we need an appropriate version of the whitening by windowing principle beyond the pointwise properties of the estimators. Neumann (1997, 1998) derived such results for nonparametric estimation of the autoregression function.

Remark. It should be mentioned, that it is possible to construct formal statistical test procedures to test on parametric model structure or, perhaps more important, on nonparametric additive structure for the conditional mean and variance function as well. As a test statistic we propose a specific distance (e.g. sup-distance or L_2-distance) of a parametric and a nonparametric estimator and rejecting the hypothesis of the parametric or the nonparametric additive model when this distance is large. The interested reader is referred to Kreiß, Neumann and Yao (1998) and Neumann and Kreiß (1998).
In a paper of Ghysels et al. (1997) it is suggested to apply nonparametric techniques to the situation where the observed bivariate time series consists of a derivative price (like an option price) and the price of the corresponding underlying (an asset or exchange rate for example). It would be interesting to see whether the proposed bootstrap tests on a specific structure of the underlying conditional mean or variance function could be extended to this situation in order to test on observed data, whether a specific nonlinear pricing formula (like the Black and Scholes formula in the easiest case) is in accordance with the observed data.

In case that we are only interested in pointwise properties of the nonparametric estimators, then a very simple resampling scheme could be applied which also is generally valid for mixing observations and does not rely on the special Markov chain structure (19). This method is just an ordinary resampling with replacement from the pairs $(R_0, R_1), (R_1, R_2), ..., (R_{T-1}, R_T)$. In so far it is rather closely related to the original idea of bootstrapping (Efron (1979)). More exactly, if we denote by $N_1, ..., N_T$ independent Laplace distributed random variables on $\{1, ..., T\}$ then this specific bootstrap resample is defined as

$$(R_{N_1-1}, R_{N_1}), ..., (R_{N_T-1}, R_{N_T})$$

and for example the bootstrap Nadaraya-Watson kernel estimator for the conditional mean reads as follows

$$\hat{m}_h(x) = \frac{\sum_t K\left(\frac{x-R_{N_t-1}}{h}\right) R_{N_t}}{\sum_t K\left(\frac{x-R_{N_t-1}}{h}\right)}.$$

Finally, it has to be mentioned that the attractive idea of subsampling also applies to heteroscedastic time series, cf. Politis et al. (1997).

References

1. Ait-Sahalia, Y., Nonparametric Pricing of Interest Rate Derivative Securities. *Econometrica* **64**, 527-560 (1996)
2. Ait-Sahalia, Y., Lo, A. W., Nonparametric Estimation of State-Price Densities Implicit in Financial Asset Prices. *J. Finance* **LIII**, 499-547 (1998)
3. Black, F., Scholes, M., The Pricing of Options and Corporate Liabilities. *J. Political Econ.* **81**, 637-659 (1973)
4. Bollerslev, T., Generalized Autoregressive Conditional Heteroskedasticity. *Journal of Econometrics* **31**, 307-327 (1986)
5. Duan, J.-C., The GARCH Option Pricing Model. *Mathematical Finance* **5**, 13-32 (1995)
6. Efron, B., Bootstrap methods: Another look at the jackknife. *Ann. Statist.* **7**, 1-26 (1979)
7. Engle, R., Autoregressive Conditional Heteroskedasticity with Estimates of the Variance of U.K. inflation. *Econometrica* **50**, 987-1008 (1982)
8. Engle, R., Lilien, D., Robins, R., Estimating Times Varying Risk Premia in the Term Structure. *Econometrica* **55**, 391-407 (1987)
9. Fan, J., Design-adaptive nonparametric regression. *J. Amer. Statist. Assoc.* **87**, 998-1004 (1992)
10. Fan, J., Local linear regression smoothers and their minimax efficiencies. *Ann. Statist.* **21**, 196-216 (1993)
11. Fan, J. and Gijbels, I., Variable bandwidth and local linear regression smoothers. *Ann. Statist.* **20**, 2008-2036 (1992)
12. Fan, J. and Gijbels, I., Local Polynomial Modeling and Its Application - Theory and Methodologies. Chapman and Hall, New York, 1995
13. Fan, J, and Yao, Q., Efficient Estimation of Conditional Variance Functions in Stochastic Regression. *Biometrika* **85**, 645-660 (1998)

14. Franke, J., Kreiss, J.-P. and Mammen, E., Bootstrap of kernel smoothing in nonlinear time series. Preprint, 1997
15. Franke, J., Kreiss, J.-P., Mammen, E. and Neumann, M. H., Properties of the nonparametric autoregressive bootstrap. Preprint, 1997
16. Ghysels, E., Patilea, V., Renault, E. and Torres, O., Nonparametric Methods and Option Pricing. Preprint, 1997
17. Härdle, W. and Tsybakov, A. B., Local polynomial estimators of the volatility function in nonparametric autoregression. *Journal of Econometrics* **81**, 223-242 (1997)
18. Hart, J. D., Some automated methods of smoothing time-dependent data. *J. Nonpar. Statist.* **6**, 115-142 (1995)
19. Korostelev, A.P. and Tsybakov, A.B., Minimax theory of image reconstruction. *Lecture Notes in Statistics* **82**, (Springer, New York, 1993)
20. Kreiss, J.-P., Neumann, M. H. and Yao, Q., Bootstrap tests for simple structures in nonparametric time series regression. Preprint, 1997
21. Mammen, E., When does the Bootstrap work? Asymptotic results and simulations. *Springer Lecture Notes in Statistics* **77**, (Springer, Heidelberg, 1992)
22. Masry, E., Multivariate local polynomial regression for time series: uniform strong consistency and rates. *J. Time Ser. Anal.* **17**, 571-599 (1996)
23. Merton, R., The Theory of Rational Option Pricing. *Bell J. Econ. Management Sci.* **4**, 141-183 (1973)
24. Neumann, M. H., On robustness of model-based bootstrap schemes in nonparametric time series analysis. Discussion Paper 88/97, SFB 373, Humboldt University, Berlin, 1997
25. Neumann, M.H., Strong approximation of density estimators from weakly dependent observations by density estimators from independent observations. *Ann. Statist.* **26**, 2014-2048 (1998)
26. Neumann, M.H., Kreiß,J.-P., Regression-Type Inference in Nonparametric Autoregression. *Ann. Statist.* **26**, 1570-1613 (1998)
27. Politis, D.N., Romano, J.P. and Wolf, M., Subsampling for Heteroskedastic Time Series. *J. Econometrics* **81**, 281-317 (1997)
28. Robinson, P. M., Nonparametric estimators for time series. *J. Time Ser. Anal.* **4**, 185-207 (1983)
29. Stone, C. J., Consistent nonparametric regression. *Ann. Statist.* **5**, 595-620 (1977)
30. Tsybakov, A. B., Robust reconstruction of functions by the local approximation method. *Problems of Information Transmission* **22**, 133-146 (1986)

COMPARISON OF TWO DISCRETIZATION METHODS FOR ESTIMATING CONTINUOUS-TIME AUTOREGRESSIVE MODELS

HENGHSIU TSAI
Department of Statistics, Tunghai University, Taichung, Taiwan 407, R.O.C.
E-mail: htsai@mail.thu.edu.tw

K. S. CHAN
Department of Statistics and Actuarial Science, The University of Iowa, Iowa City, IA 52242, USA
E-mail: kchan@stat.uiowa.edu

We have applied the trapezium method to approximate integrals in an implementation of the EM algorithm proposed by Tsai and Chan (1999b) for estimating continuous-time autoregressive models, whose original implementation was based on Euler's method for approximating integrals. It is well known that the trapezium method generally provides a second order approximation to an integral of a well-behaved functional of Wiener process, whereas the Euler method is generally of first order. Simulation results confirm that with increasing discretization frequency, the EM estimators based on the trapezium method converge to the (conditional) ML estimator at a faster rate than the EM estimators based on Euler's method. However, with an appropriate choice of discretization frequency, the EM estimator based on Euler's method outperforms both the EM estimator based on the trapezium method and the ML estimator in terms of biases and standard deviations of the estimates. An invariance property of the EM estimator based on the trapezium method is briefly discussed.

Some key words: Trapezium method, Girsanov formula, Maximum likelihood estimation, Stochastic differential equations, irregularly sampled time series, Kalman filter.

1 Introduction

Owing to the sampling procedure or the presence of missing data, many time series, say, $\{Y_{t_i}\}_{i=0,\ldots,N}$, are sampled with irregular time intervals. Irregularly sampled time series data are often analyzed by assuming that these data are sampled from an underlying continuous-time process. The underlying continuous-time process may be modeled as driven by some stochastic differential equations, for example, the linear continuous-time autoregressive moving average (ARMA) models. This linear specification results in a tractable likelihood for the observed discrete-time data. Hence this method is rather popular for analyzing irregularly sampled time-series data. See, e.g., Harvey (1989), Bergstrom (1990), Tong (1990) and Jones (1980, 1993). See Belcher et al.

(1994) for some discussions of parameterizing continuous-time autoregressive models. We also note that in many applications, for example, economics, the main interest may consist of drawing inference on the underlying stochastic differential equation even with equally spaced data; see Bergstrom (1990).

The likelihood function of a CAR(p) model can be computed via Kalman filters. Maximum likelihood estimation can be done by means of some nonlinear optimization algorithm such as the simplex method. Tsai and Chan (1999b) proposed the use of the Expectation-Maximization (EM) algorithm to derive approximate ML estimators of the CAR(p) models. The EM algorithm is based on an integral representation of the likelihood function and an approximation of the integrals by Euler's method. Simulation results reported by Tsai and Chan (1999b) suggested that with suitable choices of the discretization frequency, the EM estimators are comparable to the ML estimator in terms of bias, but with smaller standard errors than those of the ML estimator. Tsai and Chan (1999b) proved in the first order case that as the discretization frequency increases to infinity, the EM estimators converge to the ML estimator in probability. Tsai and Chan (1999b) conjectured that in the limit of no discretization error, the EM estimator becomes the ML estimator.

An alternative to Euler's method is the trapezium approximation. It is well known that the trapezium approximation converges to its limit with a faster convergence rate than Euler's method. We now briefly recall the trapezium method for approximating an integral, and some of its properties without proofs; see Milstein (1995, pp.6, 135-141) for the exact statements and their proofs. Let $\{Z(s), a \leq s \leq b\}$ be a well-behaved functional of a Wiener process; more specifically, $Z(s)$ is a smooth functional of $\{W(t), a \leq t \leq s\}$, with polynomial growth in the tail. Consider integrals of the form $\int_a^b Z(s)ds$. Let $a = t_0 < t_1 < \cdots < t_N = b$ be an equally spaced partition of the interval $[a, b]$. Denote $h = (b-a)/N$. Write $\int_a^b Z(s)ds = \sum_1^N \int_{t_{i-1}}^{t_i} Z(s)ds$. The trapezium method approximates $\int_{t_{i-1}}^{t_i} Z(s)ds$ by $(Z(t_{i-1}) + Z(t_i))h/2$ whereas the Euler method approximates the integrals by $Z(t_{i-1})h$. The trapezium method can be shown to be of order two, that is, the mean error is $O(h^2)$, but the Euler method is of order one. Moreover, it holds generically that there exists no 'simple' approximating scheme for $\int_a^b Z(s)ds$ which is of order higher than two.

In this paper, we compare the finite sample performance of the (conditional) ML estimator with those of the EM estimators when the integrals are approximated by Euler's method and the trapezium method. Simulation results confirm that the EM estimators based on the trapezium method converge to the (conditional) ML estimator at a faster rate than the EM estimators based on Euler's method. In a first order case, the EM estimators based on the trapezium method converge to a limit very close to the conditional ML estimator. In a second order case, the EM estimators based on the trapezium method seem to converge to the ML estimator generated by the simplex method in terms of biases. The standard deviations of the EM estimators based on the trapezium method are also close to those of the ML estimator, although a bit larger. It is also interesting to note that with appropriate choices of discretization frequencies, the EM estimators based on Euler's method outperform the EM estimators based on the trapezium method and the ML estimator in terms of biases and standard deviations of the estimates.

The organization of this paper is as follows. Section 2 provides a review of ML estimation of CAR(p) models. An integral representation of the likelihood, its approximation by the trapezium method and the EM algorithm are described in section 3. Simulation results comparing the finite sample performance of the EM estimators based on Euler's approximation method and the trapezium method, and that of the (conditional) ML estimator are reported in section 4. An invariance property of the EM estimator based on the trapezium method is briefly discussed. We conclude in section 5.

2 Review of Maximum Likelihood Estimation of CAR Models

Nonlinear optimization is the most popular method in the literature for doing maximum likelihood estimation of the CAR(p) models, which we briefly review below. For further discussions of the CAR(p) processes, see, e.g., Brockwell (1993) and Brockwell and Stramer (1995). A CAR(p) process is defined to be a solution of the p-th order differential equation:

$$dY_t^{(p-1)} = (\alpha_0 + \alpha_1 Y_t + \cdots + \alpha_p Y_t^{(p-1)})dt + \sigma dW_t, \tag{1}$$

where the superscript $^{(j)}$ denotes j-fold differentiation with respect to t; $\{W_t,\ t \geq 0\}$ is the standard Brownian motion, and $\alpha_0,\ldots,\alpha_p$ and σ are constants. We assume that $\sigma > 0$ and $\alpha_1 \neq 0$. The Brownian motion $\{W_t, t \geq 0\}$ is nowhere differentiable. So, the solution of equation (1) is interpreted as satisfying the integral equation:

$$Y_t^{(p-1)} - Y_0^{(p-1)} = \int_0^t (\alpha_0 + \alpha_1 Y_s + \cdots + \alpha_p Y_s^{(p-1)})ds + \sigma W_t.$$

The term $(\alpha_0 + \alpha_1 Y_s + \cdots + \alpha_p Y_s^{(p-1)})$ is referred to as the instantaneous mean of the process, and σ the instantaneous standard deviation. Equation (1) can also be equivalently cast in terms of the *observation* and *state* equations:

$$Y_t = H^{'} X_t, \qquad t \geq 0, \tag{2}$$

$$dX_t = (AX_t dt + \alpha_0 \delta_p)dt + \sigma \delta_p dW_t, \tag{3}$$

where

$$A = \begin{bmatrix} 0 & 1 & 0 & \cdots & 0 \\ 0 & 0 & 1 & \cdots & 0 \\ \vdots & \vdots & \vdots & \ddots & \vdots \\ 0 & 0 & 0 & \cdots & 1 \\ \alpha_1 & \alpha_2 & \alpha_3 & \cdots & \alpha_p \end{bmatrix}, \quad X_t = \begin{bmatrix} X_t^{(0)} \\ X_t^{(1)} \\ \vdots \\ X_t^{(p-2)} \\ X_t^{(p-1)} \end{bmatrix}, \quad \delta_p = \begin{bmatrix} 0 \\ 0 \\ \vdots \\ 0 \\ 1 \end{bmatrix}, \quad H = \begin{bmatrix} 1 \\ 0 \\ \vdots \\ 0 \\ 0 \end{bmatrix},$$

and the superscript $'$ denotes the transpose of a vector, and $X_t^{(0)} = Y_t$. Equation (3) is an Ito differential equation for the state vector X_t. We assume that X_0 is independent of $\{W_t, t \geq 0\}$ and X_0 is determined by initial conditions that could be random or deterministic. The process $\{Y_t, t \geq 0\}$ is thus said to be a CAR(p) process with parameter $(\theta, \sigma^2) = (\alpha_0, ..., \alpha_p, \sigma^2)$ if $Y_t = H' X_t$, where $\{X_t\}$ is a solution of (3). The solution of (3) can be written as

$$X_t = e^{At} X_0 + \alpha_0 \int_0^t e^{A(t-u)} \delta_p du + \sigma \int_0^t e^{A(t-u)} \delta_p dW_u,$$

where $e^{At} = I + \sum_{n=1}^{\infty} [(At)^n (n!)^{-1}]$, and I is the identity matrix.

Let the mean vector of X_t be denoted by μ_t. It satisfies the equation:

$$\mu_t = \frac{\alpha_0}{\alpha_1}(e^{At} - I)H + e^{At}\mu_0.$$

The covariance kernel of $\{X_t\}$, denoted by $\gamma_{s,t}$, equals

$$\begin{aligned} \gamma_{s,t} &= E[(X_s - \mu_s)(X_t - \mu_t)'] \\ &= e^{As} V_0 e^{A' t} + \sigma^2 \int_0^{t \wedge s} e^{A(s-u)} \delta_p \delta_p' e^{A'(t-u)} du \\ &= \begin{cases} e^{A(s-t)} V_t, & 0 \leq t \leq s < \infty \\ V_s e^{A'(t-s)}, & 0 \leq s \leq t < \infty, \end{cases} \end{aligned}$$

where $t \wedge s = \min(s, t)$ and

$$\begin{aligned} V_t &= \gamma_{t,t} \\ &= e^{At} V_0 e^{A' t} + \sigma^2 \int_0^t e^{A(t-u)} \delta_p \delta_p' e^{A'(t-u)} du. \end{aligned}$$

It follows from the above equations on μ_t and V_t that the states and the observations, X_{t_i} and Y_{t_i}, at the sampling times $t_0, t_1, ...$, satisfy the discrete-time state and observation equations:

$$X_{t_{i+1}} = \mu_{t_{i+1}} + e^{A(t_{i+1} - t_i)}(X_{t_i} - \mu_{t_i}) + Z_{t_i}, \quad i = 0, 1, ..., \tag{4}$$

$$Y_{t_i} = H' X_{t_i}, \qquad i = 0, 1, ..., \tag{5}$$

where Z_{t_i} is independent of X_{t_i}, and $\{Z_{t_i}, i = 0, 1, ...\}$ is an independent sequence of Gaussian random vectors with zero mean and covariance matrices

$$\Sigma_i = E(Z_{t_i} Z_{t_i}') = \sigma^2 \int_{t_i}^{t_{i+1}} e^{A(t_{i+1}-u)} \delta_p \delta_p' e^{A'(t_{i+1}-u)} du. \tag{6}$$

These equations are needed for applications of the Kalman recursions (see, e.g., Chapter 12 of Brockwell and Davis, 1991). Let capital letters be used for random vectors and corresponding small letters for observed vectors. Let $Y = \{Y_{t_i}\}_{i=0,\dots,N}$ and define $\hat{X}_{t|s}$ as the conditional expectation of X_t based on the observations up to time s and $P_{t|s}$ the corresponding covariance matrix, i.e., $\hat{X}_{t|s} = E(X_t | y_j, j \le s) = (\hat{X}_{t|s}^{(0)}, \cdots, \hat{X}_{t|s}^{(p-1)})$ and $P_{t|s} = \text{var}(X_t | y_j, j \le s)$. Then we can compute recursively $\tilde{y}_{t_i} = y_{t_i} - \hat{X}_{t_i|t_{i-1}}^{(0)}$, the predictive residuals, and $p_{t_i|t_{i-1}}^{(1,1)}$, the $(1,1)th$ element of $P_{t_i|t_{i-1}}$, $i \ge 1$, which are handy for computing minus twice the log-likelihood function,

$$-2l_Y(\theta, \sigma^2) = \sum_{i=0}^{N} \left[\frac{\tilde{y}_{t_i}^2}{p_{t_i|t_{i-1}}^{(1,1)}} + \log p_{t_i|t_{i-1}}^{(1,1)} \right] + (N+1) \log(2\pi). \tag{7}$$

Here, we start with a diffuse initial condition as we do not assume stationarity, i.e., let

$$\begin{aligned} \hat{X}_{t_{-1}|t_{-1}} &= [\bar{y}, 0, ..., 0]', \\ P_{t_{-1}|t_{-1}} &= \delta \, s_y^2 I, \end{aligned}$$

where $t_{-1} < t_0$ is some arbitrarily chosen time point, δ is some positive number, $\bar{y}$ and s_y^2 are the sample mean and sample variance of y, respectively. A reasonable choice of δ is, e.g., 5.

A non-linear optimization algorithm can then be used in conjunction with the expression for $-2l_Y(\theta, \sigma^2)$ to find the maximum likelihood estimate of the parameter (θ, σ^2). The calculations of e^{At} are most readily performed by first block-diagonalizing A and then applying a Padé approximation on each block. See, e.g., Ward (1977). For a faster alternative for computing Σ_i, see Tsai and Chan (1999a) .

The parameter σ^2 can be concentrated out of the likelihood (see Jones, 1980). We only need to replace $P_{t|s}$ by $P_{t|s}^* = P_{t|s}/\sigma^2$ and equation (7) becomes

$$-2l_Y(\theta, \sigma^2) = \sum_{i=0}^{N} \left[\frac{\tilde{y}_{t_i}^2}{\sigma^2 p_{t_i|t_{i-1}}^{*(1,1)}} + \log \left(\sigma^2 p_{t_i|t_{i-1}}^{*(1,1)} \right) \right] + (N+1) \log(2\pi). \tag{8}$$

Differentiating (8) with respect to σ^2 and equating to zero gives

$$\sigma^2 = \frac{1}{N+1}\sum_{i=0}^{N}\frac{\tilde{y}_{t_i}^2}{p^{*(1,1)}_{t_i|t_{i-1}}} \tag{9}$$

and substituting into (8), the objective function becomes

$$-2l_Y(\theta) = (N+1)\log\left(\sum_{i=0}^{N}\frac{\tilde{y}_{t_i}^2}{p^{*(1,1)}_{t_i|t_{i-1}}}\right) + \sum_{i=0}^{N}\log p^{*(1,1)}_{t_i|t_{i-1}} + C, \tag{10}$$

where $C = (N+1)(1-\log(N+1)+\log(2\pi))$. This function is then minimized with respect to θ to get the maximum likelihood estimate $\hat{\theta}$. The parameter estimate $\hat{\sigma}^2$ is then calculated from (9).

3 An Integral Representation of the Likelihood, Its Approximation by the Trapezium Method and the EM Algorithm

Tsai and Chan (1999b) applied the Cameron-Martin-Girsanov formula (see, e.g., Corollary 8.23 of Øksendal, 1995) to derive an integral representation of the *pdf* of $Y = \{Y_{t_0}, Y_{t_1}, ..., Y_{t_N}\}$ with respect to the Lebesgue measure. This representation will be useful for applying an Expectation-Maximization (EM) algorithm to estimate the parameters. The notations $E_\theta(\cdot|y)$, $\text{var}_\theta(\cdot|y)$ and $\text{cov}_\theta(\cdot|y)$ denote the conditional expectation, variance and covariance of the enclosed expression given $Y = y$, respectively, where θ is the true parameter. Also, the parameter θ is omitted if no confusion would occur. The cumulative distribution function of Y is

$$\begin{aligned}
&P_{\theta,\sigma^2}(Y_{t_0} \le y_{t_0}, Y_{t_1} \le y_{t_1}, \cdots, Y_{t_N} \le y_{t_N}) \\
&= E_{\theta,\sigma^2}[I(Y_{t_0} \le y_{t_0}, Y_{t_1} \le y_{t_1}, \cdots, Y_{t_N} \le y_{t_N})] \\
&= E_{0,\sigma^2}\left[I(Y_{t_0} \le y_{t_0}, Y_{t_1} \le y_{t_1}, \cdots, Y_{t_N} \le y_{t_N})\frac{dP_{\theta,\sigma^2}}{dP_{0,\sigma^2}}\{X_s^{(p-1)}, t_0 \le s \le t_N\}\right] \\
&= E_{0,\sigma^2}\left[E_{0,\sigma^2}\left[I(Y_{t_0} \le y_{t_0}, \cdots, Y_{t_N} \le y_{t_N})\frac{dP_{\theta,\sigma^2}}{dP_{0,\sigma^2}}\{X_s^{(p-1)}, t_0 \le s \le t_N\}\middle| Y\right]\right] \\
&= E_{0,\sigma^2}\left[I(Y_{t_0} \le y_{t_0}, \cdots, Y_{t_N} \le y_{t_N})E_{0,\sigma^2}\left[\frac{dP_{\theta,\sigma^2}}{dP_{0,\sigma^2}}\{X_s^{(p-1)}, t_0 \le s \le t_N\}\middle| Y\right]\right] \\
&= \int_{-\infty}^{y_{t_N}}\cdots\int_{-\infty}^{y_{t_0}} E_{0,\sigma^2}\left[\frac{dP_{\theta,\sigma^2}}{dP_{0,\sigma^2}}\{X_s^{(p-1)}, t_0 \le s \le t_N\}\middle| y\right] f_{Y;(0,\sigma^2)}(y)dy_{t_0}\cdots dy_{t_N}
\end{aligned}$$

where

$$\frac{dP_{\theta,\sigma^2}}{dP_{0,\sigma^2}}\left\{X_s^{(p-1)}; 0 \le s \le t_N\right\}$$
$$= \exp\left\{\frac{1}{\sigma^2}\int_0^{t_N}(\alpha_0 + \alpha' X_s)dX_s^{(p-1)} - \frac{1}{2\sigma^2}\int_0^{t_N}(\alpha_0 + \alpha' X_s)^2 ds\right\}.$$

Thus,

$$f_{Y;(\theta,\sigma^2)}(y) = E_{0,\sigma^2}\left[\frac{dP_{\theta,\sigma^2}}{dP_{0,\sigma^2}}\left\{X_s^{(p-1)}; 0 \le s \le t\right\}\Big| y\right] f_{Y;(0,\sigma^2)}(y),$$

where $f_{Y;(\theta,\sigma^2)}(y)$ is the joint pdf of $Y = y$ under the CAR model with (θ, σ^2) as its true parameter. Therefore, $l_Y(\theta, \sigma^2)$, the log-likelihood function of Y is

$$l_Y(\theta,\sigma^2) = log\ E_{0,\sigma^2}\left[\frac{dP_{\theta,\sigma^2}}{dP_{0,\sigma^2}}\left\{X_s^{(p-1)}; 0 \le s \le t\right\}\Big| y\right] + l_Y(0,\sigma^2).$$

Let $R = R(\theta, \sigma^2) = 1/\sigma^2 \int_0^{t_N}(\alpha_0+\alpha' X_s)dX_s^{(p-1)} - 1/(2\sigma^2)\int_0^{t_N}(\alpha_0+\alpha' X_s)^2 ds$. Note that for fixed σ^2, R is the complete-data log-likelihood for the process $\{X_s, 0 \le s \le t_N\}$. Let D be the differentiation operator with respect to θ, then

$$Dl_Y(\theta,\sigma^2) = \frac{DE_{0,\sigma^2}[\exp(R)|y]}{E_{0,\sigma^2}[\exp(R)|y]} = \frac{E_{0,\sigma^2}[\exp(R)(DR)|y]}{E_{0,\sigma^2}[\exp(R)|y]} = E_{\theta,\sigma^2}[(DR)|y] \quad (11)$$

where the last equality follows from the change of measure formula (see, e.g., Lemma 3.5.3 of Karatzas and Shreve, 1991). We remark that (11) is similar to a result due to Louis (1982) for Euclidean sample spaces. Thus,

$$\frac{\partial l_Y}{\partial \alpha_0} = \frac{1}{\sigma^2}E_{\theta,\sigma^2}\left[\int_{t_0}^{t_N} dX_s^{(p-1)} - \int_{t_0}^{t_N}(\alpha_0 + \alpha' X_s)ds\Big| y\right], \quad (12)$$
$$\frac{\partial l_Y}{\partial \alpha_r} = \frac{1}{\sigma^2}E_{\theta,\sigma^2}\left[\int_{t_0}^{t_N} X_s^{(r-1)}dX_s^{(p-1)} - \int_{t_0}^{t_N}(\alpha_0 + \alpha' X_s)X_s^{(r-1)}ds\Big| y\right], \quad (13)$$
$$r = 1, 2, ..., p.$$

The above equations can be used to estimate the parameters by an EM algorithm. The M-step is done by equating the above equations to zero and then solving the resulting linear equations. The E-step involves the computation of the conditional expectation of integrals, which is complex. Tsai and Chan (1999b) applied Euler's method to approximate the integrals in these equations. The conditional expectations are then computed by Kalman filters. An alternative to Euler's method is the trapezium method which has a faster convergence rate than the former.

The EM algorithm is a data augmentation method which augments the observed data Y with some latent data Z so that $l_{Y,Z}(\theta)$, the log-likelihood function of $X = (Y, Z)$, is tractable. Here θ is an arbitrary element of the parameter space Ω. Let $Y = y$ be the observed incomplete data. Let $X = \{X_0, X_{1/m}, \cdots, X_{k_N/m}\}$ be the unobserved complete data of which Y is a measurable function, where $Y = \{Y_{t_j}\}_{0,\ldots,N}$ and m is chosen to be some moderately large integer such that, for each $0 \le j \le N$, $t_j = k_j/m$ (approximately) for some integer k_j. Note that the preceding condition that the observed times $t_j = k_j/m$ (approximately) can be lifted by employing irregularly spaced partitions at the expense of more complex notations. To simplify notations, we write X_k for $X_{k/m}$, Y_k for $Y_{k/m}$ and q for k_N in this section.

The EM algorithm consists of two steps:

$$\begin{aligned} E \text{ step: } & \text{form } Q(\theta'|\theta) = E_\theta(l_X(\theta')|y); \\ M \text{ step: } & \text{maximize } Q(\cdot|\theta). \end{aligned}$$

Suppose that for all θ, $Q(\cdot|\theta)$ has a unique global maximizer at $M(\theta)$ and that $M(\theta)$ is continuous in θ. Then an EM sequence $\{\theta_k\}$ is obtained from $\{\theta_{k+1}\} = M(\theta_k)$, and $\{\theta_k\}$ is a Markov chain which converges to a stationary point of $l_Y(\theta)$. An important property of the EM algorithm is that the likelihood of the observed data always increases along an EM sequence. See, e.g., Tanner (1991). Dempster, Laird and Rubin (1977) showed that the EM algorithm converges at a linear rate, with the rate depending on the proportion of information about θ in $l_Y(\theta)$ which is observed. For other convergence properties of the EM algorithm, see Dempster, Laird and Rubin (1977), and Wu (1983).

We now describe some formulas useful for the E-step. The conditional distribution of X given $Y = y$ is Gaussian. The computation of the means and the variances of the conditional distribution of X given $Y = y$ can be carried out by a forward Kalman filtering sequence, followed by backward iterations. See, e.g., page 189 of Anderson and Moore (1979). We now outline the Kalman computation below. For $0 \le k \le q$, $\hat{X}_{k|k}, \hat{X}_{k+1|k}, P_{k|k}, P_{k+1|k}$, can be computed via a forward Kalman filter as follows.

First, let $\hat{X}_{-1|-1} = [\bar{y}, 0, \ldots, 0]'$ and $P_{-1|-1} = \delta\, s_y^2 I$ as in section 2. Then, for $0 \le k \le q$, compute $P_{k|k-1}$ and $P_{k|k}$ iteratively via equations (14) and (15):

$$P_{k|k-1} = e^{\frac{A}{m}} P_{k-1|k-1} e^{\frac{A'}{m}} + \Sigma, \tag{14}$$

$$P_{k|k} = \begin{cases} P_{k|k-1} - \frac{1}{p^{(1,1)}_{k|k-1}} P_{k|k-1} H H' P_{k|k-1}, & if \;\; k \in \{k_0, ..., k_N\} \\ P_{k|k-1}, & if \;\; k \notin \{k_0, ..., k_N\}, \end{cases} \tag{15}$$

where $p^{(i,j)}_{k|k-1}$ is the $(i, j)th$ element of $P_{k|k-1}$; $\Sigma = V - e^{\frac{A}{m}} V e^{\frac{A'}{m}}$ and V is the solution of the matrix equation $AV + VA' = -\sigma^2 \delta_p \delta_p'$. The preceding result on Σ is well-known for the stationary case but it also holds for the non-stationary case (see Tsai and Chan, 1999a, for a proof).

For $0 \leq k \leq q$, compute $\hat{X}_{k|k-1}$ and $\hat{X}_{k|k}$ iteratively via equations (16) and (17).

$$\hat{X}_{k|k-1} = \mu + e^{\frac{A}{m}} (\hat{X}_{k-1|k-1} - \mu) \tag{16}$$

$$\hat{X}_{k|k} = \begin{cases} \hat{X}_{k|k-1} + \frac{1}{p^{(1,1)}_{k|k-1}} P_{k|k-1} H (y_k - \hat{X}_{k|k-1}), & if \;\; k \in \{k_0, ..., k_N\} \\ \hat{X}_{k|k-1}, & if \;\; k \notin \{k_0, ..., k_N\}, \end{cases} \tag{17}$$

where $\mu = -\alpha_0 H / \alpha_1$. Next, go through the Kalman filter backward, we can compute the conditional means and the variances of X_k's given all observed data via the following recursive equations:

$$\hat{X}_{k|q} = \hat{X}_{k|k} + B_k (\hat{X}_{k+1|q} - \hat{X}_{k+1|k}), \tag{18}$$

$$P_{k|q} = P_{k|k} + B_k (P_{k+1|q} - P_{k+1|k}) B_k', \tag{19}$$

where $B_k = P_{k|k} e^{\frac{A'}{m}} P^{-1}_{k+1|k}, \quad k = q-1, ..., 0.$

The M-step can be carried out by solving the equations obtained by equating the right hand sides of equations (12) and (13) to zero. The integrals in equations (12) and (13) can be approximated by Euler's method (see Tsai and Chan, 1999b). Alternatively, we may approximate the integrals by the trapezium method. Using the trapezium method and the fact that for any $a < b$

$$\begin{aligned} \int_a^b X_s^{(r-1)} dX_s^{(p-1)} &= X_b^{(r-1)} X_b^{(p-1)} - X_a^{(r-1)} X_a^{(p-1)} - \int_a^b X_s^{(p-1)} dX_s^{(r-1)} \\ &= X_b^{(r-1)} X_b^{(p-1)} - X_a^{(r-1)} X_a^{(p-1)} - \int_a^b X_s^{(p-1)} X_s^{(r)} ds, \\ & \quad r = 1, ..., p-1, \\ \int_a^b X_s^{(p-1)} dX_s^{(p-1)} &= \frac{1}{2} \left[(X_b^{(p-1)})^2 - (X_a^{(p-1)})^2 - \sigma^2 (b-a) \right], \end{aligned}$$

equations (12) and (13) become

$$\frac{\partial l_Y}{\partial \alpha_0} = \frac{1}{\sigma^2}\left[\delta_p'(\hat{X}_{q|q} - \hat{X}_{0|q}) - \frac{1}{2m}\sum_{j=1}^{q}\left\{(\alpha_0 + \alpha' X_{j-1}) + (\alpha_0 + \alpha' X_j)\right\}\middle| y\right]$$

$$= \frac{1}{\sigma^2}\left[\delta_p'(\hat{X}_{q|q} - \hat{X}_{0|q}) - \frac{q}{m}\alpha_0 - \frac{\alpha'}{2m}\sum_{j=1}^{q}(\hat{X}_{j-1|q} + \hat{X}_{j|q})\right], \quad (20)$$

$$\frac{\partial l_Y}{\partial \alpha_r} = \frac{1}{\sigma^2}\delta_r'\left[E(X_q X_q'|y) - E(X_0 X_0'|y)\right]\delta_p$$

$$-\frac{1}{2m\sigma^2}\sum_{j=1}^{q}\delta_p'\left[E(X_{j-1}X_{j-1}'|y) + E(X_j X_j'|y)\right]\delta_{r+1}$$

$$-\frac{1}{2m\sigma^2}\sum_{j=1}^{q}E\left[(\alpha_0 + \alpha' X_{j-1})X_{j-1}'\delta_r + (\alpha_0 + \alpha' X_j)X_j'\delta_r\middle| y\right], \quad (21)$$

$r = 1, 2, ..., p-1,$

$$\frac{\partial l_Y}{\partial \alpha_p} = \frac{1}{2\sigma^2}\delta_p'\left[E(X_q X_q'|y) - E(X_0 X_0'|y)\right]\delta_p - \frac{1}{2}(t_N - t_0)$$

$$-\frac{1}{2m\sigma^2}\sum_{j=1}^{q}E\left[(\alpha_0 + \alpha' X_{j-1})X_{j-1}'\delta_p + (\alpha_0 + \alpha' X_j)X_j'\delta_p\middle| y\right]. \quad (22)$$

Setting equations (20), (21) and (22) to zero yields the approximate maximum likelihood estimate $\hat{\theta} = (\hat{\alpha}_0, ..., \hat{\alpha}_p) = (\hat{\alpha}_0, \hat{\alpha})$ as the solution of the linear equations:

$$\begin{bmatrix} 2q & \sum_{j=1}^{q}(\hat{X}_{j-1|q}' + \hat{X}_{j|q}') \\ \sum_{j=1}^{q}(\hat{X}_{j-1|q} + \hat{X}_{j|q}) & \sum_{j=1}^{q}\Gamma_{1,j} \end{bmatrix}\begin{bmatrix} \hat{\alpha}_0 \\ \hat{\alpha} \end{bmatrix}$$

$$= \left[2m(\hat{X}_{q|q} - \hat{X}_{0|q})'\delta_p \; b_1 \cdots b_p\right]', \quad (23)$$

where

$$\Gamma_{1,j} = E(X_{j-1}X_{j-1}'|y) + E(X_j X_j'|y)$$

$$= P_{j-1|q} + \hat{X}_{j-1|q}\hat{X}_{j-1|q}' + P_{j|q} + \hat{X}_{j|q}\hat{X}_{j|q}', \quad (24)$$

and

$$b_r = 2m\delta_r'\left[E(X_q X_q'|y) - E(X_0 X_0'|y)\right]\delta_p - \delta_{r+1}'\sum_{j=1}^{q}\Gamma_{1,j}\delta_p,$$

$$r = 1, ..., p-1,$$
$$b_p = m\delta_p^{'}\left[E(X_q X_q^{'}|y) - E(X_0 X_0^{'}|y)\right]\delta_p - m\sigma^2(t_N - t_0).$$

Given $\hat{\theta} = (\hat{\alpha}_0, \hat{\alpha})$, the (approximate) maximum likelihood estimate $\hat{\sigma}^2$ is computed by (9). Thus, a sequence of EM iterates of (θ, σ^2) can now be constructed as follows. Given the jth iterate $(\hat{\theta}_j, \hat{\sigma}_j^2)$, compute $\hat{X}_{k|q}$ and $P_{k|q}$, for $0 \le k \le q$, using equations (14) to (19). Then solve equation (23) to get a new set of estimates $\hat{\theta}_{j+1}$. Let $P^*_{t_{-1}|t_{-1}} = P_{t_{-1}|t_{-1}}/\hat{\sigma}_j^2 = \delta s_y^2 I/\hat{\sigma}_j^2$. Then with $\hat{\theta}_{j+1}$ and making use of equation (9), we can get $\hat{\sigma}_{j+1}^2$. It remains to specify the initial estimates, which will be described below.

- **Initial estimates**

Here, we describe how to get a set of initial estimates of the parameters.
(1) Note that $X_{t_j}^{(0)} = Y_{t_j}$, for all j.
(2) If $p > 1$, then for i from 1 to $p-1$, estimate $X_{t_j}^{(i)}$ by $(X_{t_j}^{(i-1)} - X_{t_{j-1}}^{(i-1)})/\Delta t_j$, for $i \le j \le N$, iteratively, where $\Delta t_j = t_j - t_{j-1}$.

Now, use the trapezium method again, but set the discretization interval to be Δ_{t_j} between t_{j-1} and t_j, and then we can approximate equations (12) and (13) by the following equations.

$$\frac{\partial l_X}{\partial \alpha_0} = \frac{1}{\sigma^2}\left[X_{t_N}^{(p-1)} - X_{t_{p-1}}^{(p-1)} - \frac{1}{2}\sum_{j=p}^{N}\left\{(\alpha_0 + \alpha^{'} X_{t_{j-1}}) + (\alpha_0 + \alpha^{'} X_{t_j})\right\}\Delta t_j\right]$$

$$\frac{\partial l_X}{\partial \alpha_r} = \frac{1}{\sigma^2}\left(X_{t_N}^{(r-1)} X_{t_N}^{(p-1)} - X_{t_{p-1}}^{(r-1)} X_{t_{p-1}}^{(p-1)}\right)$$
$$-\frac{1}{2\sigma^2}\sum_{j=p}^{N}\left(X_{t_{j-1}}^{(p-1)} X_{t_{j-1}}^{(r)} + X_{t_j}^{(p-1)} X_{t_j}^{(r)}\right)\Delta t_j$$
$$-\frac{1}{2\sigma^2}\sum_{j=p}^{N}\left\{(\alpha_0 + \alpha^{'} X_{t_{j-1}})X_{t_{j-1}}^{(r-1)} + (\alpha_0 + \alpha^{'} X_{t_j})X_{t_j}^{(r-1)}\right\}\Delta t_j,$$
$$r = 1, 2, ..., p-1,$$
$$\frac{\partial l_X}{\partial \alpha_p} = \frac{1}{2\sigma^2}\left[(X_{t_N}^{(p-1)})^2 - (X_{t_{p-1}}^{(p-1)})^2 - \sigma^2(t_N - t_{p-1})\right]$$
$$-\frac{1}{2\sigma^2}\sum_{j=p}^{N}\left\{(\alpha_0 + \alpha^{'} X_{t_{j-1}})X_{t_{j-1}}^{(p-1)} + (\alpha_0 + \alpha^{'} X_{t_j})X_{t_j}^{(p-1)}\right\}\Delta t_j.$$

Setting the above equations to zero yields the initial estimate $\tilde{\theta}_0 = (\tilde{\alpha}_0, \tilde{\alpha})$ as the solution of the linear equations:

$$\begin{bmatrix} 2\sum_{j=p}^{N} \Delta t_j & \sum_{j=p}^{N} (X'_{t_{j-1}} + X'_{t_j})\Delta t_j \\ \sum_{j=p}^{N} (X_{t_{j-1}} + X_{t_j})\Delta t_j & \sum_{j=p}^{N} (X_{t_{j-1}} X'_{t_{j-1}} + X_{t_j} X'_{t_j})\Delta t_j \end{bmatrix} \begin{bmatrix} \tilde{\alpha}_0 \\ \tilde{\alpha} \end{bmatrix}$$

$$= \left[2(X_{t_N}^{(p-1)} - X_{t_{p-1}}^{(p-1)}) \; d_1 \cdots d_p \right]', \tag{25}$$

where

$$\begin{aligned} d_r &= 2\left(X_{t_N}^{(r-1)} X_{t_N}^{(p-1)} - X_{t_{p-1}}^{(r-1)} X_{t_{p-1}}^{(p-1)} \right) \\ &\quad - \sum_{j=p}^{N} \left(X_{t_{j-1}}^{(p-1)} X_{t_{j-1}}^{(r)} + X_{t_j}^{(p-1)} X_{t_j}^{(r)} \right) \Delta t_j, \quad r = 1, ..., p-1, \\ d_p &= (X_{t_N}^{(p-1)})^2 - (X_{t_{p-1}}^{(p-1)})^2 - \sigma^2 (t_N - t_{p-1}), \end{aligned}$$

and the initial estimate of σ^2 is again given by $\tilde{\sigma}^2 = (N+1)^{-1} \sum_{i=0}^{N} \tilde{y}_{t_i}^2 / p_{t_i|t_{i-1}}^{*(1,1)}$, where $P_{t_i|t_{i-1}}^{*}$ is defined as before, while $P_{t_{-1}|t_{-1}}^{*} = \delta s_y^2 I / \bar{\sigma}^2$ and $\bar{\sigma}^2 = N^{-1} \sum_{i=1}^{N} (X_{t_i} - X_{t_{i-1}})^2 / (t_i - t_{i-1})$.

4 Simulations

Regularly spaced time series data, $Y_{t_i} = X_i, i = 0, 1, 2, ..., 100$, are simulated from the following stationary CAR(p) processes via equations (4) and (5):

$$\begin{aligned} &\text{Model 1, } dX_t = -X_t dt + dW_t, \\ &\text{Model 2, } dX_t^{(1)} = (-2X_t - X_t^{(1)})dt + dW_t. \end{aligned}$$

For model 1, the conditional maximum likelihood estimator has a closed form solution. See Tsai and Chan (1999b) for details. The averages, the standard deviations and the *mean squared errors* (MSEs) of 1000 replications of the conditional maximum likelihood estimators, the EM estimators with both Euler's method and the trapezium approximation method are given in Table 1.

From Table 1, we see that with increasing discretization frequency, the EM estimators based on the trapezium approximation method converge to the conditional maximum likelihood estimator much faster than the EM estimators based on Euler's approximation method. In fact, with the trapezium method, the EM estimators remain the same for the discretization frequency

Table 1: Averages, (standard deviations) and (MSEs) of 1000 simulations of the conditional ML and EM estimators of Model 1 (m = 1, 5, 10, 20, 40 and 80)

value of m		True Value	Euler's method	trapezium method
initial estimate	α_0	0	-0.0001(0.0690)(0.0048)	-0.0003(0.0692)(0.0048)
	α_1	-1	-0.6515(0.0886)(0.1293)	-0.6515(0.0894)(0.1294)
	σ^2	1	0.7825(0.1317)(0.0647)	0.7825(0.1319)(0.0647)
EM (m=1)	α_0	0	-0.0001(0.0690)(0.0048)	-0.0001(0.1167)(0.0136)
	α_1	-1	-0.6515(0.0886)(0.1293)	-1.0740(0.2843)(0.0863)
	σ^2	1	0.7830(0.1318)(0.0645)	1.0327(0.2479)(0.0625)
EM (m=5)	α_0	0	-0.0002(0.0906)(0.0082)	-0.0001(0.1176)(0.0138)
	α_1	-1	-0.8519(0.1421)(0.0421)	-1.0825(0.2929)(0.0926)
	σ^2	1	0.8889(0.1620)(0.0386)	1.0401(0.2565)(0.0674)
EM (m=10)	α_0	0	-0.0003(0.0989)(0.0098)	-0.0001(0.1176)(0.0138)
	α_1	-1	-0.9268(0.1741)(0.0357)	-1.0827(0.2931)(0.0927)
	σ^2	1	0.9346(0.1806)(0.0369)	1.0402(0.2566)(0.0675)
EM (m=20)	α_0	0	-0.0003(0.1055)(0.0111)	-0.0001(0.1176)(0.0138)
	α_1	-1	-0.9849(0.2062)(0.0427)	-1.0827(0.2931)(0.0927)
	σ^2	1	0.9723(0.2003)(0.0409)	1.0402(0.2566)(0.0675)
EM (m=40)	α_0	0	-0.0004(0.1103)(0.0122)	-0.0001(0.1176)(0.0138)
	α_1	-1	-1.0257(0.2353)(0.0560)	-1.0827(0.2931)(0.0927)
	σ^2	1	1.0001(0.2191)(0.0480)	1.0402(0.2566)(0.0675)
EM (m=80)	α_0	0	-0.0005(0.1135)(0.0129)	
	α_1	-1	-1.0523(0.2594)(0.0700)	
	σ^2	1	1.0188(0.2354)(0.0558)	
conditional MLE	α_0	0	-0.0005(0.1185)(0.0140)	-0.0005(0.1185)(0.0140)
	α_1	-1	-1.0936(0.3050)(0.1018)	-1.0936(0.3050)(0.1018)
	σ^2	1	1.0509(0.2639)(0.0722)	1.0509(0.2639)(0.0722)

m larger than 10 in terms of the averages and the standard deviations of 1000 replications. For m larger than 10, there is a minor difference between the EM estimators based on the trapezium method and the conditional maximum likelihood estimator. This can be explained as follows: The objective function, $-2l_Y(\boldsymbol{\theta}, \sigma^2)$, is the same for these two methods up to contributions from the initial conditions.

We repeat the same simulation study for model 2. The results are given in Table 2. The conditional maximum likelihood estimator of model 2 does not have a closed form solution. So, the conditional maximum likelihood estimators are replaced by the maximum likelihood estimators obtained from the simplex nonlinear optimization method. The simplex method is executed by applying the DUMPOL subroutine of the Microsoft Fortran PowerStation IMSL package to minimize the right hand side of equation (10). For a detailed description of the implementation of the DUMPOL subroutine for estimating continuous-time autoregressive models, see Tsai and Chan (1999b).

From Table 2, we also see that the EM estimators based on the trapezium approximation method converge to the maximum likelihood estimator much faster than the EM estimators based on Euler's method. With the trapezium method, the EM estimators seem to converge to a limit soon after $m = 20$, but this is not the case with Euler's method. The EM estimators based on the trapezium method seem to converge to the ML estimator generated by the simplex method in terms of biases. The standard deviations of the EM estimators based on the trapezium method are also close to those of the ML estimator, although the standard deviations of the former seem to be larger than those of the latter a bit. It is also interesting to note that with an appropriate discretization frequency, the EM estimator based on Euler's method outperforms the EM estimator based on the trapezium method and the ML estimator in terms of biases and standard deviations of the estimates.

Tsai and Chan (1999b) proved an invariance property of the EM estimators under linear transformations when Euler's approximation method is used. It can be shown that the invariance property also holds when the trapezium approximation method is used. The proof is similar to that of Tsai and Chan (1999b), although it is more tedious. Now, we state the results in the following theorem without proof.

Theorem 4.1 *(Invariance property of the EM estimators under linear transformations)*

Let $\{Y_{t_j}\}_{j=0,1,\ldots,N}$ be a series of discrete-time data sampled from a CAR(p) model defined by

$$dX_t^{(p-1)} = (\alpha_0 + \alpha' X_t)dt + \sigma dW_t.$$

Table 2: Averages, (standard deviations) and (MSEs) of 1000 simulations of the ML and EM estimators of Model 2 (m = 1, 2, 5, 10 and 20)

value of m		True Value	Euler's method	trapezium method
initial estimate	α_0	0	0.0009(0.0475)(0.0023)	0.0011(0.0722)(0.0052)
	α_1	-2	-0.9314(0.1108)(1.1542)	-1.4092(0.1995)(0.3888)
	α_2	-1	-0.5389(0.0813)(0.2192)	0.2045(0.1012)(1.4611)
	σ^2	1	0.7549(0.1385)(0.0793)	0.5764(0.0980)(0.1890)
EM (m=1)	α_0	0	0.0009(0.0448)(0.0020)	0.0021(0.1043)(0.0109)
	α_1	-2	-0.8807(0.0806)(1.2593)	-2.0429(0.2486)(0.0636)
	α_2	-1	-1.0443(0.0476)(0.0042)	-1.0192(0.2658)(0.0710)
	σ^2	1	1.0717(0.1829)(0.0386)	1.0045(0.2897)(0.0839)
EM (m=2)	α_0	0	0.0015(0.0737)(0.0054)	0.0021(0.1040)(0.0108)
	α_1	-2	-1.4462(0.1467)(0.3282)	-2.0381(0.2422)(0.0601)
	α_2	-1	-1.1348(0.0929)(0.0268)	-1.0130(0.2616)(0.0686)
	σ^2	1	1.0365(0.1816)(0.0343)	0.9980(0.2841)(0.0807)
EM (m=5)	α_0	0	0.0019(0.0923)(0.0085)	0.0021(0.1041)(0.0108)
	α_1	-2	-1.8090(0.1971)(0.0753)	-2.0391(0.2428)(0.0605)
	α_2	-1	-1.1050(0.1475)(0.0328)	-1.0137(0.2625)(0.0691)
	σ^2	1	1.0228(0.2037)(0.0420)	0.9989(0.2854)(0.0815)
EM (m=10)	α_0	0	0.0020(0.0984)(0.0097)	0.0021(0.1041)(0.0108)
	α_1	-2	-1.9277(0.2164)(0.0521)	-2.0400(0.2434)(0.0608)
	α_2	-1	-1.0704(0.1861)(0.0396)	-1.0148(0.2638)(0.0698)
	σ^2	1	1.0135(0.2262)(0.0513)	1.0002(0.2871)(0.0824)
EM (m=20)	α_0	0	0.0021(0.1013)(0.0103)	0.0021(0.1042)(0.0109)
	α_1	-2	-1.9848(0.2276)(0.0520)	-2.0405(0.2438)(0.0611)
	α_2	-1	-1.0452(0.2153)(0.0484)	-1.0156(0.2647)(0.0703)
	σ^2	1	1.0065(0.2461)(0.0606)	1.0011(0.2882)(0.0831)
simplex method	α_0	0	0.0021(0.1041)(0.0108)	0.0021(0.1041)(0.0108)
	α_1	-2	-2.0425(0.2409)(0.0598)	-2.0425(0.2409)(0.0598)
	α_2	-1	-1.0194(0.2557)(0.0658)	-1.0194(0.2557)(0.0658)
	σ^2	1	1.0010(0.2793)(0.0780)	1.0010(0.2793)(0.0780)

Consider $\{Y^*_{t_j}\}_{j=0,1,\ldots,N}$, *where* $Y^*_{t_j} = aY_{t_j} + b$, $\forall j$, *and* $a \neq 0$, *a series of discrete-time data sampled from the corresponding CAR(p) model defined by*

$$dX_t^{*(p-1)} = (\alpha_0^* + \alpha^{*'} X_t^*)dt + \sigma^* dW_t.$$

Then, the two CAR models are related as follows: $(\alpha_0^*, \alpha^*, \sigma^*) = (a\alpha_0 - b\alpha_1, \alpha, a\sigma)$. *Furthermore, for a fixed* m, *let* $\{\hat{\alpha}_j\}_{j=0,\ldots,p}$ *and* $\hat{\sigma}^2$ *be the EM estimators of* $\{\alpha_j\}_{j=0,\ldots,p}$ *and* σ^2 *based on* $\{Y_{t_j}\}$; *similarly,* $\{\hat{\alpha}^*_j\}_{j=0,\ldots,p}$ *and* $\hat{\sigma}^{*2}$ *denote the EM estimators of* $\{\alpha^*_j\}_{j=0,\ldots,p}$ *and* σ^{*2} *based on* $\{Y^*_{t_j}\}$. *Then,* $(\hat{\alpha}_0^*, \hat{\alpha}^*, \hat{\sigma}^*) = (a\hat{\alpha}_0 - b\hat{\alpha}_1, \hat{\alpha}, a\hat{\sigma})$.

The above invariance properties imply that the choice of m is independent of the scale of the data.

5 Conclusion

We have applied the trapezium approximation method in an implementation of the EM algorithm proposed by Tsai and Chan (1999b) for estimating continuous-time autoregressive models. Simulation studies confirm that the EM estimators converge much faster to the ML estimator with the trapezium method than with Euler's method. Tsai and Chan (1999c) proposed a likelihood based approach for testing for nonlinearity with irregularly sampled data. Tsai and Chan (1999c) used Euler's method for approximating the integrals involved in calculating the score and the information matrix. It is interesting to investigate whether or not the power of their test may be enhanced by an alternative implementation of the test via the trapezium method.

Acknowledgments

The work of Tsai was partially supported by the Institute of Statistical Science, Academia Sinica, Taipei, Taiwan. Part of this work was done during Tsai's visit at the Institute of Statistical Science, Academia Sinica, Taipei, Taiwan and he gratefully thanks the institute for providing a very productive environment and computer facilities.

References

1. Anderson, B. D. O., and Moore, J. (1979), *Optimal Filtering*, Englewood Cliffs, N.J. : Prentice-Hall.

2. Belcher, J., Hampton, J. S. and Tunnicliffe Wilson, G. (1994), Parameterization of continuous time autoregressive models for irregularly sampled time series data, *Journal of the Royal Statistical Society*, Series B **56**, 141-155.
3. Bergstrom, A. R. (1990), *Continuous Time Econometric Modeling*, Oxford University Press, New York, NY.
4. Brockwell, P. J. (1993), Threshold ARMA processes in continuous time, 170-90 in *Dimension Estimation and Models*, ed. by H. Tong, World Scientific Publishing, River Edge, NJ.
5. Brockwell, P. J. and Davis, R. A. (1991), *Time Series: Theory and Methods*, Springer-Verlag, New York.
6. Brockwell, P. J. and Stramer, O. (1995), On the approximation of continuous time threshold ARMA processes, *Annals of the Institute of Statistical Mathematics* **47**, no. 1, 1–20.
7. Dempster, A. P., Laird, N. M. and Rubin, D. B. (1977), Maximum Likelihood from Incomplete Data via the EM Algorithm (with discussion), *Journal of the Royal Statistical Society*, Series B **39**, 1-38.
8. Harvey, A. (1989), *Forecasting, Structural Time Series Models and the Kalman Filter*, New York : Cambridge University Press.
9. Jones, R. H. (1980), Maximum Likelihood Fitting of ARMA Models to Time Series with Missing Observations, *Technometrics* **22**, 389-395.
10. Jones, R. H. (1993), *Longitudinal Data with Serial Correlation: A State-space Approach*, Chapman and Hall, London.
11. Karatzas, I. and Shreve S. E. (1991), *Brownian Motion and Stochastic Calculus*, Springer-Verlag, New York.
12. Louis, T. A. (1982), Finding the Observed Information Matrix when Using the EM Algorithm, *Journal of the Royal Statistical Society, Ser. B*, **44**, 226-233.
13. Milstein, G.N. (1995), *Numerical Integration of Stochastic Differential Equations*, Kluwer Academic Publishers: Dordrecht.
14. Øksendal, B. K. (1995), *Stochastic Differential Equations : An Introduction with Applications*, New York : Springer.
15. Press, W. H., Flannery, B. P., Teukolsky, S. A., and Vetterling, W. T. (1996), *Numerical recipes in Fortran 90 : the art of parallel scientific computing*, Cambridge, U. K.: Cambridge University Press.
16. Tanner, Martin A. (1991), *Tools for Statistical Inference : Observed Data and Data Augmentation Methods*, Springer-Verlag, New York.
17. Tong, H. (1990), *Non-linear Time Series : A Dynamical System Approach*, Oxford University Press, Oxford.
18. Tsai, H. and Chan, K. S. (1999a), A note on the covariance structure of

a continuous-time ARMA process, *To appear in Statistica Sinica.*

19. Tsai, H. and Chan, K. S. (1999b), A new EM method for estimating continuous-time autoregressive models, *Technical Report 285*, Department of Statistics and Actuarial Science, University of Iowa.
20. Tsai, H. and Chan, K. S. (1999c), Testing for nonlinearity with partially observed time series, *Technical Report 284*, Department of Statistics and Actuarial Science, University of Iowa.
21. Ward, R. C. (1977), Numerical computation of the matrix exponential with accuracy estimate, *SIAM J. Num. Anal.* **14**, 600-610.
22. Wu, J. C. F. (1983), On the Convergence Properties of the EM Algorithm, *The Annals of Statistics*, **11**, 95-103.

A NOTE ON KERNEL ESTIMATION IN INTEGRATED TIME SERIES

YINGCUN XIA, W.K. LI, HOWELL TONG
Department of Statistics and Actuarial Science,
The University of Hong Kong, Pokfulam Road, Hong Kong
E-mail: ycxia@hkucc.hku.hk, hrntlwk@hku.hk and htong@hku.hk

The kernel estimation method for ***integrated*** time series has recently received attention in the literature. See e.g. Xia, Li and Tong (1998), Karlsen and Tjøstheim (1998) and Phillips and Park (1999). Specifically, two relevant observations are discussed: (1) in contrast to a linear time series regression set-up, which has a convergence rate much faster than the case of stationary data, the convergence rate of kernel estimators is much slower in this case than in the case of stationary data and (2) the optimal bandwidth is $O(n^{-1/10})$ rather than $O(n^{-1/5})$ which holds for stationary data, thus suggesting that the procedures developed by Granger *et al.* (1997) have to be modified. This note gives some intuition behind these results and explores the problem of bandwidth selection for integrated time series.

1 Introduction

The kernel smoother method is a powerful tool to explore relationships between variables. See, for example, Härdle (1990). In recent years, the method has been extended to the analysis of time series. See, for example, Härdle and Vieu (1990), Masry and Tjøstheim (1995) and Fan and Gijbels (1996). A basic assumption for the extension is that the time series is strongly mixing. This condition holds for many time series models and extensive study of kernel smoother for strongly mixing data can be found in e.g. Andrews (1995) and Masry (1995). It is generally accepted that there is no material difference in the asymptotic properties of kernel smoothers between strongly mixing data and *i.i.d.* data.

A number of authors have applied the kernel smoother to integrated time series data (see, e.g., Granger *et al.*, 1997). Justification has been given in Karlsen and Tjøstheim (1998) and Phillips and Park (1999). An earlier work is Yakowitz (1993) about the nearest neighbor regression estimation for such a set-up. Integrated time series appear frequently in the econometric literature. It is well known that this kind of data has a much different performance from standard results in parametric analysis. See e.g. Hargreaves (1996). In nonparametrics, it is found that the convergence rate for the kernel smoother depends on the design of explanatory variables. See for example Fan and Truong (1993) and Hall *et al.* (1997). A natural question is whether the kernel estimation method can be justified for integrated time series. If so, what is the corresponding convergence rate? Our answer to the first question is in the affirmative. However, interestingly, the optimal bandwidth is of $O(n^{-1/10})$ rather than $O(n^{-1/5})$. Consequently, the convergence rate is much slower than that for *i.i.d.* data. The bandwidth selection for the kernel estimation of integrated time series is also investigated by simulations in this note. Some bandwidth selectors such as cross-validation selectors are totally data driven, whilst others such as the plug-in selector are not. We suggest that some totally data driven bandwidth selectors can still be used here.

Recently, much attention has been paid to the kernel smoother of sparse data. See for example Hall *et al.* (1997), Hall and Turlach (1997a,b). They have suggested that the bandwidth should be of $O(n^{-1/(5+\delta)})$, where $\delta > 0$, when the design is sparse. Our note not only lends support to their suggestion but also exhibits interesting models which generate sparse data.

Xia *et al.* (1998), Karlsen and Tjøstheim (1998) and Phillips and Park (1999) have independently considered similar problems. The last two papers have paid their attention to the detailed theoretical aspects. On the other hand, Xia *et al.* (1998) have employed more elementary methods (giving weaker theoretical results) and addressed additionally the choice of bandwidth for integrated data.

2 Convergence of Kernel Smothers in Integrated Time Series

Consider a simple nonparametric model

$$y_t = g(x_t) + \varepsilon_t,$$

where $g(\cdot)$ is an unknown function, ε_t's are *i.i.d.* random variables and independent of $\{x_t\}$. Here, $\{x_t\}$ is an integrated time series with order 1 and follows the process

$$x_t = c + \sum_{j=1}^{t} \xi_j,$$

where ξ_j's are *i.i.d.* random variables, as considered by Granger *et al.* (1997). It is easy to see that

$$g(x) = E(y_t | x_t = x).$$

Therefore, a natural kernel estimator of $g(x)$ is

$$\hat{g}(x) = \sum_{t=1}^{n} K_h(x_t - x) y_t \Big/ \Big[\sum_{t=1}^{n} K_h(x_t - x) + \delta_n \Big], \tag{1}$$

where $K(\cdot)$ is a kernel function, $K_h(\cdot) = K(\cdot/h)$, h is a bandwidth and δ_n is a positive constant which tends to zero as $n \to \infty$ so that the problem of 'zero-divisor' is avoided. See e.g. Hall *et al.* (1997).

It is not difficult to see that if $E(\xi_j) \neq 0$, then $\hat{g}(x)$ does not converge unless $g(x)$ is a linear function. In order to investigate the asymptotic properties of $\hat{g}(x)$, we further make the following assumptions.

(A1) ε_t's are *i.i.d.* random variables and $E\varepsilon_t = 0$, $E\varepsilon_t^2 = \sigma^2$.

(A2) ξ_j's are *i.i.d.* random variables with a density function, $E\xi_j = 0$, $E\xi_j^2 = 1$ and $E\xi_j^3 < \infty$. The characteristic function $\phi(x)$ of ξ_j satisfies $\phi(x) = (1 - |x|^2)(1 + o(1))$ as $x \to 0$ and $|\phi(x)| = O(|x|^{-\delta})$ as $|x| \to \infty$ for some $\delta > 0$.

(A3) $g(x)$ has a bounded third derivative.

(A4) The kernel function $K(\cdot)$ is a symmetric density function with compact support and $\int u^2 K(u)du = 1$.

(A1), (A3) and (A4) are very common assumptions for kernel smoothers. (A2) is assumed in order to facilitate the use of some properties of sums of *i.i.d.* random variables (see, Petrov, 1975, p.175) and the occupation times theorems of Markov chain (see, Darling and Kac, 1957). Distributions such as normal, uniform, double exponential and many others satisfy (A2). An interesting problem is that if $E\xi_j^2 = \infty$, then the data set is more sparse than the case of (A2). We shall discuss this problem later.

Notice that if we replace x_t by z_t, which is a sequence of *i.i.d.* random variables with density $f(x)$, then

$$E\sum_{t=1}^{n} K_h(z_t - x) = nhf(x)(1 + o(1)). \tag{2}$$

and

$$\frac{1}{nh}\sum_{t=1}^{n} K_h(z_t - x) = f(x) + o_p(1).$$

However, in our case, the distribution of x_t changes with t. We have the following two theorems (Karlsen and Tjøstheim, 1998, Theorem 4.3; Phillips and Park, 1999, Theorem 3.7 and Xia, Li and Tong, 1998)

Theorem 1. Suppose (A2) holds, $h \to 0$ and $\sqrt{n}h \to \infty$. Then for any x

$$E\sum_{t=1}^{n} K_h(x_t - x) = \frac{1}{\sqrt{2\pi}}\sqrt{n}h(1 + o(1))$$

and

$$\sum_{t=1}^{n} K_h(x_t - x) = O_p(\sqrt{n}h).$$

In fact, $\sum_{t=1}^{n} K_h(x_t - x)/(\sqrt{n}h)$ tends to a nondegenerated random variable. Comparing the first part of Theorem 1 with (2), we can expect that the local linear smoother and the Nadaraya-Watson kernel smoother are asymptotically equivalent for the integrated time series by analogy with the case of uniform design density. See Fan and Gijbels (1996, p.17). Although $\sum_{t=1}^{n} K_h(x_t - x)/(\sqrt{n}h)$ does not converge to a nonrandom constant, we can still prove that the kernel smoother is a consistent estimator of the unknown regression function.

Theorem 2. Suppose assumptions (A1)-(A4) hold, $h \to 0$ *and* $\sqrt{n}h \to \infty$. *Then for any* x

$$\hat{g}(x) - g(x) = O_p(h^2 + (\sqrt{n}h)^{-1/2}).$$

From Theorem 2, if we choose the bandwidth $h_{int} \propto n^{-1/10}$, where '$\propto$' means that both sides have the same order, then $\hat{g}(x)$ achieves the optimal convergence rate $O_p(n^{-1/5})$, which is slower than $O_p(n^{-2/5})$, the convergence rate of kernel smoother of *i.i.d.* data. It is known that the optimal bandwidth in the sense of convergence rate for *i.i.d.* data is $h_{iid} \propto n^{-1/5}$ and therefore h_{int} is asymptotically much wider than h_{iid}.

We can give an intuitive explanation for the difference between h_{int} and h_{iid}. In the case of a scalar integrated time series, x_t is for each fixed t a sum of independent random variables. By Theorem 7 of Petrov (1975, p.175) we have

$$E\#\{x_t;\ |x_t - x| \le \Delta,\ t = 1,\ \cdots,\ n\} \propto \Delta\sqrt{n}, \tag{3}$$

where $\#A$ denotes the number of elements in set A and Δ is a small positive number. In contrast, for the *i.i.d.* data z_t, $t = 1, \cdots, n$, if the common density is positive at x, we have

$$E\#\{z_t;\ |z_t - x| \le \Delta,\ t = 1,\ \cdots,\ n\} \propto \Delta n. \tag{4}$$

Comparing (3) with (4), we describe the situation by saying that the integrated time series data are much more sparse in the neighborhood of x than the *i.i.d.* data. Alternatively, the integrated series is an example of a null recurrent Markov chain, and it is therefore obvious that the rate of filling up a fixed neighborhood with sufficiently many points is slower due to the slow recurrence rate of the chain. For the integrated time series data, there are much fewer observations which can usefully contribute to the estimation of $g(x)$ than the case with *i.i.d.* data. Therefore, it is not surprising that the convergence rate is slower and the bandwidth wider. Furthermore, if $E\xi_j^2 = \infty$ and $E|\xi_j|^{1+\delta} < \infty$ for some $0 < \delta < 1$ (but $E|\xi_j|^{1+\delta'} = \infty$ for $\delta' > \delta$), then we have

$$E\#\{x_t;\ |x_t - x| \le \Delta,\ t = 1,\ \cdots,\ n\} \propto \Delta n^{\delta/(1+\delta)},$$

which implies that x_t's are more sparse than the case under (A2). The optimal bandwidth is

$$h \propto n^{-\delta/(5+5\delta)}, \qquad (0 < \delta < 1).$$

Granger *et al.* (1997) suggested that the bandwidth for the integrated time series is of $O(n^{-1/5})$. This suggestion is invalid under our assumptions.

To see the performance of the bandwidth and the convergence rate for kernel smoothers for integrated time series, we carry out a simulation study. Suppose that $x_t = 0.1 + \sum_{j=1}^{t} \xi_j$ with $\xi_j \sim N(0,\ 0.01)$,

$$y_t = 1 - x_t^2 + \varepsilon_t, \qquad \varepsilon_t \overset{i.i.d.}{\sim} N(0,\ 0.01). \tag{5}$$

In order to compare the difference between the integrated time series data and the *i.i.d.* data, we consider another set-up

$$y_t = 1 - z_t^2 + \varepsilon_t, \quad z_t \overset{i.i.d.}{\sim} N(0.1,\ 4), \quad \varepsilon_t \overset{i.i.d.}{\sim} N(0,\ 0.01). \tag{6}$$

We generate 500 pairs of samples from models (5) and (6) for $n = 400, 800, 2000, 5000$ and 10000. Note that models (5) and (6) have the same innovation variance. We estimate the regression functions $g(x) = 1 - x^2$ at $x = 0.1$. The sample means of squared errors of $\hat{g}(0.1)$ against the bandwidths are shown in Figure 1. Comparing Figure 1(a) with Figure 1(b), the optimal bandwidths for the integrated time series data change much more slowly than those for the *i.i.d.* data as n changes and the convergence rate for the integrated time series data is slower than that for the *i.i.d.* data.

3 Bandwidth Selection

Bandwidth selection always plays an important role in kernel smoother. A sparse design entails a wider bandwidth and a slower convergence rate. Since the cross-validation selector is totally data-driven, it changes according to the

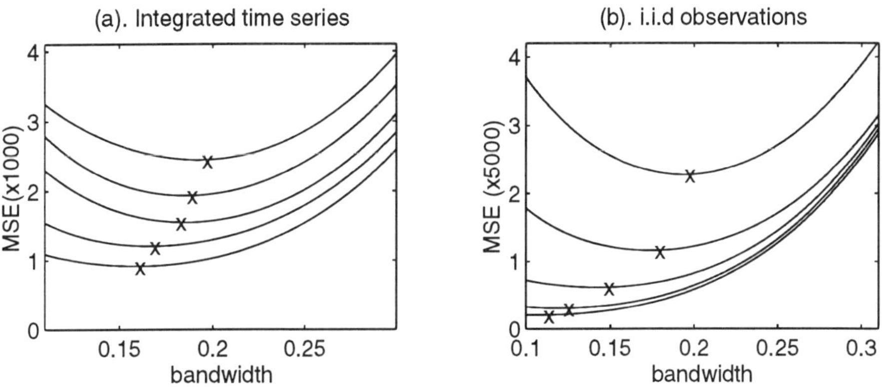

Figures 1: (a) and (b) are simulation results for model (5) and model (6) respectively. The solid curves are the relations of means of squared errors (MSE) against the bandwidth. The sample sizes are 400, 800, 2000, 5000 and 10000 as we move from the uppermost solid curve downwards. '×' is the optimal choice in each case.

data design automatically. We conjecture that the cross-validation can also work here. However, theoretical results are unavailable as we have been unable to overcome the technical difficulties pertinent to nonstationary data. Instead, we use simulations to gain some insight into the feasibility of the cross-validation method.

Suppose that we are interested in an interval $[a,\ b]$ and a constant bandwidth for the kernel smoother on this interval. Let $\hat{g}_t(x)$ be the estimator of $g(x)$ as in (1) using data $\{(x_s,\ y_s) :\ s \neq t\}$. Then the bandwidth selected by the cross-validation is

$$h_C = \arg\min_h \sum_{x_t \in [a,\ b]} \left\{ \hat{g}_t(x_t) - y_t \right\}^2 .$$

In the case of *i.i.d.* observations, it is well known that h_C is a consistent estimator of the minimizer, h_{ISE}, of $ISE(h)$, where

$$\mathrm{ISE}(h) = \int_a^b \left\{ \hat{g}(x) - g(x) \right\}^2 f(x) dx.$$

See e.g. Jones (1991).

Now, consider model (5) and the interval $[-0.5,\ 0.5]$. We generate 100 samples from (5) for total sample sizes n=200, 400, 800, 1500 and 3000 respectively. Table 1 lists the simulation results. It shows that the cross-validation bandwidth is close to h_{ISE}. To see the relation between the bandwidth h_{ISE} and sample size n, we list the ratios of $h_{ISE}/n^{-1/10}$ and $h_{ISE}/n^{-1/5}$ in Table 1. We can see that the bandwidth is proportional to $n^{-1/10}$, which therefore lends support to our theoretical results. Consequently, it is natural that $h_{ISE}/n^{-1/5}$ increases as n goes large.

TABLE 1: The mean of cross-validation bandwidth and the mean of optimal bandwidth in the sense of ISE for model (5)

n	h_C	h_{ISE}	$h_{ISE}/n^{-1/10}$	$h_{ISE}/n^{-1/5}$	mean of $\lvert h_C - h_{ISE}\rvert$
200	0.1748	0.1788	0.3037	0.5159	0.0476
400	0.1635	0.1710	0.3113	0.5668	0.0436
800	0.1612	0.1550	0.3024	0.5901	0.0394
1500	0.1448	0.1438	0.2988	0.6208	0.0359
3000	0.1298	0.1307	0.2911	0.6482	0.0322

References

1. D.W.K. Andrews, Nonparametric kernel estimation for semiparametric models. *Econometric Theory* **11**, 560–596 (1995).
2. D.A. Darling and M. Kac (1957). On occupation times for Markoff processes. *Trans. Amer. math. Soc.* **84**, 444–458 (1957).
3. J. Fan. and I. Gijbels, *Local Polynomial Modeling and Its Applications.* (Chapman & Hall, London, 1996).
4. J. Fan and Y.K. Truong, Nonparametric regression with errors in variables. *Ann. Statist.* **21**, 1900–1925 (1993).
5. C.W.J. Granger, T. Inoue and N. Morin, Nonlinear stochastic trends. *J. of Econometrics* **81**, 65–92 (1997).
6. P. Hall and B.A. Turlach, Enhancing convolution and interpolation methods for nonparametric regression. *Biometrika* **84**, 779–790 (1997a).
7. P. Hall and B.A. Turlach, Interpolation method for adapting to sparse design in nonparametric regression (with discussion). *J. Amer. Stat. Ass.* **92**, 466–476 (1997b).
8. P. Hall, J.S. Marron, M.H. Neumann and D.M. Titterington, Curve estimation when design density is low. *Ann. Stat.* **25**, 756–770 (1997).
9. C.P. Hargreaves, *Nonstationary Time Series Analysis and Cointegration.* (Oxford University Press, 1996).
10. W. Härdle, *Applied Nonparametric Regression.* (Cambridge University Press, Boston, 1990)
11. W. Härdle and P. Vieu, Kernel regression smoothing of time series. *J. Time Ser. Anal.* **13**, 209–224 (1992).

12. M.C. Jones, The roles of ISE and MISE in density estimation. *Statistics & Probability Letters* **12**, 51–56 (1991).
13. H. Karlsen and D. Tjøstheim, *Nonparametric estimation in null recurrent time series.* Discussion paper 50, Sonderforschungsbereich 373, Wirtschaftswissensshaftliche Fakultät, Humboldt-Universität zu Berlin, Spandauer Strasse 1, 10178 Berlin, Germany, (1998).
14. E. Masry, Multivariate local polynomial regression for time series: uniform strong consistency and rates. *J. Time Ser. Anal.* **17**, 571–599 (1996).
15. E. Masry and D. Tjøstheim, Nonparametric estimation and identification of nonlinear ARCH time series: strong convergence and asymptotic normality, *Econometric Theory* **11**, 258–289 (1995).
16. V.V. Petrov, *Sums of Independent Random Variables.* (Springer-verlag, New York, 1975).
17. P.C.B. Phillips and J. Park, Nonstationary density estimation and kernel autoregression. *Econometric Theory* **15**, 269–298 (1999).
18. Y. Xia, W.K. Li and H. Tong, *A note of kernel estimation in integrated time series.* Technical Report #189, (1998), Department of Statistics and Actuarial Science, the University of Hong Kong.
19. S. Yakowitz, Nearest neighbor regression estimation for null-recurrent Markov time series. *Stoc. Proc. Appl.* **48**, 311–318 (1993).

Part II: Long Memory and Value at Risk

STYLIZED FACTS ON THE TEMPORAL AND DISTRIBUTIONAL PROPERTIES OF ABSOLUTE RETURNS: AN UPDATE

CLIVE W.J. GRANGER, SCOTT SPEAR
Department of Economics, University of California, San Diego, La Jolla, CA 92093-0508, USA
E-mail: cgranger@weber.ucsd.edu

ZHUANXIN DING
Frank Russell Company, Tacoma, WA 98402, USA

The possibility of specific long-memory temporal properties exponential marginal distributionals of absolute returns are considered for daily data for a number of markets and similar results are found in each case. Possible explanations are considered but no complete explanation is found. A fractionally integrated model is considered, found to require an unusual distribution for its inputs, has a poor forecasting performance, and its properties may be explained by a regime-switching process.

1 Introduction

If r_t is the return derived from a price series P_t by $r_t = \log P_t - \log P_{t-1}$, a simple decomposition is, given the mean return m,

$$r_t - m = \text{sign}\ (r_t - m) \cdot |r_t - m| \tag{1}$$

where

$$\text{sign}\ (x) = \begin{cases} 1 & \text{if } x > 0 \\ 0 & \text{if } x = 0 \\ -1 & \text{if } x < 0 \end{cases}$$

and $|r_1 - m|$ is the absolute return about its mean. Two earlier papers, Ding, Granger, and Engle (1993) (henceforth DGE) and Granger and Ding (1995) (henceforth GD) considered various aspects of the components of the decomposition using a long series of daily data for the Standard and Poors Index of U.S. stocks. These papers concentrated on the temporal properties of $|r_t - m|^\theta$ for various values of θ, but particularly for $\theta = 1$ and 2 and also on the marginal distributional property of $|r_t|$. In the second of these papers it was pointed out the Luce (1980) had shown from an axiomatic argument that $|r_t - m|^\theta$ was an appropriate class of risk-measures to consider, with the value of θ at the choice of the individual investor. This possibility is not followed up here.

In this paper the essential empirical facts found in the first two papers mentioned above are summarized and the question is then asked if similar properties are found elsewhere in speculative markets, for stock indices from other exchanges, for individual shares, for commodity prices, exchange rates and for interest rates. In all cases the answer appears to be yes, very similar stylized facts are found in these other markets. The paper then turns to asking the question – why does this occur? Although several plausible reasons can be suggested, the similarity in all of the markets still requires an explanation.

The data used in the first two papers was the S&P 500 stock market daily closing prices for the period January 3, 1928 to August 30, 1991, giving altogether 17,055 observations (the data was kindly supplied by William Schwert). Various parts of the analysis was also repeated for ten subsamples of equal size, this division being arbitrary. The mean return for the whole period was 0.00018 and so autocorrelations for powers of absolute returns changed very little whether or not a mean was first subtracted from returns. Some of the analysis used an "outlier reduced" series in which any value outside $\pm 4\hat{\sigma}$ was replaced by $4\hat{\sigma}$ or $-4\hat{\sigma}$ as appropriate, where $\hat{\sigma}$ is estimated from the raw data. This reduction was selected in an arbitrary way but some experimentation suggested that if the value 4 was changed somewhat the results were not substantially altered.

Some of the temporal properties established are:

TP1: returns are not autocorrelated (except possibly at lag one) (DGE Table 3.1, Figure 3.1), consistent with a form of the efficient market theory.

TP2: $|r_t|$, r_t^2 are "long-memory" with $\text{corr}(|r_t|, |r_{t-k}|) > \text{corr}(r_t^2, r_{t-k}^2)$.

The autocorrelations of $|r_t|$, r_t^2 die out slowly, much slower than the exponential rate of a typical stationary ARMA model, for example (DGE, Table 3.1, Figure 3.1). An important component of this "fact" is that the estimated autocorrelations stay positive for many lags, the first $\hat{\rho}_k$ that is negative is at $k = 2598$ for r_t^2 and at $k = 2705$ for $|r_t|$ using the complete S&P daily sample.

TP3: $\text{corr}(|r_t|, |r_{t-k}|) > \text{corr}(|r_t|^\theta, |r_{t-k}|^\theta)$, $\theta \neq 1$.

Autocorrelations of powers of absolute returns are highest at power one, which clearly agrees with TP2 (DGE Table 3.2, Figures 3.2 to 3.7). This is called the "Taylor Effect" in GD as it was partly observed in Taylor (1986).

A second effect, called the "Machina Effect", that $\text{corr}(|r_t|^\delta, |r_t|) > \text{corr}(|r_t|^\delta, |r_t|^\theta)$ for any δ and any $\theta \neq 1$ that is discussed in GD and found to be approximately true will not be considered further here.

TP4: The autocorrelations of sign (r_t) are all insignificant.

According to Table 4 of GD, this is, at best, only approximately true, as some low lag autocrrelations appear to be significant, particularly in the sub-samples. However, the series sign (r_t) is clearly short-memory.

The "long-memory" property of the autocorrelation of absolute returns may indicate that the series are fractionally integrated, or $\mathrm{I}(d)$, processes where $0 < d < 1$, although other processes also have the property, as discussed in Granger and Ding (1996). In DG, d is estimated for the full sample, given $\hat{d} = 0.474$ (standard error $= 0.058$) but for the ten sub-samples, $\hat{d}$ is usually in the range 0.336 to 0.446 with the exception of a value of $\hat{d} = 0.156(.106)$ in the early 1960's and of $\hat{d} = 0.714(.105)$ in the late 1970's. It thus seems that d is neither zero nor 1 and is probably not constant.

A difficulty with this finding is that most of the standard tests available in statistics either assume the series is iid or, at least, is short memory. The long-memory property can substantially change the basic properties of familiar tests. Results on the properties of tests using $\mathrm{I}(d)$ processes are just being developed, for example Beran and Ghosh (1991), Beran (1994), but many gaps remain.

Turning to the distributional properties,

DP1: $|r_t|$ and sign (r_t) are independent (contemporaneously).

If the conditional distribution of $|r_t| \mid$ sign $(r_t) = 1$ is the same as the distribution of $|r_t| \mid$ sign $(r_t) = -1$ (assuming prob (sign $(r_t) = 0) = 0$) this is equivalent to DP1. Using histograms for these two conditional distributions, their equality can be tested using a Kolmogorov-Smirnov test. GD rejects DP1 for the full sample but not for all sub-samples and they present evidence that it is approximately correct.

DP2: mean $|r_t|$ = standard deviation $|r_t|$.

DP3: marginal distribution $|r_t|$ is exponential (after outlier reduction).

Outlier reduction as defined above does not greatly impact estimates of the mean and standard deviation of the absolute return. For the full sample of over 17,000 terms, a total of 138 terms were outlier reduced (64 positive, 74 negative). The same procedure applied to subsamples produced an average of about 8 per subsample, the most being 11, the least 4 out of samples of 1700 terms. Naturally, outlier reduction produces substantial reduction on estimates of higher moments. It may be noted that an exponential distribution with parameter $-\ \alpha$ has:

mean = standard deviation $= \alpha$
skewness $\left(= \mu_3/\mu_2^{3/2} \right) = 2$
kurtosis $\left(= \mu_4/\mu_2^2 \right) = 9$

regardless of the value of α. DGE, Table 2A shows that DP2 holds approximately for the full sample and all ten sub-samples for both the original data and also outlier reduced data, and further that DP3 holds quite closely for the outlier reduced data, except possibly for sub-samples 8 and 9. For the full sample skewness is 2.4 and sub-sample values range from 1.5 to 2.3 and kurtosis is 10.3 for the full sample and ranges over 6.0 to 10.3 for the sub-samples. Naturally, much larger kurtosis values are usually achieved if outlier are not removed, with a range from 7.2 to 207 (thanks to October, 1987). The estimate of α varies across subsamples.

The next sections examine if some of these properties are found in other series.

2 New Data Sets

The analysis will now be extended to cover the following series:

2.1 Stock Markets

(1) CAC-40 stock price index, Paris, 07/02/88 - 09/30/91. $N = 1382$.

(2) Nikkei stock average, Tokyo, 01/03/71 - 09/30/91. $N = 5594$.

(3) Chevron share prices, NYSE, 07/02/82 - 12/31/91. $N = 2365$.

(4) Ten individual share prices, every third stock in the Dow-Jones Thirty, 07/02/62 - 12/31/91. $N = 7900$.

2.2 Commodity Markets (prices supplied by Dr. Walter Labys)

(5) Silver prices, Jan 1968 - March 1989, $N = 5287$.

(6) Copper prices, Jan 1968 - March 1989, $N = 5308$.

(7) Sugar prices, Jan 1971 - March 1989, $N = 4981$.

(8) Coffee prices, Oct 1972 - March 1989, $N = 4087$.

2.3 Interest Rates (provided by Dr. Pierre Siklos)

(9) Daily U.S. Treasury Bill rate for the period 01/01/85 - 10/07/93, $N =$ 2180.

3 Correlograms of Powers of Absolute Returns

This section will discuss estimates of $\text{corr}(|r_t|^\theta, |r_{t-k}|^\theta)$ plotted against k, for the series mentioned in the previous section (results for individual share prices are discussed later). These correlograms can take a variety of shapes, some of which are illustrated in Figures 1 to 2. Figures 1 and 2 show autocorrelations up to lag 200 for copper and coffee prices respectively. The top figure in each case superimposes autocorrelations for r_t, $|r_t|$ and r_t^2 and it is seen that those for r_t generally lie within the 95% confidence interval for the hypothesis that the true autocorrelation is zero, and the autocorrelations for $|r_t|$ are above those for r_t^2, in agreement with stylized fact TP2. The middle figure shows the correlograms for $|r_t|^\theta$, for $\theta = 1$, 0.75, 0.5 and 0.25 and the lower figure shows the correlograms for $\theta = 1$, 1.25, 1.50, 1.75 and 2. For copper, in Figure 1b, the correlations are generally larger for $\theta = 1$ than other values and then declines as θ declines, but this distinction is not clear for higher lag values, say $k > 100$. For coffee in Figure 2b it is seen that the correlograms are indistinguishable for θ values between 0.25 and 1, whereas in Figure 2c a clear ordering occurs as θ increases from 1 to 2, with declining memory.

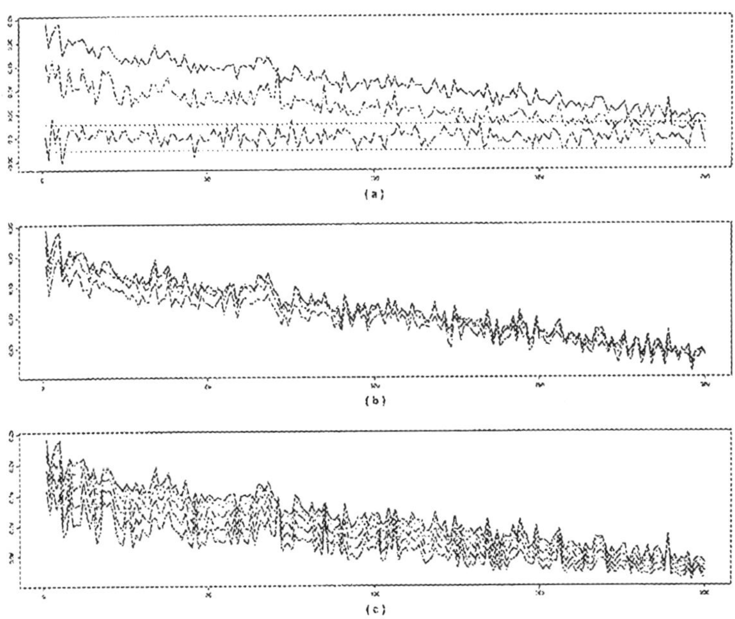

Figure 1: Autocorrelations of r_t and $|r_t|^\theta$ for returns of Copper prices

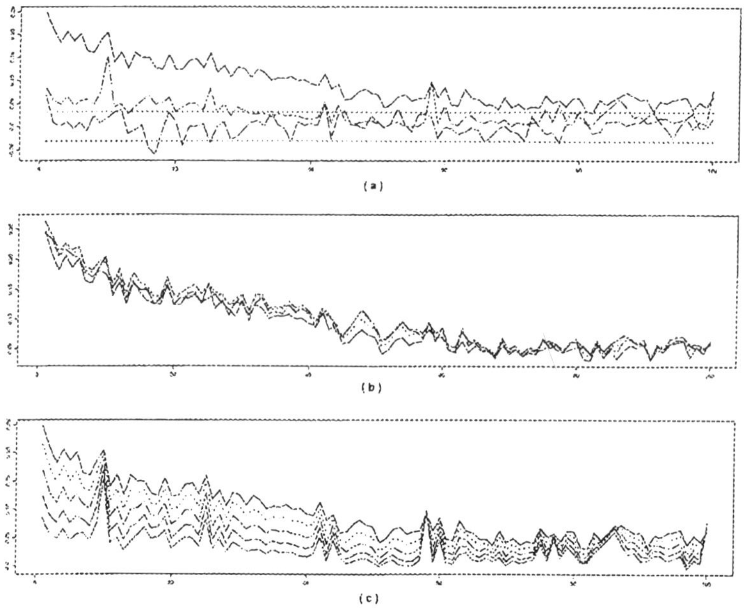

Figure 2: Autocorrelations of r_t and $|r_t|^\theta$ for returns of Coffee prices

In Figure 3, the correlograms for $\theta = 1$ (absolute returns) and $\theta = 2$ (variances) for daily Coca-Cola prices is shown for roughly 300 lags. The case for $\theta = 1$ is seen to have longer, positive memory than for $\theta = 2$. This is a typical finding for all the series analyzed.

These figures illustrate groups of correlograms that do not change as θ changes (described as "flat") and those that steadily increase or decrease as θ increases. These patterns may hold for all of the lags shown in a diagram of just for some sub-set.

There is clearly too much information generated by the series considered to be fully presented in the paper, so the results are summarized in Table 1:

Table 1: Major Properties of Correlograms of $|r_t|^\theta$

Series	$\theta < 0.25$	$0.25 \leq \theta \leq 1$	$1 \leq \theta \leq 1$
Stock Prices			
NYSE Price Indices (S&P)	Clear Increase	Increase to $\theta = .5$ Flat .5 to 1 as $\theta \uparrow$	Clear decrease as $\theta \uparrow$
Paris Index (CAC-40)		Clear increase as $\theta \uparrow$ up to $k = 10$, flat $k > 15$ to 50	Flat from $k \leq 5$ Clear decrease as $\theta \uparrow$ for $k > 5$
Tokyo Index (Nikkei)		Increase .125 to .5 lower lags flat .5 to 1 all lags, and all θ value for $k > 50$.	Clear decrease as $\theta \uparrow$
Chevron		Clear increase as θ increases for $k < 100$, less clear for $k \geq 100$	Clear decrease as $\theta \uparrow$
Commodity Prices			
Silver		Flat	Clear decrease
Sugar		Very Flat	Clear decrease until $k = 100$
Copper		Some increase	Clear decrease
Coffee		Flat	Clear decrease
Interest Rates			
Treasury 3 month		Clear increase	Clear decrease
Ten Individual Stocks			
Memory clearly increased as θ went from .1 to.5. Then flat from .7 to 1.0; no change in memory as θ changed from 1 to 1.5 Small but distinct reduction in memory as θ goes from 1.7 to 2.0.			

For each series correlograms were estimated for θ values 0.25, 0.5, 0.75, 1.0, 1.25, 1.5, 1.75 and 2 and for a few series the value 0.125 was also used. The typical result has the correlogram for $\theta = 1$ higher than for higher θ values and also often for lower values, but $\theta = 0.75$, $\theta = 0.5$ may have similar correlograms. Thus, for the stock prices, commodity prices and interest rate clear evidence is found for the temporal properties TP1 and TP2, mentioned in the first section, and some evidence for TP3, the Taylor effect. The common properties within a group of series might be explained by some common factor, but the similarities across groups is perhaps more surprising.

The remaining temporal property, TP4, concerns the auto-correlations of the sign (r_t). For virtually every series some short memory was observed, with one or a few significant correlations, but no evidence of any memory beyond that. The evidence is not presented to save space.

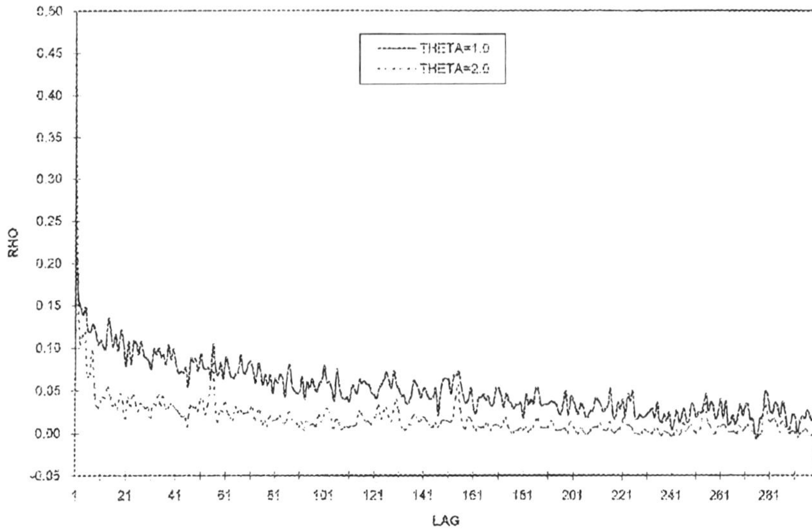

Figure 3: Correlograms for $|r_t|^\theta$ for daily Coca-Cola prices

4 Distributional Properties

In Table 2, the mean and standard deviation of the original absolute return series and this ratio is shown, and then these same quantities plus the skewness and kurtosis measures for the outlier reduced data. For each series these quantities are shown conditionally for $r_t > 0$ and $r_t < 0$ (denoted just +, −) together with the relative frequencies (called probabilities) of those signs occurring. As pointed out in above, if the two conditional distributions are the same this implies that the variables $|r_t|$ and sign (r_t) are independent. This is tested using the Kolmogorov-Smirnov test statistic D. In Column 10, "out" indicates the number of outliers reduced in each sub-sample. Table 2A shows the distributional properties for the outlier adjusted absolute returns for ten individual stock prices, plus the standard deviation/mean ratio for the unadjusted series.

The table contains a great deal of information. Some of the obvious "facts" can be summarized as:

(1) The marginal distribution of r_t is approximately symmetric, with Prob $(r_t > 0) = 1/2$, but not exactly so, according to the figures in column (1). If the r_t were i.i.d., the standard deviation for the estimate of Prob$(r_t >$ $0)$ would be 0.004 for the NYSE series. It is not clear how the long-

memory property of $|r_t|$ affects this value.

(2) In all cases, the estimated mean and standard deviation values are approximately equal both for the unadjusted series, according to column 4, and the adjusted series, column 7. Mean daily returns are, of course, very small, see columns 2 and 5, so outlier reduction has little affect, but standard deviation is naturally somewhat reduced by the reduction. There appears to be a slight tendency for the mean/standard deviation ratio to be nearer to unity for the adjusted series but the effect is not overwhelming. Thus, the proposition DP2, introduced in section 1, appears at least visually to be plausible, even though no formal test has been applied. The same can be said for DP3, that the marginal distributions of the outlier adjusted series have the appearance of being at least fairly well approximated by an exponential distribution, according to their first four moments. The skewness coefficients are all positive, are near the hypothesized value of two, with a range of 1.591 (silver +), to 2.487 (NYSE +). Similarly the kurtosis coefficients are generally quite near the hypothesized value of 9, with a range of 6.26 (silver +) to 11.44 (Tokyo stock index +). Of course, one cannot take these details very seriously as the series have been outlier reduced in an arbitrary fashion and are not independent either through time or between themselves. What matters is that the suggestive proposition DP2, DP3 that originally were suggested by the NYSE data have been found to hold in a very similar way in various series from speculative markets.

(3) The other stylized fact, DP1, that a $|r_t|$ and sign (r_t) are independent, which corresponds here to the Kolmogorov-Smirnov Statistic D being insignificant, or at least small, is found to hold for about half of the series and to be very strongly rejected for interest rates. It is acceptable for the NYSE, silver and coffee prices if the usual rule of allowing wider critical intervals with very large samples is followed. Apart from interest rates DP1 would be a reasonable approximation to make when building a simple, parsimonious model but consideration of relaxing it would be necessary for some series. However, the long-memory property of absolute returns affects the Kolmogorov-Smirnov test in ways that have yet to be determined.

The table just considers the marginal distributions for $|r_t|$, and could be repeated for $|r_t|^\theta$, for different values of θ, but these calculations have not been performed. In general terms, the exponential distribution values for skewness and kurtosis occurs in the range $\theta = 0.9$ to 1.1 for outlier adjusted data.

Table 2. Properties of Marginal Distributions of Series

		Not Adjusted			Outlier Adjusted						
	1	2	3	4	5	6	7	8	9	10	11
	Prob	Mean ×100	St.Dev ×100	Mean/S:D	Mean ×100	St.Dev ×100	Mean/S:D	Skew	Kurt	Out	D
Stock Prices											
S&P 500	+0.52	0.70	0.80	0.875	0.70	0.70	1.000	2.487	11.12	64	0.19
	−0.46	9.80	0.90	0.889	0.80	0.80	1.000	2.303	09.45	74	($\rho = 11$)
Tokyo	+0.52	0.56	0.64	0.877	0.55	0.55	0.989	2.463	11.44	19	0.41
	−0.48	0.61	0.75	0.818	0.48	0.60	0.942	2.235	09.19	22	
Paris	+0.52	0.84	0.88	0.952	0.82	0.78	1.054	2.113	09.84	4	0.88
	−0.48	0.89	1.02	0.875	0.87	0.93	0.942	2.220	09.18	9	
Chevron	+0.46	1.26	1.13	1.118	1.25	1.09	1.146	1.806	06.44	16	0.48
	−0.44	1.19	1.04	1.146	1.18	0.98	1.208	1.811	07.23	9	
*Commodity Prices**											
Silver	+0.51	1.47	1.43	1.025	1.45	1.32	1.099	1.591	06.26	3	.036
	−0.49	1.50	1.80	0.833	1.47	1.41	1.048	1.725	07.04	8	
Sugar	+0.49	2.50	2.61	0.958	2.46	2.33	1.053	2.083	08.79	16	.058
	−0.51	2.36	2.37	0.996	2.34	2.21	1.059	1.827	07.68	8	
Copper	+0.50	1.29	1.11	1.157	1.29	1.10	1.171	1.743	07.31	9	0.69
	−0.50	1.27	1.29	0.981	1.25	1.12	1.042	1.778	06.44	14	
Coffee	+ 0.53	1.41	1.39	1.011	1.40	1.35	1.040	2.010	08.39	8	0.48
	−0.47	1.42	1.61	0.885	1.40	1.35	0.987	2.049	08.52	13	
Interest Rate											
TB3M	+ 0.53	0.71	0.81	0.884	0.71	0.76	0.930	2.070	08.70	6	.150
	−0.47	0.80	0.94	0.854	0.78	0.77	1.007	2.310	09.47	8	

* For commodity prices the Critical Values for D are 1%=0.45, 5%=0.38

Table 2A. Marginal Distributions of Individual Share Series

	Mean ×100	S.D. ×100	Outlier Adjusted Ratio S.D./Mean	Skew-ness	Kurt-osis	Non Adjusted Ratio S.D./Mean
Bethehem	1.45	1.43	98	1.90	7.9	107
Coca-Cola	1.05	.99	95	1.89	7.9	102
Disney	1.44	1.35	94	1.81	7.4	100
Dupont	1.02	.94	92	1.68	7.7	95
Exxon	.89	.80	90	1.66	7.9	98
Gen. Elect	1.03	.94	91	1.78	7.3	95
G.M.	1.08	.99	92	1.75	7.1	97
Goodyear	1.24	1.15	92	1.77	7.4	98
Merckk	1.07	.96	90	1.59	7.3	92
3M	.99	.91	92	1.75	7.1	98
Average			93	1.76	7.5	98

As an illustration Table 3 shows estimated means, standard deviations and skewness measures for $|r_t - \text{mean}|^\theta$ for two stock price indices, the S&P500 (NYSE) and the Nikkei (Tokyo) for various values of θ. If $|r_t|^\theta$ are used the results are very similar. The table also shows the ratios of mean over standard deviation. The data were not outlier adjusted. If a fractionally integrated I(d) model is considered, d can be estimated for each θ. The pattern observed is that d initially increased (up to .508 as $\theta = .2$ for S&P index and up to .479 at $\theta = 3$ for the Nikkei index) and then steadily decreases as θ decreased, taking values .486 as $\theta = .7$, corresponding values are .421, .357 and .251 for the Nikkei. For each series, the kurtosis is below 10 up to $\theta = 0.6$ and steadily or even rapidly increases as θ increases thereafter. For both series, the Jaques-Bera statistics, used to test for normality, reached a minimum value of about 48 near $\theta = .2$.

The sample sizes for the two series are very different but the numbers for the two indices are generally remarkably similar, especially the plot of the ratio of mean/standard deviation against theta. Generally both mean and standard deviation decline with θ, the mean declines quicker so that their ratio also declines and takes a value near one around $\theta = 1$ or a little less. The skewness measure increases fairly steadily, and the two series produce similar values for many values of θ.

Table 3. Distributional Properties as θ Varies

	S&P Index				Nikkei Index			
	1	2	3	4	5	6	7	8
θ	Mean	St.Dev	$\frac{\text{Mean}}{\text{St.Dev}}$	Skew	Mean	St.Dev	$\frac{\text{Mean}}{\text{St.Dev}}$	Skew
.10	.580	.068	8.53	-.379	.570	.066	8.64	-.538
.20	.342	.078	4.38	.025	.329	.073	4.51	-.085
.30	.204	.068	3.00	.415	.192	.062	3.10	.333
.40	.123	.054	2.78	.812	.113	.048	2.35	.754
.50	.075	.041	1.83	.124	.068	.036	1.89	1.210
.60	.046	.031	1.48	1.710	.041	.026	1.58	1.730
.70	.029	.022	1.32	2.250	.025	.019	1.32	2.360
.80	.018	.016	1.125	2.900	.015	.013	1.15	3.160
.90	.011	.012	0.92	3.680	.009	.010	0.90	4.180
.94	.010	.011	0.91	4.040	.008	.008	1.00	4.670
.98	.008	.009	0.89	4.430	.006	.007	0.86	5.220
1.00	.007	.009	0.78	4.640	.006	.007	0.86	5.510
1.02	.007	.008	0.88	4.860	.005	.007	0.71	5.820
1.06	.006	.007	0.86	5.340	.004	.006	0.67	6.490
1.10	.005	.007	0.71	5.850	.004	.005	0.80	7.220
1.20	.003	.005	0.60	7.370	.002	.004	0.50	9.370
1.30	.002	.004	0.50	9.270	.001	.003	0.33	11.990
1.50	.002	.002	0.50	14.500	.001	.002	0.50	18.500
1.80	.000	.001	–	26.800	.000	.001	–	29.900
2.00	.000	.001	–	37.800	.000	.000	–	37.200

5 Models To Fit The Facts

It is obviously imperative that econometricians suggest models, or at least one model, that can produce the facts that appear to have been discovered in the series as discussed above. The facts that need an explanation for a series of returns r_t and absolute returns $a_t = |r_t|$ or $a_t = |r_t - \text{mean}|$ are

(i) r_t are uncorrelated (or at least short-memory)

(ii) a_t is long-memory

(iii) The marginal distribution of a_t has mean = standard deviation.

Fact (ii) can be extended to the Taylor effect or something similar and fact (iii) can be extended to the marginal distribution being exponential, after outlier reduction.

It is quite easy to find a simple model that fits some of the facts but it is more difficult to find a model that reproduces all of the facts. For example

$r_t = e_t a_t$ where e_t is zero mean, iid and a_t has the form

$$\begin{aligned} a_t &= 0, \; t = 1, \ldots, T/2 \\ &= a t = T/2 + 1, \ldots, T \quad a > 0 \end{aligned} \tag{2}$$

then facts (iii) is satisfied and also an approximation of (ii). However, the facts in extended (iii) and in Table 3 are not reproduced. Similarly, fact (iii) does not hold if the jump is anywhere but near the half-way point. Other simple models, such as if a_t is a long cycle or a linear tend produce none of the facts.

A more sophisticated model is a stochastic volatility model of the form $r_t = a_t e_t$ where e_t is iid, $E[e_t] = 0$, $E[|e_t|] = 0$, e_t, a_t are independent

$$a_t = a \exp(\nu_t) \tag{3}$$

where ν_t has mean 0, variance V is fractionally integrated I(d) with $|d| < 1/2$, with an unconditional distribution $\nu_t \sim N(0, V)$. Lima, Breidt and Crato (1994) have recently considered this model as an alternative to the Fractionally Integrated GARCH model, discussed in Baillie, Bollerslev and Mikkelsen (1994), which consider long memory properties of $|r_t|^\theta$ for $\theta = 2$. It should be noted that there are two sources of random shocks into this model, the e_t and those that generate ν_t which makes estimation difficult. However the unconditional moments of ν_t can be found after some algebra and, with the assumptions above, mean = standard deviation of $|r_t|$ if V is given by $\log(4/\pi) = .2416$. Using this value it is found that the population skewness and kurtosis measure are

$$\begin{aligned} \text{skewness} &= \frac{64}{\pi^2} - 4 = 2.485 \\ \text{and kurtosis} &= 12 \times \frac{4^4}{\pi^4} - \frac{4^4}{\pi^2} + 9 = 14.6 \; . \end{aligned} \tag{4}$$

A simulation shows that the estimated values of the quantities closely approximate these theoretical values. Following Lima, Briedt and Crato (1994) it is easily shown that

$$\operatorname{corr}(|y_t|^\theta, |y_{t-k}|^\theta) = \frac{\exp V\theta^2 \rho_k - 1}{\exp V\theta^2 - 1} \tag{5}$$

where $\rho_k = \operatorname{corr}(\nu_t, \nu_{t-k})$ so that $\operatorname{corr}(|y_t|, |y_{t-k}|) = (V\rho_k)/(e^\nu - 1)$ for k large $= .88\rho_k$ using the value of V given above. Thus, $|r_t|$ will have the same long-memory property as ν_t. With this model property (ii) is captured, property (iii) can be produced if the coefficient V is chosen correctly, although the value required has little intuitive appeal, and the extended (iii) does not hold. Further, $\text{corr}(|y_t|^\theta, |y_{t-k}|^\theta)$ will decrease as θ decreases and so the Taylor effect is not captured by this simple model.

Although the marginal distribution being exponential has not been formally established it is important from an interpretational point of view. If $|r_t|$ is exponential then it is reasonable to expect that the pair $|r_t|$, $|r_{t-k}|$ will be jointly exponential. This joint distribution, which is discussed in Granger and Ding (1995), has the properties that the marginals are each univariate exponential and $E[|r_t| \mid |r_{t-k}|] = \alpha_1 + \alpha_2 |r_{t-k}|$ and so the conditional mean is linear. It is shown in the same paper that when $|r_t|$, $|r_{t-k}|$ have a joint exponential distribution then the Machina effect, TP3 and the property TP2, as described in the first section are automatically explained. This is obviously useful in modeling the properties of the series being considered except the exchange rates.

A model that can provide properties (i), (ii), and (iii) extended is the long memory ARCH type model proposed by Ding and Granger (see Ding (1994) and Ding and Granger (1996)). Here $r_t = \sigma_t e_t$, $e_t \sim$ i.i.d. with

$$E[e_t] = 0, \;\; E[e_t^2] = 1$$

$$\sigma_t = \alpha\sigma + (1-\alpha)\sum_{k=1}^{n} a_k |\varepsilon_{t-k}|/\lambda \tag{6}$$

where $0 \le \alpha \le 1$, $\lambda = e|e_t|$ and

$$a_k = \frac{q\Gamma(p+q)\Gamma(p+k-1)}{\Gamma(p)\Gamma(p+q+k)} \tag{7}$$

$= Ak^{-1-q}$ when k is large. The Fractionally Integrated ARCH Model of Granger and Ding [see Granger and Ding (1995)] is a special case of the above model when $p + q = 1$ and $\alpha = 0$. It should be noted that it is not necessary for the coefficient of $|\varepsilon_{t-k}|/\lambda$ to sum to 1 in order to have the long memory property desired. In the above model, $(1-\alpha)\sum_{k=1}^{\infty} a_k = 1 - \alpha$ which is less than 1 when $\alpha > 0$, but $|\varepsilon_{t-k}|$ still has the long memory property as far as a_k decreases hyperbolically as in the above formula. A detailed discussion concerning the simulation and estimation of this class of models is provided in Ding (1994) and Ding and Granger (1996). Their estimation and simulation results show that this class of models can produce the empirical stylized facts found in the financial time series quite well, particularly if e_t has a double exponential distribution.

We have made an investigation of the effects of temporal aggregation on the stylized facts, using only the S&P index series. In DGE, averages of five adjacent returns were formed, and then every fifth term used to form a new series $\bar{r}_{t,s}$. The correlograms are presented in their Tables 3, 4 for the power $|\bar{r}_{t,s}|^\theta$. In the following Table 4, corresponding values for $\rho_{5k}(|r|)$ and $\rho_k(|\bar{r}_5|)$ are shown in adjacent columns for three different values of k. Most columns peak at $\theta = 1$ or perhaps at $\theta = .75$ but are generally much larger for $\theta > 1$ for the aggregated series.

Table 5 compares the mean, standard deviation and their ratio for the daily S&P absolute returns as reported in Table 2, and the five period aggregates $|\bar{r}_{t,5}|$.

Table 4

θ	$\rho_{10}(\|r\|)$	$\rho_2(\|\bar{r}_5\|)$	$\rho_{20}(\|r\|)$	$\rho_4(\|\bar{r}_5\|)$	$\rho_{100}(\|r\|)$	$\rho_{20}(\|\bar{r}_5\|)$
0.125	.121	.109	.100	.149	.089	.129
0.25	.191	.155	.164	.184	.131	.158
0.50	.259	.213	.224	.227	.165	.188
0.75	.291	.255	.241	.255	.173	.203
1.00	.303	.279	.237	.267	.162	.205
1.25	.295	.286	.211	.263	.138	.197
1.50	.270	.277	.170	.245	.106	.180
1.75	.234	.255	.125	.217	.073	.160
2.00	.193	.227	.083	.186	.045	.138
3.00	.072	.109	.009	.075	.003	.068

Table 5

		Mean ×100	St. Dev ×100	Ratio Mean/St. Dev
No Aggregation	+	.70	.80	.875
	–	.80	.90	.889
5 Day Aggregation	+	.34	-.36	.947
	–	.40	.41	.963

The aggregate values are seen to be supportive of DP2, with means nearly equal to standard deviations.

A rather different result occurs with a data set that has not been previously mentioned in this paper, being the share prices for the Ajinomoto for every transaction, on the Tokyo exchange for the period 04/03/89 to 03/30/92, with a sample size $N = 28,099$. They are clearly measured at a higher frequency than the other series used here, involving several values with a day (an average of about 34). Table 6 shows the same quantities as above the original data and after returns are aggregated over ten adjacent terms, both with and without outlier reduction.

Before aggregation the marginal distribution does not display properties DP2 or DP3 but after aggregation mean and standard deviation are approximately equal (DP2) and after outlier reduction the moments approximate those of an exponential distribution (DP3).

Table 6

		Mean ×100	St. Dev ×100	Ratio Mean St. Dev	Skew	Kurt
Original Data						
No aggregations	+	.71	.43	1.675	11.15	35.2
	−	.70	.37	1.879	4.07	37.7
10 term aggregations	+	.12	.12	.972	5.48	63.0
	−	.11	.10	1.11	2.19	10
Outlier Reduced						
No aggregations	+	.71	.35	2.011	2.32	11.4
	−	.69	.34	2.021	2.561	13.6
10 term aggregations	+	.11	.10	1.163	1.86	7.68
	−	.11	.10	1.140	2.00	8.43

Whether or not temporal aggregation can explain the observed properties DP2, DP3 for the daily series clearly deserves further investigation.

Apart form aggregation, there are two other simple explanations that have been proposed and have been investigated. The first is that the long-memory property can be occurring because of the dividend component of stock prices and the second is that it can be explained by a common factor, which is itself long-memory, as in the CAPM formulation. Both of these explanations would not explain the stylized facts in the commodity prices or interest rates but are easily examined. Using both prices that contain dividends and are dividend corrected makes no discernible difference to the results. For each of the ten individual shares, CAPM models were estimated for returns, residuals obtained ε_t and then the question asked if the absolute values of these residuals $|\varepsilon_t|$ had the main stylized facts, the long-memory in the correlogram and the first four moments similar to the exponential distribution after outlier reduction. Somewhat surprisingly, the answer in most cases was yes. The following table shows the averages over the ten shares:

	S/DMean	Skewness	Kurtosis
$\lvert r \rvert$	93	1.76	7.5
$\lvert \varepsilon \rvert$	90	1.73	7.1

and we observe that the correlogram for $|\varepsilon_t|$ and $|\varepsilon_t|^2$ for Bethlehem Steel, which is typical for the group, with $\theta = 1$ being most persistent and more persistent than $\theta = 2$. These results suggest that the CAPM residuals have absolute values with the main stylized facts and that it then follows that the stylized facts for absolute returns cannot be explained by a common factor linking the various returns derived from prices on speculative markets.

Suppose that one accepts that a reasonable model for absolute returns is $(1-B)^d a_t = \varepsilon_t$ with a_t having an exponential distribution, after removal of a few outliers. It is interesting to ask what distribution the shocks ε_t should have to produce the stylized facts, ignoring the outliers. Granger and Jeon (1999) (GJ) have considered this question and start with the moving average representation

$$a_t = (1-B)^{-d}\varepsilon_t \approx \sum_{j=0}^{m} b_j(d)\varepsilon_{t-j} \tag{8}$$

where

$$b_j(d) = \Gamma(j+d)/(\Gamma(d)\Gamma(j+1)) \quad j \geq 1 \tag{9}$$

which is approximately

$$b_j(d) \approx \frac{1}{\Gamma(d)} j^{d-1} \text{ for large } j\ . \tag{10}$$

Using (8) and (9) and a given, large m, and with a_t having an exponential distribution normalized to have variance one, it is possible to derive the first four moments of a_t. Details are in GJ. Using an affine transformation of a_t will give any pair of mean variances and so attention is concentrated on the third and fourth central moments, which produce the standard measures of skewness (ξ_3) and of kurtosis (ξ_4), which takes the values 2 and 9 for the exponential distribution. Using $m = 150$, some examples of the skewness and kurtosis for ε_t required to produce a_t which has an exponential distribution are:

d	Skewness	Kurtosis	Chi-squared ν
0	2	9	
0.16	2.16	9.68	1.72
0.26	2.52	11.40	1.265
0.36	3.33	15.98	0.72
0.46	5.16	29.41	0.30
0.56	8.90	76.63	
0.66	11.74	214.60	
0.76	20.36	514.00	
0.86	23.50	797.00	

For $d < 1/2$ these quantities will converge as m increases, but not for $d \geq 1/2$ as the process then has a variance which depends on t, the time since it started. GJ considers fitting several standard distributions to these numbers, such as the inverse Gaussian and the Lognormal, but the most satisfactory was the χ^2_ν distribution, although the fit was far from perfect. The "degrees of freedom" is seen to be fractional and is under one for d greater or equal to 0.31. It must be emphasized this is just an exercise in matching higher moments, but it does illustrate the strange distribution from which the shocks have been drawn (even after removal of outliers) to produce the "stylized facts."

6 Long-Memory Models

A simple model that provides the stylized facts is

$$r_t = s_t|r_t| \tag{11}$$

(ignoring means for ease of exposition), where s_t is -1 or 1 with (roughly) equal probability, of 1/2, and independent of $|r_t|$. The absolute return $|r_t|$ is fractionally integrated I(d), with d near 1/2 but just under this value and has a marginal exponential distribution $E_\alpha(x)$, with mean = standard deviation of α. The fractionally integrated processes are explored in detail in the special issue of the *Journal of Econometrics* vol. 73, No. 1 (July 1996), edited by R.T. Bailey and M.L. King. This model involves just two parameters, d and α, and can generate data with the stylized facts on most days, i.e. on the 99.7% of days that are not outlier reduced. It cannot capture the marginal distribution on these exceptional days when perhaps an ARCH formulation will perform better – that possibility has not been investigated. Although this model captures the stylized facts it does not follow that it is the only model that can do so. A variety of alternatives have been suggested. Rydán, Teräsvirta, and Åsbrink (1998) (RTA) propose a Markov switching model with a mixture of normals for the error distribution, which approximates the exponential distribution and captures some aspects of the long-memory property. Harvey (1993) proposes embedding the I(d) property into a stochastic volatility model and Granger and Ding (1996), Baillie, Bollerslev, and Mikkelsen (1996) and others have proposed and studied generalized ARCH models with this property. A difficulty with these alternative models is that they are less parsimonious, appear to fit not better than the simple model – although no real comparative study is available – and all of the models share the problem that they are not based on a firm theoretical foundation.

A further possible is that nonlinear dynamic models will have similar linear properties to a fractional I(d) process. An example is the simple nonlinear AR(1) process

$$X_t = \text{sign}(X_{t-1}) + \varepsilon_t \tag{12}$$

where ε_t is zero mean, iid $N(0, \sigma^2)$, as discussed by Granger and Teräsvirta

(1999). If ρ_k is the kth autocorrelation of the process, then theoretically

$$\rho_k = (1-2p)^k \tag{13}$$

where

$$p = \text{Prob}(\varepsilon_t) < 1 = \text{Prob}(\varepsilon_t) > 1 \tag{14}$$

but simulations show that for p small estimated autocorrelations will decline very slowly in appearance like those of an I(d) process. Further, using a generated series of sample size $N = 20,000$ estimated d values (using the Geweke, Porter-Hudak method) are:

$$\begin{array}{ll} p = 0.01 & \hat{d} = 0.28 \\ p = 0.005 & \hat{d} = 0.45 \\ p = 0.001 & \hat{d} = 0.75 \end{array} \tag{15}$$

Visually this NLAR(1) looks like a sequence with structural breaks, with values around 1, then with a break and values around −1, then a break back to 1, and so forth.

There is one further stylized fact, only slightly discussed previously in this paper but introduced in Granger and Ding (1996), which is that the long-memory parameter d in the I(d) model does not appear to be constant. They divided the original 17 thousand points of the daily S&P index series into ten equal sub-samples, estimated d for each and found apparently significant variation through time, with values going from 0.156 (in period 1954-60) to 0.714 (in period 1973-79). To illustrate these movements further the Figures 4, 5, and 6 show d estimated using the Geweke and Porter-Hudak (1983) techniques using a running window length of 2500 days, which is approximately ten years, the window being moved one day at a time and d re-estimated. Figure 4 shows the plot of estimated d are using the U.S. S&P 500 index data and Figure 5 shows the same plot using the London FT index. Figure 6 superimposes the two plots of the estimated d for the NYSE and the London exchange against time. The years are for the start of the (approx) ten year window and the shaded periods are the NBER designated business-cycle downturns for the U.S. economy, and so are not strictly relevant for the London market. There appears to be some common movement in the markets. $\hat{d}$ is basically moving around a constant value from the start of the period until 1953, and then increases in both markets but starting from a lower level in London. The $\hat{d}$ for both markets contain a dramatic drop in 1964 or so followed by a sharp rise and in recent periods move less closely together. There appears to be no clear, visual relationship between rolling window mean returns or mean absolute returns and the changing $\hat{d}$ values for either market.

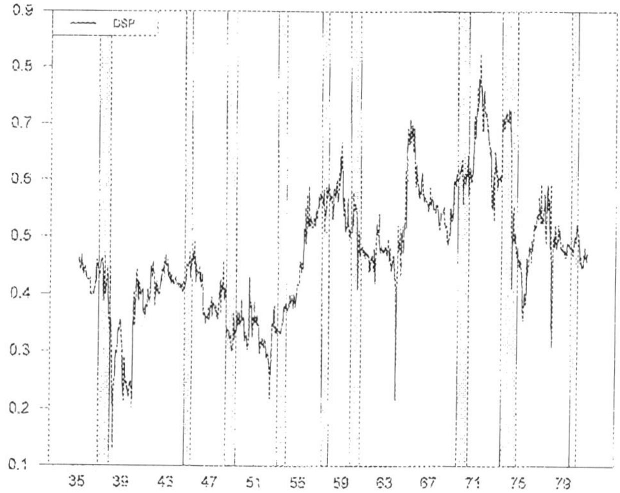

Figure 4: Estimated d for the S&P 500 index

Figure 5: Estimated d for the FT 30 index

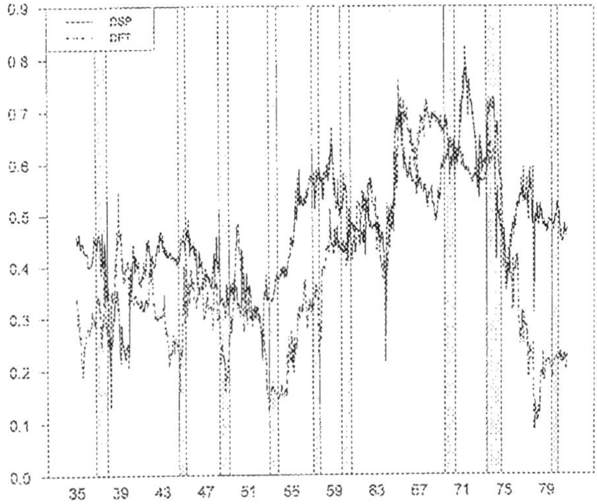

Figure 6: Overlay plot of the estimated d for the NYSE and the London exchange

Mills (1997) has recently considered two London indices based on the 100 largest companies and on 250 mid-to-large companies for the period November 1987 to November 1994. He finds all the stylized facts for these indices except long-memory, which fits in with the results above a moving estimate of $\hat{d}$ as it was found that the estimated values were exceptionally low at the end of the sample, which is around 1992.

When individual shares on the NYSE were considered in a similar fashion several, but not all, had $\hat{d}$ plots that were similar in shape to that for the whole market, but simple regressions did not reveal any "market factor" relationship.

7 Experience With Forecasting Using the I(d) Model

If X_t is a pure I(d) model, and so obeys

$$(1-B)^d X_t = \varepsilon_t \tag{16}$$

where ε_t is zero mean, white noise, there will be high-order autoregressive models which provide good approximation, so that

$$X_{t+1} = \sum_{j=0}^{N} a_j X_{t-j} + \varepsilon_{t+1} \tag{17}$$

giving a one step forecast

$$f_{t,1}(N) = \sum_{j=0}^{N} a_t X_{t-j} \ . \tag{18}$$

Looking head two steps, using (17) gives

$$X_{t+2} = a_0 X_{t+1} + \sum_{j=0}^{N-1} a_{j+1} + \varepsilon_{t+2} \tag{19}$$

and so the two step forecast $f_{t,2}$ will given by

$$f_{t,2} = a_0 f_{t,1} + \sum_{j=1}^{N-1} a_{j+1} X_{t+j} \ . \tag{20}$$

By iteration one obtains the h-step forecast $f_{t,h}$. The theoretical quality of these forecasts, with d known, are given in Beran (1994b) and depend on d, or N, the number of lags used, and on h, the forecast horizon. For example, if $d = 0.4$ and $N = 100$ the following theoretical equations are found:

h	$\mathrm{corr}(x_{t+h}, f_{t,h})$
1	0.72
10	0.55
20	0.51
100	0.41
500	0.35

While these values appear promising, the question now arises whether such levels of forecastability can be achieved in practice. The experiences using daily absolute returns of the S&P 500 were varied. For example, if the period 1928 – 1978 is used, giving a sample of approximately 9,000 terms, and forecasts are evaluated over the following year, the estimated d is 0.45, and compared to an AR(12) model, whose estimated coefficient happens to add to 0.83, the forecasts based on the I(d) model prove to be superior to the AR(12) model, the superiority increasing with horizon h. The correlation (actual, forecast) changes with horizon as:

h	corr (act, forecast)
1	0.43
25	0.26
50	0.28

These correlations seem to be promising but are much lower than the theoretical values. However, considering other data sets, that is different ending dates and sample lengths, usually produced less satisfactory forecasting results, particularly over longer horizons where fractionally integrated processes should have a clear advantage. There are several possible reasons for this poor or mixed performance, including a bad estimate of d, other model mis-specifications and changing values of d through time. There is certainly empirical evidence for the latter, and its effect on forecasting needs further study.

8 Conclusions

The stylized facts presented are that absolute returns of daily prices have long-memory, and have, approximately, an exponential distribution after outlier reduction. These facts are found for stock price indices, for individual shares, and for the residuals of CAPM model as well as commodity prices and an interest rate. A simple parsimonious model using a fractional difference I(d) model finds that the d coefficient is not constant through time and may vary similarly in New York and London. It is a challenge to stock market observers and theorists to explain these "fact", in the same way that the efficient market theory explained the random walk hypothesis, which might once have been considered a stylized fact. An implication of the results is that many of the statistical tests being used with data are inappropriate because of the long-memory property.

Acknowledgments

The institute for Quantitative Investment Research for financial support and to Dr. Walter Labys, Dr. Pierre Siklos, Dr. William Schwert, Dr. Frank Diebold, Dr. Robert Engle, and Prof. Terry Mills for supplying the commodity prices, interest rates stock prices, and other data. Partial support was provided by NSF grant SBR-9708615.

References

1. R.T. Baillie, T. Bollerslev and H.O. Mikkelsen, *Fractionally integrated generalized autoregressive conditional heterskedasticity.* (Department of Finance, Kellogg Graduate School of Management, Northwestern University, manuscript, 1994).
2. J. Beran and S. Ghosh, Slowly decaying correlations, testing normality, nuisance parameters. *Journal of the American Statistical Association* **86**, 785–791 (1991).
3. J. Beran, On a class of m-estimator for Gaussian long-memory models. *Biometrika* **81**, 755–766 (1994a).

4. J. Beran, *Statistics for long memory processes.* (Chapman-Hall, London, 1994b).
5. P. de Lima, F. Jay Breidt, N. Crato, *Modeling long-memory stochastic volatility.* (Working paper in Economics No. 323, Johns Hopkins University 1994).
6. Z. Ding, *Time series analysis of speculative returns.* (Ph.D. dissertation, Department of Economics, University of California, San Diego 1994).
7. Z. Ding and C.W.J. Granger, Modeling volatility persistence of speculative returns. A new approach. *Journal of Econometrics* **73**, 185–216 (1996).
8. Z. Ding, C.W.J. Granger and R.F. Engle, A long memory property of stock market returns and a new model. *Journal of Empirical Finance* **1**, 83–106 (1993).
9. J. Geweke and S. Porter-Hudak, The estimation and application of long-memory time series models. *Journal of Time Series Analysis* **4**, 221–238 (1983).
10. C.W.J. Granger and Z. Ding, Some properties of absolute return; an alternative measure of risk. *Annales d'Economic et de Statistique* **40**, 67–92 (1995).
11. C.W.J. Granger and Z. Ding, Varieties of long memory models. *Journal of Econometrics* **73**, 61–78 (1996).
12. C.W.J. Granger and Y. Jeon, The distributional properties of shocks to a fractional I(d) process having a marginal exponential distribution. *Applied Financial Economics*, (forthcoming, 1998).
13. C.W.J. Granger and T. Teräsvirta, A simple nonlinear time series model with misleading linear properties. *Economic Letters* **62**, 161–165 (1999).
14. R.D. Luce, Several possible measures of risk. *Theory and Decision* **12**, 217–228 (1980).
15. T.C. Mills, Stylized facts on the temporal and distributional properties of daily FT-SE Returns. *Applied Financial Economics* **7**, 597–604 (1997).
16. T. Rysén, T. Teräsvirta and S. Åsbrink, Stylized facts of daily return series and the hidden Markov model. *Journal of Applied Econometrics* **13**, 217–244 (1998).
17. S. Taylor, *Modelling financial time series.* (John Wiley & Sons: New York 1986).

VOLATILITY COMPUTED BY TIME SERIES OPERATORS AT HIGH FREQUENCY

ULRICH A. MÜLLER
Olsen & Associates,
Seefeldstr. 233, CH - 8008 Zürich, Switzerland
E-mail: ulrichm@olsen.ch

Most financial markets produce inhomogeneous (i. e. unequally spaced) tick-by-tick data at high frequency. Recently developed time series operators can be used to directly compute statistical variables such as volatility from inhomogeneous data. This is not possible with traditional time series methods. Value-at-Risk computations require measurements of current volatility, but the conventional calculation from daily data, sampled at a certain daytime, is strongly sensitive to the choice of this daytime, revealing a high amount of stochastic noise. An alternative calculation from high-frequency, tick-by-tick data with time series operators is shown to have similar results, except for two advantages: distinctly reduced noise and up-to-date results at each tick. The time series operator method is flexible and computationally efficient. It can be used to express generating process equations and to compute the Value-at-Risk in real time.

1 Introduction

For banks and other financial institutions, risk measurement plays a central role. Risk levels must conform to the capital adequacy rule. An error in the computed risk level may thus affect a bank's investment strategy. The state of the art is measuring risk by analyzing daily data: using one market price per working day and per financial instrument. In this paper, the stochastic error of such a risk measure is demonstrated in a new way, concluding that using only daily data is insufficient.

The challenge for statisticians is to analyze the limitations of risk measures based on daily data and to develop better methods based on high-frequency data. This paper meets this challenge by introducing the time series operator method, applying it to risk measurement and showing its superiority when compared to a traditional method based on daily data.

Intra-day, high-frequency data is available from many financial markets nowadays. Many time series can be obtained at tick-by-tick frequency, including every quote or transaction price of the market. These time series are inhomogeneous because market ticks arrive at random times. Irregularly spaced series are called inhomogeneous, regularly spaced series are homogeneous. An example of a homogeneous time series is a series of daily data, where the data points are separated by one day (on a business time scale which omits the weekends and holidays).

Inhomogeneous time series by themselves are conceptually simple; the difficulty lies in efficiently extracting and computing information from them. In most standard books [3,11,4] on time series analysis, the field of time series is restricted to homogeneous time series already in the introduction. This restriction induces numerous simplifications, both conceptually and computationally, and was almost inevitable before cheap computers and high-frequency time series were available.

In a recent paper, Zumbach and Müller [12] have presented a new time series operator technique, together with a computationally efficient toolbox, to directly analyze and model inhomogeneous as well as homogeneous time series. This method has many applications, among them volatility or Value-at-Risk (VaR) computations tick by tick. Traditional methods such as RiskMetrics™ [6] can be shown to be special cases of the operator technique.

A comparison is made between VaR results based on daily data, sampled at a certain daytime, and results based on tick-by-tick data and the new time series operator technique. If using daily data, a surprising and (for practitioners) alarming sensitivity against the choice of the sampling daytime is observed. The stochastic noise seems higher than acceptable to risk managers. An alternative VaR computation based on tick-by-tick data and a new time series operator technique is shown to have similar properties, except for two advantages: distinctly reduced noise and availability of up-to-date results at each tick.

These results may encourage both researchers and practitioners to study and use high-frequency tick-by-tick data and, thereby, to apply methods based on time series operators. There exist some applications of the operator technology, such as the "Scale of Market Shocks" by Zumbach et al. [13] and some indicators of automated technical trading algorithms as described by Pictet et al. [10]. An obvious step would be the implementation of real-time, tick-by-tick VaR computations in practice.

The time series operators can also be used in the formulation of old and new generating processes of time series. This opens new ways to develop process equations with new properties, also for inhomogeneous time series.

2 The time series operator technique

In this paper, only a minimum of a description is given, so the applications of the following sections can be understood. The theory of the time series operators has been derived and explained by Zumbach and Müller [12].

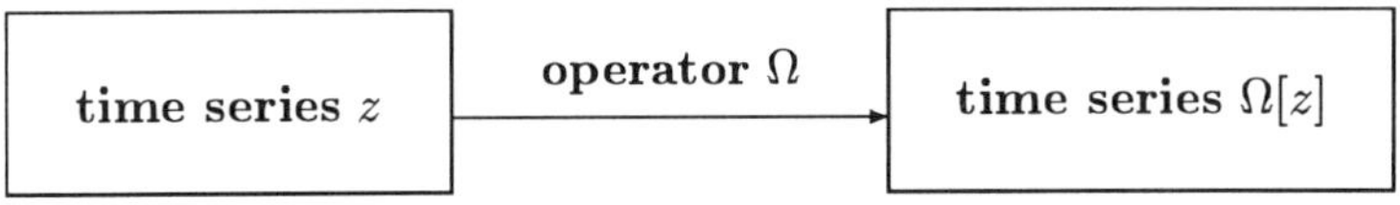

Figure 1: Time series operator Ω

2.1 Inhomogeneous time series

A time series z consists of elements or ticks z_i at times t_i. The sequence of these time points is required to be growing, $t_i > t_{i-1}$.

A general time series is inhomogeneous, meaning that the sampling times t_i are irregular. For a homogeneous time series, the sampling times are regularly spaced, $t_i - t_{i-1} = \delta t$.

For some discussions and derivations, a continuous-time version of z has to be assumed: $z(t)$. However, the operator methods that are eventually applied only need the discrete time series (t_i, z_i).

The letter x is used to represent the time series of logarithmic middle prices, $x = (\ln p_{\text{bid}} + \ln p_{\text{ask}})/2$. This quantity is used in the applications.

2.2 Operators

An operator Ω, mapping from the space of time series into itself, is visualized in Figure 1. The resulting time series $\Omega[z]$ has a value of $\Omega[z](t)$ at time t. Important examples are moving average operators and more complex operators that construct a time series of volatility from a time series of prices.

Linear and translation-invariant operators are equivalent to a *convolution* with a kernel ω:

$$\Omega[z](t) \;=\; \int_{-\infty}^{t} \omega(t-t')\, z(t')\, \mathrm{d}t' \tag{1}$$

A *causal* kernel has $\omega(t) = 0$ for all $t < 0$. No information from the "future" is used. If $\omega(t)$ is non-negative, $\Omega[z]$ is a weighted *moving average* of z whose kernel should be normalized:

$$\int_{-\infty}^{t} \omega(t-t')\, \mathrm{d}t' \;=\; \int_{0}^{\infty} \omega(t)\, \mathrm{d}t \;=\; 1 \tag{2}$$

The kernel $\omega(t)$ is the weighting function of the past. Figure 2 gives an example.

The range of an operator is the first moment of its kernel:

$$r \;=\; \int_{-\infty}^{\infty} \omega(t)\, t\, \mathrm{d}t \tag{3}$$

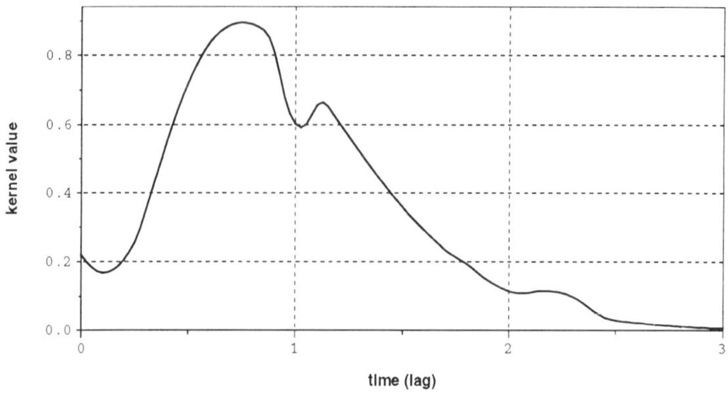

Figure 2: Example of the causal kernel $\omega(t)$ of a moving average

This indicates the characteristic depth of the past covered by the kernel.

Operators are useful for several reasons, as to be shown. One important aspect is to replace individual ticks from the market by local short-term averages of ticks. This mirrors the view of traders who consider not only the most recent tick but also the prices offered by other market makers within a short time interval.

2.3 The simple EMA operator

The Exponential Moving Average (EMA) operator is a simple example of an operator. It is written $\text{EMA}[\tau; z]$ and has an exponentially decaying kernel, as shown in Figure 3:

$$\omega(t) = \text{ema}(t) = \frac{e^{-t/\tau}}{\tau} \tag{4}$$

According to eqs. (3) and (4), the range of the operator $\text{EMA}[\tau; z]$ and its kernel is

$$r = \tau \tag{5}$$

The variable τ thus characterizes the depth of the past of the EMA.

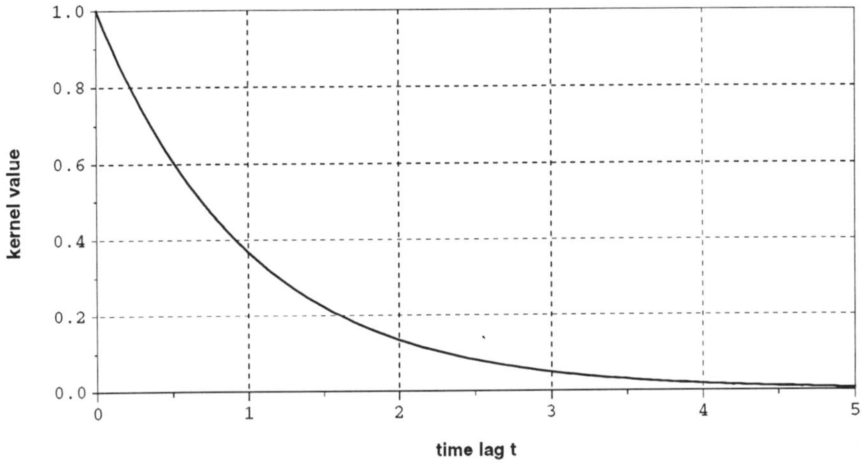

Figure 3: Kernel of the simple EMA operator EMA$[\tau; z]$. Here: $\tau = 1$.

The values of EMA$[\tau; z](t)$ can be computed by the convolution of eq. (1), if $z(t)$ is known in continuous time. This implies an integration whose numerical computation for many time points t is costly. Fortunately, there is an *iteration formula* that makes this computation much more efficient and, at the same time, solves the problem of discrete data. This means that we do not need to know the whole function $z(t)$; we just need the discrete time series values $z_i = z(t_i)$ at irregularly spaced time points t_i. The EMAs are calculated by the following iteration formula:

$$\mathrm{EMA}[\tau; z](t_n) \;=\; \mu\, \mathrm{EMA}[\tau; z](t_{n-1}) \;+\; (1-\mu)\, z(t_n) \;+\; (\mu-\nu)\, [z(t_n) - z(t_{n-1})] \tag{6}$$

with

$$\mu \;=\; \mathrm{e}^{-\alpha} \quad , \quad \alpha \;=\; \frac{t_n - t_{n-1}}{\tau} \tag{7}$$

and

$$\nu \;=\; \frac{1-\mu}{\alpha} \tag{8}$$

This variable ν is related to the problem of using discrete data in a convolution defined in continuous time. We need an assumption on the behavior of $z(t)$ between the discrete time points t_i. Eq. (8) is based on the assumption of linear interpolation between points; other formulas for ν are implied by other interpolation assumptions, as explained by Zumbach and Müller[12]. In the case of assuming the value of the old tick for the whole interval before the new tick, the correct formula is

$$\nu = 1 \tag{9}$$

For a homogeneous time series, μ and ν are constants. A homogeneous time series can alternatively be regarded as a truly discrete time series to which interpolation does not apply. This is mentioned here because it is a popular approach used by traders (in their "Technical Analysis") and by RiskMetrics. For such a discrete time series, $t_n - t_{n-1}$ is defined to be 1, and the following definition is appropriate:

$$\mu = \nu = \frac{1}{1+\alpha} = \frac{\tau}{\tau + t_n - t_{n-1}} = \frac{\tau}{\tau + 1} \tag{10}$$

as shown by Müller (1991)[7]. The range of an operator for a genuine discrete time series has a new definition: $r = \sum_{i=0}^{\infty} \omega_i i$. For EMA, this means $r = \mu/(1-\mu) = \tau$ with $\omega_i = (1-\mu)\mu^i$. The μ and ν values resulting from eq. (10) are very similar to those of eqs. (7) and (8) as long as α is small.

The iteration equation (6) is computationally efficient, extremely so when compared to a numerical convolution based on eq. (1). No other operator can be computed as efficiently as the simple EMA operator. However, there are means to use the iteration equation (6) as a tool to efficiently compute operators with other kernels, as shown below.

An iteration formula is not enough. We have to initialize $\text{EMA}[\tau; z](t_0)$ at the start of the time series. For this, we can take $z_0 = z(t_0)$ or another typical value of z. This choice introduces an initial error of $\text{EMA}[\tau; z](t_0)$ which decreases exponentially with time. Therefore, we also need a build-up period for $\text{EMA}[\tau; z]$: a time period over which the values of $\text{EMA}[\tau; z]$ should not yet be applied because of their initial error. Build-up periods should be multiples of τ, e. g. 5τ. The choice of a large enough build-up period is discussed by Zumbach and Müller[12].

2.4 The operator $\text{EMA}[\tau, n; z]$

Time series operators can be convoluted: a time series resulting from an operator can be mapped by another operator. This is a powerful method to generate new operators with different kernels.

The $\text{EMA}[\tau, n; z]$ operator results from the repeated application of the same simple EMA operator, as defined by Müller (1991)[7] for the first time. The following recursive definition applies:

$$\text{EMA}[\tau, k; z] = \text{EMA}[\tau; \text{EMA}[\tau, k-1; z]] \tag{11}$$

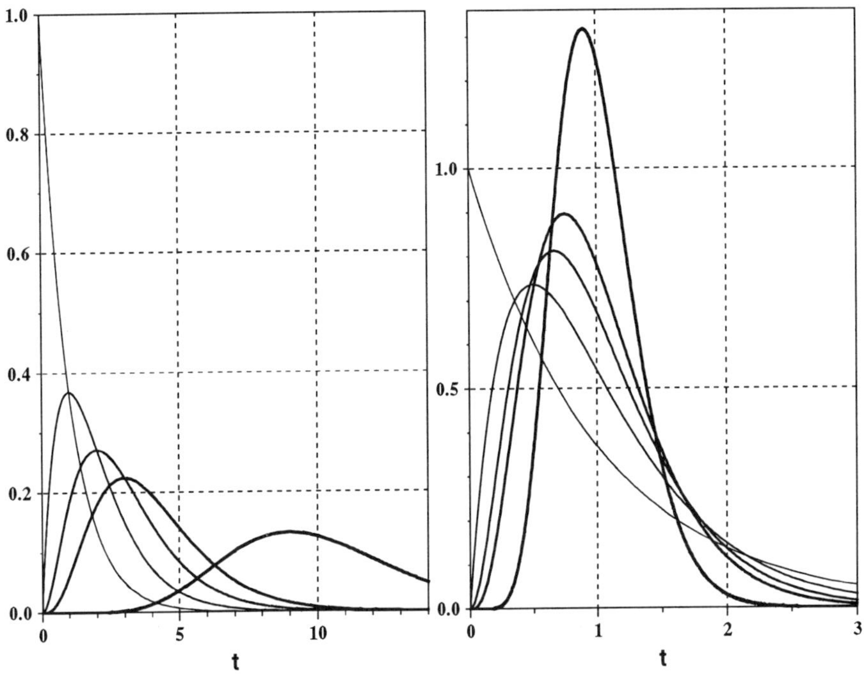

Figure 4: Kernels of EMA$[\tau, n; z]$. Curves for $n = 1$ (thin), 2, 3, 4, 10 (bold). Left graph: for $\tau = 1$; right graph: $r = n\tau = 1$.

with EMA$[\tau, 1; z] =$ EMA$[\tau; z]$. The computationally efficient iteration formula of the simple EMA, eq. (6), can again be used; we have to apply it recursively (n times) for each new tick (t_i, z_i). For ν, we insert eq. (8) which is based on a linear interpolation assumption between ticks. (This assumption is just a good approximation in some cases, as discussed by Zumbach and Müller [12]).

The operator EMA$[\tau, n; z]$ has the following kernel:

$$\mathrm{ema}[\tau, n](t) \;=\; \frac{1}{(n-1)!} \left(\frac{t}{\tau}\right)^{n-1} \frac{\mathrm{e}^{-t/\tau}}{\tau} \tag{12}$$

This kernel is plotted in Figure 4, for several n. For large n (e. g. $n = 10$ in Figure 4), the mass of the kernel is concentrated in a relatively narrow region around a time lag of $n\tau$. The corresponding operator can thus be seen as a smoothed backshift operator.

The family of functions of eq. 12 is related to Laguerre polynomials which

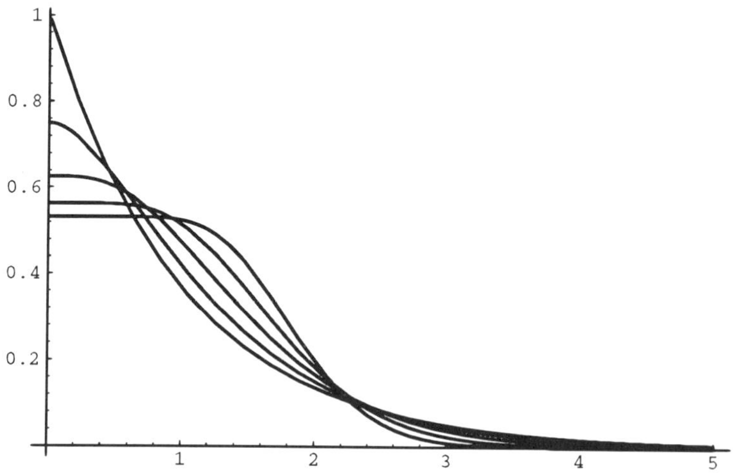

Figure 5: Kernels of $\mathrm{MA}[\tau, n; z]$; $n = 1, 2, 4, 8, 16$, for $\tau = 1$.

are orthogonal with respect to the measure e^{-t} (for $\tau = 1$).

Operators, i. e. their kernels, can be linearly combined. This is a powerful method to generate more operators. Linear combinations of $\mathrm{EMA}[\tau, n; z]$ operators with different n but identical τ values have kernels that correspond to expansions in Laguerre polynomials. This means that any kernel can be expressed as such a linear combination. The convergence of the Laguerre expansion may however be slow.

In practice, a small set of useful operators can be prepared with all the kernels needed. Aside from the discussed expansion, it is also possible to linearly combine kernels with different τ values. Some useful types of combined operators are presented by Zumbach and Müller [12].

2.5 The operator $\mathrm{MA}[\tau, n; z]$

The moving average (MA) operator has kernels with useful properties as shown in Figure 5. It is constructed as a sum of $\mathrm{EMA}[\tau, n; z]$ operators:

$$\mathrm{MA}[\tau, n] = \frac{1}{n} \sum_{k=1}^{n} \mathrm{EMA}[\tau', k] \qquad \text{with } \tau' = \frac{2\,\tau}{n+1} \tag{13}$$

The variable τ' is chosen such that the range of $\mathrm{MA}[\tau, n]$ is $r = \tau$, independent of n. The kernel form depends on the choice of n. For $n = 1$, we obtain a simple

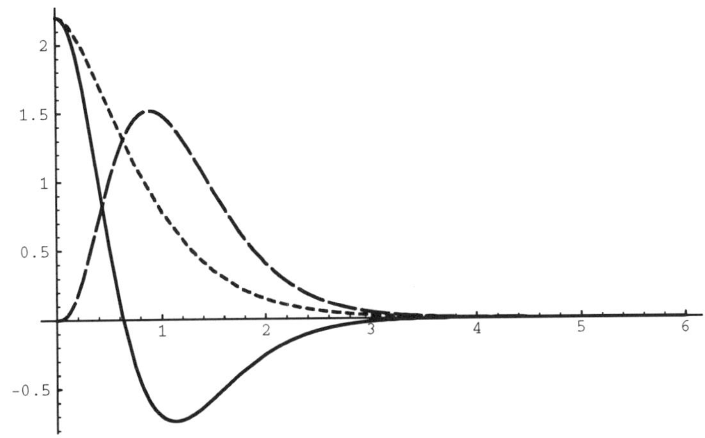

Figure 6: Kernel of the differential operator $\Delta[\tau]$: full curve. Here: $\tau = 1$; dotted curve: first two terms of the operator; dashed curve: the last term $2\,\mathrm{EMA}[\alpha\beta\tau, 4]$.

EMA operator, for $n = \infty$ the rectangularly shaped kernel of a simple moving average with constant weight up to a limit of 2τ. This simple rectangular moving average has a serious disadvantage in its dynamic behavior: additional noise when old observations are abruptly dismissed from the rectangular kernel area. Kernels with finite n are better because of their smoothness; the memory of old observations fades gradually rather than abruptly.

The formula for the $\mathrm{MA}[\tau, n]$ kernel is

$$\mathrm{ma}[\tau, n](t) \;=\; \frac{n+1}{n}\,\frac{\mathrm{e}^{-t/\tau'}}{2\,\tau}\sum_{k=0}^{n-1}\frac{1}{k!}\left(\frac{t}{\tau'}\right)^{k} \tag{14}$$

Many other kernel forms can be constructed through different linear combinations of $\mathrm{EMA}[\tau, n; z]$ and other operators.

2.6 From returns to differentials

Most statistics in finance is based on returns: price changes rather than prices. Simple returns have a rather noisy behavior over time; we often want differences between local averages of x: *smoothed* returns.

Smoothed returns are computed by differential operators. Examples:

- $x - \mathrm{EMA}[\tau, n; x]$, where the EMA replaces $x(t-\tau)$. This is used by the application of section 3.2.

- $\text{EMA}[\tau_1, n_1] - \text{EMA}[\tau_2, n_2]$, with $\tau_1 < \tau_2$ or $n_1 < n_2$.
- $\Delta[\tau] \;=\; \gamma\,\{\text{EMA}[\alpha\tau, 1] + \text{EMA}[\alpha\tau, 2] - 2\,\text{EMA}[\alpha\beta\tau, 4]\}$, with $\gamma = 1.22208$, $\beta = 0.65$ and $\alpha^{-1} = \gamma(8\beta - 3)$. This is normalized, so $\Delta[\tau, 1] = 0$, $\Delta[\tau, t] = 1$. The kernel of this differential operator proposed by Zumbach and Müller [12] is plotted in Figure 6.

The expectation value of squared smoothed returns may differ from that of the corresponding simple returns. This has to be accounted for when comparing the two concepts, for example in terms of a factor c in eq. (20).

2.7 Volatility measured by operators

Volatility is a central term in risk measurement and finance in general, but there is no unique, universally accepted definition. There are volatilities derived from option market prices and volatilities computed from diverse model assumptions. In this paper, the focus is on historical volatility: a volatility computed from recent data of the underlying instrument with a minimum of parameters and model assumptions.

For computing the time series of such a volatility, a time series operator is again the suitable tool. We first define the nonlinear moving norm operator:

$$\text{MNorm}[\tau, p; z] \;=\; \{\text{MA}[\tau; |z|^p]\}^{1/p} \tag{15}$$

This operator is based on a linear MA operator (where we are free to choose any positive, causal kernel), it is nonlinear only because a nonlinear function of the basic time series variable z is used. $\text{MNorm}[\tau, p; z]$ is homogeneous of degree 1.

The volatility of a time series x can now be computed with the help of the moving norm operator:

$$\text{Volatility}[\tau_1, \tau_2, p;\, x] = \text{MNorm}\left[\frac{\tau_1}{2},\, p;\, \Delta[\tau_2; x]\right] \tag{16}$$

$$= \left\{\text{MA}\left[\frac{\tau_1}{2};\, |\Delta[\tau_2; x]|^p\right]\right\}^{1/p} \tag{17}$$

This is the moving norm of (smoothed) returns. With $p = 2$, it is a particular version of the frequently used RMS value. However, some researchers had and have good reasons to choose a lower value such as $p = 1$ in their special studies.

Eq. (17) is based on a moving average (MA) and a differential (Δ) operator. In principle, we may choose any MA and Δ operator according to our preference. In the applications of section 3, this choice is made explicit.

The volatility definition of eq. (17), as any definition of historical volatility, necessarily has *two* timing parameters:

1. the size of the return measurement intervals: τ_2;
2. the size of the total moving sample: τ_1, often $\gg \tau_2$; defined as the double range of the used MA. The MA operator has a range (center of gravity of the kernel) of $\tau_1/2$.

3 Application: volatility computation in risk management

Computing recent volatility is a central ingredient of risk assessment in risk management. Here it serves as an example to demonstrate the usefulness and superiority of time series operators.

The RiskMetrics™ method[6] is chosen as a well-known example. First it is shown to be a special application of the time series operator technique. Then a better volatility computation method, also based on the time series operator technique, is proposed. Thus two approaches are compared:

1. The RiskMetrics method, based on an IGARCH model with *working-daily data.*
2. A *tick-by-tick* alternative, following RiskMetrics as closely as possible, based on time series operators.

In both cases, squared volatility is defined as the expectation σ^2 of squared, working-daily changes of the logarithmic middle price x. This is for the sake of a meaningful comparison; it does not imply that using squared returns is necessarily the best choice for an optimal volatility definition.

3.1 Conventional computation based on daily data

The RiskMetrics method is based on an IGARCH model. Its volatility formula gives the conditional expectation of the squared return assuming IGARCH:

$$\sigma^2(t) \;=\; \mu\,\sigma^2(t-1\,\mathrm{wday}) \;+\; (1-\mu)\,[x(t)-x(t-1\,\mathrm{wday})]^2 \qquad (18)$$

with $\mu = 0.94$. This is just an EMA iteration which can also be written in our operator notation:

$$\sigma^2(t) \;=\; \mathrm{EMA}\left[\tau = 15.67\,\mathrm{wdays};\ [x(t)-x(t-1\,\mathrm{wday})]^2\right] \qquad (19)$$

evaluated at discrete time points separated by 1 working day (= 1 wday); with EMA range $\tau = \mu/(1-\mu)$ in working days, following eq. (10).

Thanks to the regularity of the underlying homogeneous time series, $\mu = 0.94$ is a constant. In general, the constancy of μ makes the operator technique particularly efficient for homogeneous time series.

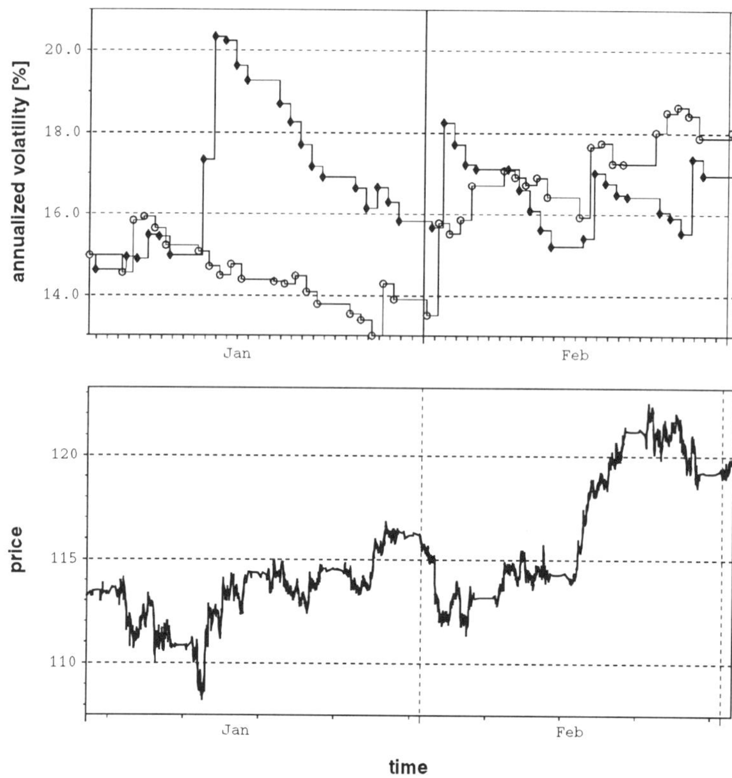

Figure 7: Standard RiskMetrics volatility. Circles: data sampled at 7 am GMT; diamonds: data sampled at 5 pm GMT. Independent computations. FX rate: USD/JPY ; period: Jan 1999 - Feb 1999. The price is plotted against time on the lower graph.

Figure 7 shows the resulting volatility as a function of time, in an empirical example from the foreign exchange (FX) market: USD/JPY data in January and February 1999. The volatility is computed only once per working day, at a given daytime; the resulting volatility value is valid until it is replaced by a new one, one working day later.

In Figure 7, two such volatilities are plotted. The difference between the two curves solely originates from the choice of daytime when the raw data x is sampled and the volatility is computed by eq. (18) or (19). One curve is sampled at 7 am GMT which is a time in the late afternoon of East Asian time zones – a suitable daytime for the daily risk calculations of an East Asian risk manager. The other curve is sampled at 5 pm GMT – a suitable daytime for a risk manager in London.

The differences between the two curves are surprisingly large: up to 25%, an alarming uncertainty for risk managers. Risk levels are linked to a bank's capital through the capital adequacy rule, so differences in risk measurements have a major impact on banking! In our case, two risk managers measure very different volatility and thus risk levels for the *same* financial instrument, just because they live in different time zones. A difference can persist over weeks, as shown in Figure 7. This figure is just an example. The same surprisingly strong effect can be found also for other financial instruments, sampling periods and choices of daytime.

Both deviating volatility values cannot be right at the same time; there must be an error in these values. This error is of stochastic nature; there is no systematic bias dependent on the daytime. In Figure 7, the difference between the two curves is neither always positive nor negative; it changes its sign.

Figure 7 demonstrates the large stochastic error of the RiskMetrics method. The large size of this error has two main reasons:

1. The rather small range of the kernel of some 16 working days. The number of independent observations is limited. We cannot essentially change this fact, because the choice of a short range is also motivated by the goal of fast adaptivity to new market events.

2. The results depend on only one observation per day, taken at a certain daytime. All the other information on prices of the day is thrown away. The value at that daytime may be little representative for the full day: it may be located on top of a short-lived local peak of the price curve. This is indeed the reason for the large deviations of the two curves in Figure 7. The effect is exacerbated by the known fact that returns have a heavy-tailed distribution function: extreme (intra-day) events dominate the statistics.

The focus of this study is not so much the behavior of RiskMetrics (IGARCH), but the problems of using homogeneous, daily data in general, no matter which GARCH-type or other model is investigated. The significance of most results can be improved by using all the available information, tick by tick, as shown in the next section.

3.2 Tick-by-tick volatility computation

For the sake of a fair comparison, a tick-by-tick volatility computation is introduced that follows RiskMetrics as closely as possible. There are only two innovative modifications:

- The squared volatility $\sigma^2(t)$ is computed at every available tick, not only once per working day.
- Simple returns are replaced by operator-based, smoothed returns.

Nothing is changed otherwise; the sampling range of 15.67 working days and the working-daily nature of (smoothed) returns are preserved.

The new volatility measure is again defined in operator notation (where "wdays" stands for working days):

$$\sigma^2 = c\,\mathrm{EMA}\left[\tau = 15.67\,\mathrm{wdays};\ (x - \mathrm{EMA}[\tau = 1\,\mathrm{wday},\ 4;\ x])^2\right] \tag{20}$$

This is just a special case of eq. (16). The computation is efficiently done at every new tick, repeatedly using the iteration formula (6). This works not only for the simple EMA but also for $\mathrm{EMA}[\tau, 4; x]$ as explained in section 2.4.

The constant c compensates for the fact that we use smoothed returns $x - \mathrm{EMA}[\tau, 4; x]$ as introduced in section 2.6 instead of the simple returns of section 3.1. In the case of x following a Gaussian random walk, the theoretically correct value is $c = 128/93$. Using this factor eliminates a systematic bias of the tick-by-tick volatility as compared to the RiskMetrics volatility.

Eq. (20) is computed on a special business time scale defined as follows. The 49 weekend hours from Friday 8 pm GMT to Sunday 9 pm GMT are compressed to the equivalent of only 1 hour outside the weekend. This fully corresponds to the time scale of RiskMetrics which omits the weekend days. A more sophisticated and appropriate choice of the business time scale would be the ϑ-time of Dacorogna et al. (1992) [1], but this is avoided here in order to keep the approach as close to RiskMetrics as possible.

Figure 8 shows the resulting volatility as a function of time. The same financial instrument and sampling period is studied as in Figure 7. High-frequency data is available here. Now, the large differences between values at 7 am GMT and 5 pm GMT have vanished. The observations at these daytimes appear as points on one continuous, consistent curve. In fact, we can obtain volatility values at any daytime now, not just once or twice a day. A risk manager in London essentially measures the same risk of the instrument as a risk manager in East Asia, as should be expected in normal situations. Only if a dramatic event between the two daytimes of measurement happens, the risk levels deviate. This is natural; the operator-based volatility quickly reacts to dramatic events, as can be seen in Figure 8.

The variations of the volatility level over time are moderate in Figure 8. The extreme volatility minima and maxima of Figure 7 which are mostly due

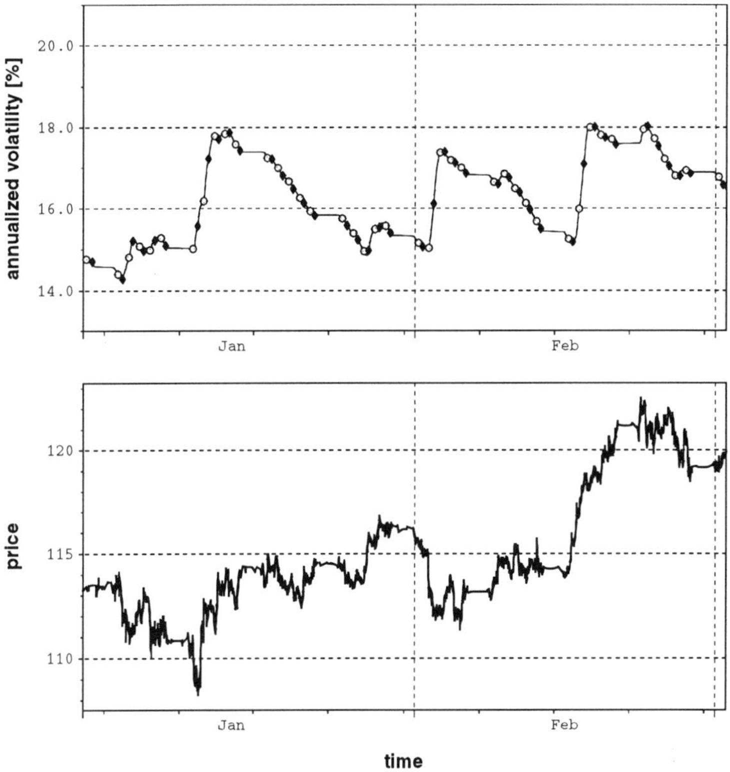

Figure 8: Operator-based tick-by-tick volatility. Circles: values at 7 am GMT (7:00); diamonds: values at 5 pm GMT (17:00). FX rate: USD/JPY ; period: Jan 1999 - Feb 1999. The price is plotted against time on the lower graph.

to stochastic noise have vanished. The new tick-by-tick volatility has less stochastic noise than the RiskMetrics volatility, although the moving sample range of 15.67 working days is the same.

The curves of Figures 7 and 8 are combined in Figure 9. Here, we can see that the tick-by-tick volatility (bold curve) has a similar dynamic behavior as the RiskMetrics volatilities while avoiding extreme oscillations due to stochastic noise.

The lower noise level of the tick-by-tick volatility is now plausible, but we need scientific evidence for this. In the general case, such evidence can be gained through Monte-Carlo studies based on a certain process assumption,

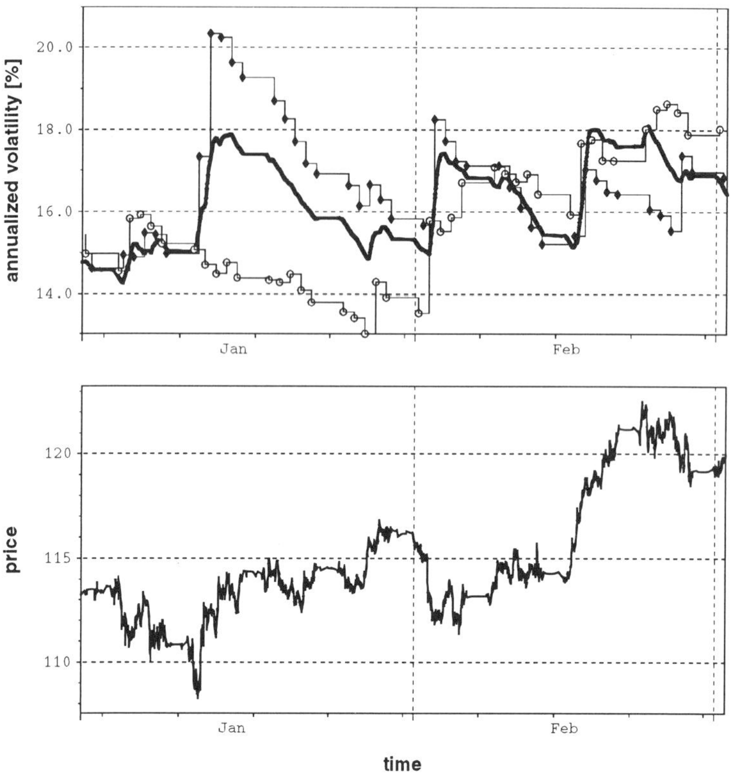

Figure 9: Comparing RiskMetrics to operator-based volatility. Bold curve: operator-based; circles: RiskMetrics 7 am GMT; diamonds: RiskMetrics 5 pm GMT. FX rate: USD/JPY ; period: Jan 1999 - Feb 1999. The price is plotted below.

comparing the error variances of the RiskMetrics volatility and the tick-by-tick volatility. In the case of Gaussian random walk, we have even stronger evidence: an exact result by Müller (1993)[8]. By using continuously overlapping returns instead of non-overlapping returns, the error variance of the empirically determined σ^2 is reduced to 2/3 of the original value.

The tick-by-tick operator is indeed using (almost) continuously overlapping returns. In addition to this, it is based on smoothed rather than simple returns, which also leads to a reduction of stochastic noise.

Other advantages of tick-by-tick, operator-based methods are the efficient computation based on iterations and the updating at every new tick. Thanks to fast updates, the volatility measure can quickly react to new market events such as shocks, at any daytime.

3.3 VaR computation in real-time

Conventional Value-at-Risk (VaR) computations are done once a day, usually in the evening. Traders and portfolio managers typically do not know their current risk; they just know yesterday evening's risk. What they would really need is a *real-time* VaR computation, updated as quickly as they can change their positions.

The tick-by-tick operator proposed in section 3.2 and eq. (20) is a tool to make a real-time VaR possible. In addition to this, a corresponding tool to compute moving covariances or correlations is needed. This is possible with similar operators; some operators are proposed by Zumbach and Müller [12].

A real-time VaR computed according to these guidelines would still be rather similar to RiskMetrics, except for lower noise and a higher updating frequency. There are many criticisms of RiskMetrics that would apply to it, too. Some researchers, for example, replace the IGARCH model by another GARCH-type model. Other researchers such as McNeil et al. [5] focus on the behavior of extreme price changes which may follow other laws than average-size changes. Moreover, return observations over intervals other than daily, for example hourly or weekly returns, contain valuable information that should also be used in a VaR computation.

All these possible refinements of risk measurement are not discussed in this paper. They have to be addressed in future research studies that will eventually lead to a more mature solution of real-time VaR computation. But it seems clear that this optimal solution will also be based on tick-by-tick operators.

4 Process equations based on operators

Processes of the ARMA and GARCH families can be expressed in terms of time series operators, as we have seen for IGARCH in eq. (19). The squared conditional volatility of a GARCH(1,1) process, for example, can be written as follows:

$$\sigma^2(t) \;=\; c + a\,\sigma^2(t') + b\,[x(t') - x(t' - \Delta t)]^2 \tag{21}$$

where $t' = t - \Delta t$. The following alternative notation is based on a simple EMA operator:

$$\sigma^2(t) \;=\; \frac{c}{1-a} + \frac{b}{1-a}\,\mathrm{EMA}\left[\frac{a}{1-a};\, [x(t') - x(t' - \Delta t)]^2\right] \tag{22}$$

This rephrased form of GARCH(1,1) is a starting point of interesting new developments. Initially, it applies to a discrete, homogeneous time series in the sense of eq. (10), but it allows for a direct and efficient computation of the GARCH(1,1) volatility from inhomogeneous data, since the operator technique is also suited to inhomogeneous time series. This is not possible with conventional methods.

Moreover, eq. (22) can be modified to obtain other processes. The kernel of the EMA operator can, for example, be replaced by other kernels. The return $x(t') - x(t' - \Delta t)$ can be replaced by a smoothed return computed by a differential operator which reflects the perception of market participants better that the simple return.

Dacorogna et al. (1998) [2] have introduced the EMA-HARCH process to model some empirical facts of high-frequency data in finance: the long memory of volatility, the fat tails of the return distribution and the asymmetric causality between fine-grained (high-resolution) and coarse-grained (low-resolution) volatilities as found by Müller et al. [9]. This is one of the first processes whose equation is written with the help of a time series operator:

$$\sigma^2(t) \;=\; c + \sum_{j=1}^{n} C_j \, \sigma_j^2(t) \tag{23}$$

The "partial volatilities" σ_j^2 correspond to market segments and are written in terms of the EMA operator:

$$\sigma_j^2(t) \;=\; \mathrm{EMA}\left[\frac{k_{j+1} - k_j}{2}\Delta t;\; [x(t') - x(t' - k_j \Delta t)]^2\right] \tag{24}$$

with $k_1 \;=\; 1$ and $k_j \;=\; 4^{j-2} + 1$ for $j > 1$.

There are many more possibilities and good reasons to introduce time series operators in process equations to a much larger extent in the future.

5 Conclusions

Most financial markets produce inhomogeneous data, irregularly spaced in time. The time series operators presented in this paper are able to directly use inhomogeneous data to estimate statistical variables such as volatilities. This computation is efficient thanks to simple iteration formulas. The operator technique efficiently works also for homogeneous, equally spaced data.

Starting from the simple exponential moving average (EMA) operator, large families of operators with different kernels and different purposes can be constructed. A wider overview of these more complex operators, which are still computationally efficient, is given by Zumbach and Müller [12]. One example is a tick-by-tick Fourier analysis on a moving sample.

Thanks to averaging, the operator technique often produces results with less noise (lower stochastic errors) than conventional methods based on homogeneous time series. This is also the case for the main application of this paper: volatility of daily returns as needed for Value-at-Risk (VaR) computations. The conventional RiskMetrics method has a rather high stochastic error which is demonstrated in a new way: volatility is computed twice with daily data. In one case, the data is always sampled at 7 am GMT (late afternoon in East Asia), in the other case at 5 pm (late afternoon in London). The results of the two computations can differ by some 25% for many days in a row - an embarrassing fact for risk managers. The proposed tick-by-tick alternative based on time series operators does not have this sensitivity against the choice of the daytime and has less noise, while keeping the essential characteristics of the RiskMetrics method: it is still based on daily returns and still has a range (center of gravity of the kernel) of around 16 working days.

The logical next step will be to design a complete market risk analysis based on the time series operator technique. Its result will be a real-time VaR, updated with every new incoming tick from a market, with less noise than the corresponding results from conventional methods.

Finally, the operator technique can be used to formulate time series generation process equations. This is possible for well-known processes such as GARCH and new, more complex processes such as EMA-HARCH. The formulation in terms of operators has many advantages: flexibility and applicability to irregularly spaced time series.

Acknowledgments

This paper resulted from my presentation at the Hong Kong International Workshop on Statistics in Finance, 5 - 9 July 1999. I would like to thank the organizing committee of this successful workshop for the excellent organization and for having given me the opportunity to present the time series operator method. The presented work is not mine alone; I am grateful for the help and the contributions of Gilles Zumbach and my other colleagues at Olsen & Associates.

References

1. Dacorogna M. M., Müller U. A., Nagler R. J., Olsen R. B., and Pictet O. V., 1993, *A geographical model for the daily and weekly seasonal*

volatility in the FX market, Journal of International Money and Finance, **12**(4), 413–438.
2. Dacorogna M. M., Müller U. A., Olsen R. B., and Pictet O. V., 1998, *Modelling short-term volatility with GARCH and HARCH models*, published in "Nonlinear Modelling of High Frequency Financial Time Series" edited by Christian Dunis and Bin Zhou, John Wiley, Chichester, 161–176.
3. Granger C. W. J. and Newbold P., 1977, *Forecasting economic time series*, Academic Press, London.
4. Hamilton J. D., 1994, *Time Series Analysis*, Princeton University Press, Princeton, New Jersey.
5. McNeil A. J. and Frey R., 1998, *Estimation of tail-related risk measures for heteroscedastic financial time series: an extreme value approach*, Preprint from the ETH Zürich, August 27, 1–28.
6. J. P. Morgan, 1996, *RiskMetrics – technical document*, Technical report, J. P. Morgan and International marketing – Reuters Ltd.
7. Müller U. A., 1991, *Specially weighted moving averages with repeated application of the EMA operator*, Internal document UAM.1991-10-14, Olsen & Associates, Seefeldstrasse 233, 8008 Zürich, Switzerland.
8. Müller U. A., 1993, *Statistics of variables observed over overlapping intervals*, Internal document UAM.1993-06-18, Olsen & Associates, Seefeldstrasse 233, 8008 Zürich, Switzerland.
9. Müller U. A., Dacorogna M. M., Davé R. D., Olsen R. B., Pictet O. V., and von Weizsäcker J. E., 1997, *Volatilities of different time resolutions – analyzing the dynamics of market components*, Journal of Empirical Finance, **4**(2-3), 213–239.
10. Pictet O. V., Dacorogna M. M., Müller U. A., Olsen R. B., and Ward J. R., 1992, *Real-time trading models for foreign exchange rates*, Neural Network World, **2**(6), 713–744.
11. Priestley M. B., 1989, *Non-linear and non-stationary time series analysis*, Academic Press, London.
12. Zumbach G. O. and Müller U. A., 1998, *Operators on inhomogeneous time series*, submitted for publication; internal document GOZ.1998-11-01, Olsen & Associates, Seefeldstrasse 233, 8008 Zürich, Switzerland.
13. Zumbach G. O., Dacorogna M. M., Olsen J. L., and Olsen R. B., 1998, *Introducing a scale of market shocks*, submitted for publication; internal document GOZ.1998-10-01, Olsen & Associates, Seefeldstrasse 233, 8008 Zürich, Switzerland.

MISSING VALUES IN ARFIMA MODELS

W. PALMA
Universidad Católica de Chile, Casilla 306, Santiago 22, CHILE
E-mail: wilfredo@mat.puc.cl

This paper discusses state space techniques to estimate and forecast long–memory models with missing observations. A Kalman filter approach to compute maximum likelihood estimates is studied and the behavior of short and long term forecasts is analyzed. As an illustration, this methodology is applied to the analysis of foreign exchange rates.

1 Introduction

An ARFIMA(p,d,q) process, $\{y_t\}$, is defined by the equation:

$$\Phi(B)(1-B)^d y_t = \Lambda(B)\epsilon_t, \tag{1}$$

where $d < 1/2$, $\{\epsilon_t\}$ is a white noise sequence, B is the backshift operator $By_t = y_{t-1}$, $\Phi(B) = 1+\phi_1 B+...+\phi_p B^p$ is the AR operator, $\Lambda(B) = 1+\lambda_1 B+...+\lambda_q B^q$ is the MA operator and $(1-B)^d = \sum_{k=0}^{\infty} \pi_k(-B)^k$ is the fractional difference operator, with $\pi_k = \frac{\Gamma(k-d)}{\Gamma(k+1)\Gamma(-d)}$, where $\Gamma(\cdot)$ denotes the Gamma function. If $d \in (0, 1/2)$, the resulting is a *long-memory* process, with non-summable correlations, that is, $\sum_{k=0}^{\infty} |\gamma_k| = \infty$, where γ_k is the autocorrelation of order k. If $d < 0$, the result is an *intermediate-memory* process, with zero spectral density at frequency zero and summable correlations, $\sum_{k=0}^{\infty} |\gamma_k| < \infty$, cf., Brockwell and Davis [2]. Estimation of these long-range dependent models is discussed by Dahlhaus [5], Li and McLeod [9], Sowell [15], Robinson [14], and Chan and Palma [3], among others. Adaptive procedures for forecasting long memory processes were proposed by Tiao and Tsay [16]. Further discussion on adaptive methods in time series in given in Tong [17]. Missing values in time series has been addressed by many authors, see for example Jones [6], Kitagawa [7], Kohn and Ansley [8] and Palma and Chan [12].

The main purpose of this paper is to discuss the issue of estimating and forecasting ARFIMA models in the presence of data gaps. Since missing values is a common problem in financial and economical time series, we apply these state space techniques to a statistical analysis of foreign exchange data.

This article is structured as follows. Section 2 addresses the application of Kalman filter techniques to the estimation of long memory processes with missing observations. Section 3 discusses the performance of short and long term ARFIMA predictions. An application of these techniques to the study of exchange rates is presented in Section 4. Final remarks are given in Section 5.

2 State Space Representations of ARFIMA models

A $MA(\infty)$ representation of an ARFIMA(p, d, q) process is given by:

$$y_t = \frac{\Lambda(B)}{\Phi(B)}(1-B)^{-d}\epsilon_t = \sum_{j=0}^{\infty} \psi_j \epsilon_{t-j}, \tag{2}$$

where ψ_j are the coefficients of:

$$\psi(z) = \sum_{j=0}^{\infty} \psi_j z^j = \frac{\Lambda(z)}{\Phi(z)}(1-z)^{-d}.$$

¿From equation (2), an infinite-dimensional state-space representation may be written as, cf. Palma and Chan [12]:

$$X_{t+1} = FX_t + H\epsilon_t. \tag{3}$$

$$y_t = GX_t + \epsilon_t, \tag{4}$$

where the state vector is defined as

$$X_t = \left[y(t|t-1) \; y(t+1|t-1) \; y(t+2|t-1) \; \cdots \right]', \qquad y(t|j) = E[y_t|y_j, y_{j-1}, \ldots],$$

and the system matrices are given by

$$F = \begin{bmatrix} 0 & 1 & 0 & 0 & \cdots \\ 0 & 0 & 1 & 0 & \cdots \\ \vdots & \vdots & & \ddots & \end{bmatrix}, \quad H = \left[\psi_1 \; \psi_2 \; \psi_3 \cdots \right]' \text{ and } G = \left[1 \; 0 \; 0 \cdots \right].$$

The calculation of this likelihood function may be carried out recursively by means of the Kalman filter equations. Let $Q = Var(H\epsilon_t) = \sigma_\epsilon^2 HH'$, $\hat{X}_t$ and $\hat{y}_t$ be the best linear predictors based on the observed past, and

$$\Delta_t = Var(y_t - \hat{y}_t), \tag{5}$$

$$\Omega_t = Var(X_t - \hat{X}_t). \tag{6}$$

The recursive equations are:

$$\Delta_t = G\Omega_t G' + \sigma_\epsilon^2, \tag{7}$$

$$\Theta_t = F\Omega_t G' + \sigma_\epsilon^2 H, \tag{8}$$

$$\Omega_{t+1} = \begin{cases} F\Omega_t F' + Q - \Theta_t \Delta_t^{-1} \Theta_t' & \text{if observation } t \text{ is known,} \\ F\Omega_t F' + Q & \text{if observation } t \text{ is missing.} \end{cases} \tag{9}$$

$$\hat{X}_{t+1} = \begin{cases} F\hat{X}_t + \Theta_t \Delta_t^{-1}(y_t - G\hat{X}_t) & \text{if observation } t \text{ is known,} \\ F\hat{X}_t & \text{if observation } t \text{ is missing.} \end{cases} \tag{10}$$

$$\hat{y}_t = G\hat{X}_t. \tag{11}$$

The likelihood with missing observation is given by

$$L(\theta) = (2\pi)^{-(n-m)/2} \left(\prod_{t \in I_n} \Delta_t^{-1/2} e^{-\frac{1}{2}[\sum_{t \in I_n} \frac{(y_t - \hat{y}_t)^2}{\Delta_t}]} \right. \tag{12}$$

where $\theta = (d, \phi_1, ..., \phi_p, \lambda_1, ..., \lambda_q, \sigma)$, $I_n = \{t_1, ..., t_{n-m}\}$ specifies the times at which the process is observed and m is the number of missing values. The next section addresses the problem of assessing the performance of forecasts errors of long memory time series with missing values.

3 Prediction Performance

The behavior of the *mean square prediction error* (MSPE) of an ARFIMA process is addressed in this section. For simplicity, we first study the one-step prediction error behavior with *one* missing value. Then, we extend the results to the analysis of the asymptotic behavior of the MSPE for a block of missing data.

3.1 Evolution of the MSPE after a single missing observation

Consider the one-step prediction error at time t: $e_t(1) = y_t - \hat{y}_t$, where $\hat{y}_t = P_{K_{t-1}} y_t$, P is the projection operator, and $K_{t-1} = [y_{t-1}, y_{t-2}, \cdots]$ is the space generated by the observed values up to time $t-1$. Define the one-step prediction error by $\sigma_t^2(1) = Var[e_t(1)] = Var[y_t - \hat{y}_t]$. Since the process is invertible, it can be written as an $AR(\infty)$ decomposition,

$$y_t = \epsilon_t - \sum_{j=1}^{\infty} \pi_j y_{t-j}.$$

For simplicity, assume $\sigma_\epsilon^2 = 1$. Suppose that y_t is the first missing value in the time series. Since up to time t there are no missing observations, $\hat{y}_t = \sum_{j=1}^{\infty} \pi_j y_{t-j}$, so that $y_t - \hat{y}_t = \epsilon_t$, and

$$\sigma_t^2(1) = Var[y_t - \hat{y}_t] = Var[\epsilon_t] = 1.$$

At time $t+1$, $K_t = K_{t-1}$ (no new information is available at time t). Therefore,

$$\sigma_{t+1}^2(1) = Var[y_{t+1} - \hat{y}_{t+1}] = Var[y_{t+1} - P_{K_t} y_{t+1}], \tag{13}$$

and

$$\begin{aligned} P_{K_t} y_{t+1} &= P_{K_t}[\epsilon_{t+1} - \sum_{j=1}^{\infty} \pi_j y_{t+1-j}] \\ &= -\pi_1 P_{K_t} y_t - \pi_2 y_{t-1} - \pi_3 y_{t-2} - \dots \ . \end{aligned} \tag{14}$$

Thus,

$$y_{t+1} - P_{K_t} y_{t+1} = \epsilon_{t+1} - \pi_1 (y_t - P_{K_t} y_t) = \epsilon_{t+1} - \pi_1 (y_t - P_{K_{t-1}} y_t). \tag{15}$$

Hence,

$$\begin{aligned} \sigma_{t+1}^2(1) &= \pi_1^2 Var[y_t - P_{K_{t-1}} y_t] + Var[\epsilon_{t+1}] = \pi_1^2 \sigma_t^2(1) + 1 \\ &= \pi_1^2 + 1 \geq \sigma_t^2(1). \end{aligned} \tag{16}$$

The next lemma characterizes the behavior of the MSPE after the missing value.

Lemma 1 *For* $k \geq 1$ $\sigma_{t+k}^2(1) - 1 \geq \left(\frac{\pi_k}{\pi_{k+1}}\right)^2 (\sigma_{t+k+1}^2(1) - 1)$.

Proof. Since y_t is missing, $K_{t+k} = [y_{t+k}, y_{t+k-1}, \dots, y_{t+1}, y_{t-1}, y_{t-2}, \dots]$ and, $y_{t+k+1} = \epsilon_{t+k+1} - \pi_1 y_{t+k} - \pi_2 y_{t+k-1} - \pi_3 y_{t+k-2} - \dots - \pi_{k+1} P_{K_{t+k}} y_t - \pi_{k+2} y_{t-1} \cdots$. Thus, $y_{t+k+1} - P_{K_{t+k}} y_{t+k+1} = \epsilon_{t+k+1} - \pi_{k+1}(y_t - P_{K_{t+k}} y_t)$ and $Var[y_{t+k+1} - P_{K_{t+k}} y_{t+k+1}] = 1 + \pi_{k+1}^2 Var[y_t - P_{K_{t+k}} y_t]$. Therefore, for all $k \geq 1$ $Var[y_t - P_{K_{t+k}} y_t] = \frac{\sigma_{t+k+1}^2(1) - 1}{\pi_{k+1}^2}$. Thus,

$$\frac{\sigma_{t+k}^2(1) - 1}{\pi_k^2} = Var[y_t - P_{K_{t+k-1}} y_t]. \tag{17}$$

Since $K_{t+k} = [y_{t+k}, y_{t+k-1}, \dots, y_{t+1}, y_{t-1}, y_{t-2}, \dots]$ and

$$K_{t+k-1} = [y_{t+k-1}, y_{t+k-2}, \dots, y_{t+1}, y_{t-1}, y_{t-2}, \dots], \tag{18}$$

$K_{t+k-1} \subset K_{t+k}$ which implies $Var[y_t - P_{K_{t+k-1}} y_t] \geq Var[y_t - P_{K_{t+k}} y_t]$. Therefore

$$\frac{\sigma^2_{t+k}(1) - 1}{\pi^2_k} \geq Var[y_t - P_{K_{t+k}} y_t] = \frac{\sigma^2_{t+k+1}(1) - 1}{\pi^2_{k+1}}. \tag{19}$$

Hence, $\sigma^2_{t+k}(1) - 1 \geq \left(\frac{\pi_k}{\pi_{k+1}}\right)^2 (\sigma^2_{t+k+1}(1) - 1)$. □

The next two theorems specify the convergence rate of the MSPE of a short and long memory process, respectively.

Theorem 1 *As $k \longrightarrow \infty$, the one-step MSPE of a standard ARMA process converges to 1 at a exponential rate*

$$|\sigma^2_{t+k}(1) - 1| \leq Ca^{-k},$$

with $a > 1$.

Proof. From (17), $\sigma^2_{t+k}(1) - 1 = \pi^2_k Var[y_t - P_{K_{t+k-1}} y_t]$. But, $Var[y_t - P_{K_{t+k-1}} y_t] \leq Var[y_t]$, then $\sigma^2_{t+k}(1) - 1 \leq \pi^2_k Var[y_t]$. On the other hand, $\sigma^2_{t+k}(1) - 1 \geq 0$. Therefore, $0 \leq \sigma^2_{t+k}(1) - 1 \leq \pi^2_k Var[y_t]$. Since for a standard ARMA process $\pi_k \sim |a|^{-k}$ for large k and $a > 1$,

$$|\sigma^2_{t+k}(1) - 1| \leq Ca^{-k}$$

as required. □

Theorem 2 *As $k \longrightarrow \infty$, the one-step MSPE of a fractional ARIMA process converges to 1 at a hyperbolic rate*

$$|\sigma^2_{t+k}(1) - 1| \leq Ck^{-\alpha}$$

with $\alpha = 2 - 2d > 0$.

Proof. This result can be proved analogously to the previous theorem. From (17),

$$\sigma^2_{t+k}(1) - 1 = \pi^2_k Var[y_t - P_{K_{t+k-1}} y_t]. \tag{20}$$

But, $Var[y_t - P_{K_{t+k-1}} y_t] \leq Var[y_t]$, then $\sigma^2_{t+k}(1) - 1 \leq \pi^2_k Var[y_t]$. On the other hand, $\sigma^2_{t+k}(1) - 1 \geq 0$. Therefore, $0 \leq \sigma^2_{t+k}(1) - 1 \leq \pi^2_k Var[y_t]$. Since $\pi_k \sim k^{d-1}$ for large k, $|\sigma^2_{t+k}(1) - 1| \leq Ck^{-2+2d}$, as required. □

As an illustration, Figure 1 depicts the evolution of the one-step MSPE, $\sigma^2_{t+k}(1)$, after a missing value for an ARFIMA$(1, d, 1)$ model with parameters $\phi = -.49$, $d = .13$, $\theta = -.67$ and an ARMA$(1, 1)$ process with parameters $\phi = -.49$, $\theta = -.67$ and fractional noise process with long memory parameter $d = .13$.

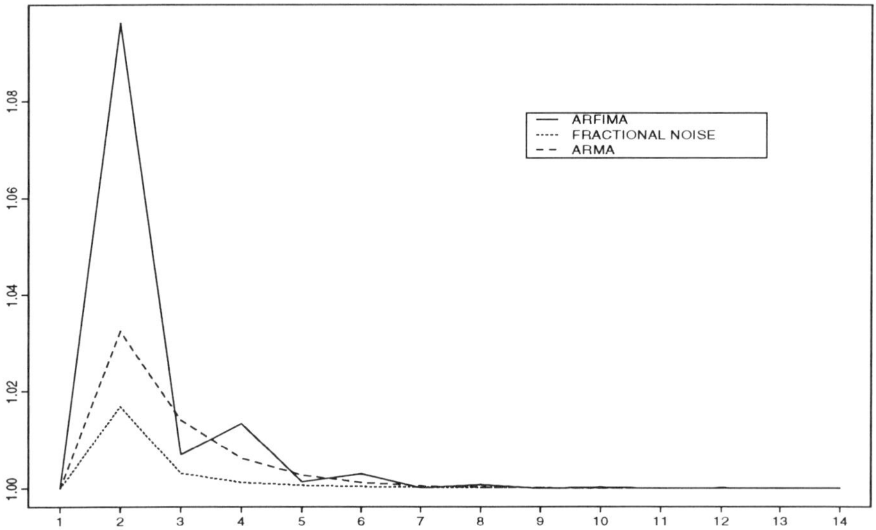

Figure 1: MSPE for an ARFIMA$(1,d,1)$ model with parameters $\phi = -.49$, $d = .13$, $\theta = -.67$ (Heavy line); an ARMA$(1,1)$ process with parameters $\phi = -.49$, $\theta = -.67$ (Broken line) and a fractional noise process with long memory parameter $d = .13$ (Dotted line).

3.2 Asymptotic Analysis of a Block of Missing Data

Suppose now that all the observations from time t_0 through $t_0 + k$ are missing. In this case, the state estimate for $t_0 + k$, $k \geq 0$, is given by $\hat{X}_{t_0+k} = F^k \hat{X}_{t_0}$, and the observation prediction is $\hat{y}_{t_0+k} = G\hat{X}_{t_0+k} = GF^k \hat{X}_{t_0}$.

The next theorem characterize the asymptotic behavior of the estimate $\hat{y}_{t_0+k}$ for short memory ARMA processes.

Theorem 3 *For standard ARMA models, (a) $\hat{y}_{t_0+k}$ converges to zero (zero-mean process) in probability, at an exponential rate, as k goes to infinity. (b) $E[\hat{y}_{t_0+k} - y_{t_0+k}]^2$ converges to the variance of the process, at an exponential rate, i.e., $|E[\hat{y}_{t_0+k} - y_{t_0+k}]^2 - \sigma_y^2| \leq Ca^{-k}$, with $a > 1$.*

Proof. To prove part (a), we observe that

$$E[\hat{y}_{t_0+k}] = GF^k E[\hat{X}_{t_0}] = 0 \tag{21}$$

and

$$Var[\hat{y}_{t_0+k}] = GF^k Var[\hat{X}_{t_0}] F'^k G' \leq \|G\|^2 \|F\|^{2k} \|Var[\hat{X}_{t_0}]\|, \tag{22}$$

where $\|\cdot\|$ is the matrix Euclidean norm. If $\{\lambda_i\}$ are the eigenvalues of F, then $\|F\| = \max_i |\lambda_i| \equiv |\lambda_0|$ Since all the eigenvalues of a causal ARMA process have absolute value smaller than 1, $|\lambda_0| < 1$, and then $Var[\hat{y}_{t_0+k}] \leq |\lambda_0|^{2k}C$ where $C = \|G\|^2 \|Var[\hat{X}_{t_0}]\|$ is a constant. Therefore,

$$Var[\hat{y}_{t_0+k}] \longrightarrow 0,$$

as $k \longrightarrow \infty$, at exponential rate. Part (b) follows directly from part (a) . □

A characterization of the asymptotic behavior of the estimate $\hat{y}_{t_0+k}$ for long memory ARFIMA processes is given in the following theorem.

Theorem 4 *For ARFIMA models, (a) $\hat{y}_{t_0+k}$ also converges to zero (zero-mean process) in probability as k goes to infinity, however, the decaying rate is hyperbolic, i.e., as $k^{-\alpha}$, $\alpha > 0$, for large k. (b) $E[\hat{y}_{t_0+k} - y_{t_0+k}]^2$ converges to the variance of the process, at a hyperbolic rate, i.e., $|E[\hat{y}_{t_0+k} - y_{t_0+k}]^2 - \sigma_y^2| \leq Ck^{-\alpha}$, $\alpha > 0$.*

Proof. Recall that $E[\hat{y}_{t_0+k}] = GF^k E[\hat{X}_{t_0}] = 0$ and

$$Var[\hat{y}_{t_0+k}] = GF^k Var[\hat{X}_{t_0}] F'^k G'. \tag{23}$$

From Brockwell and Davis [2], we can write:

$$Var[\hat{X}_{t_0}] = Var[X_{t_0}] - \Omega_{t_0}. \tag{24}$$

Hence, $Var[\hat{y}_{t_0+k}] = GF^k Var[X_{t_0}] F'^k G' - GF^k \Omega_{t_0} F'^k G'$.
Let $A_k = GF^k Var[X_{t_0}] F'^k G'$ and $B_k = GF^k \Omega_{t_0} F'^k G'$, then

$$Var[\hat{y}_{t_0+k}] = A_k - B_k. \tag{25}$$

The terms A_k and B_k converge to zero as k goes to infinity, at a hyperbolic rate, i.e., as $k^{-\alpha}$, $\alpha > 0$, for large k: Let $\Psi_{t_0} = Var[X_{t_0}]$, then

$$\Psi_{t_0} = F^{t_0} \Psi_0 F'^{t_0} + \sum_{i=0}^{t_0-1} F^i Q F'^i \tag{26}$$

therefore

$$\begin{aligned} A_k &= GF^{t_0+k} \Psi_0 F'^{t_0+k} G' + \sum_{i=0}^{t_0-1} F^i Q F'^i \\ &= \Psi_0(t_0+k+1, t_0+k+1) + \sum_{i=0}^{t_0-1} \psi_{i+k+1}^2 \\ &= \sum_{i=0}^{\infty} \psi_{i+t_0+k+1}^2 + \sum_{i=0}^{t_0-1} \psi_{i+k+1}^2 = \sum_{i=k+1}^{\infty} \psi_i^2, \end{aligned} \tag{27}$$

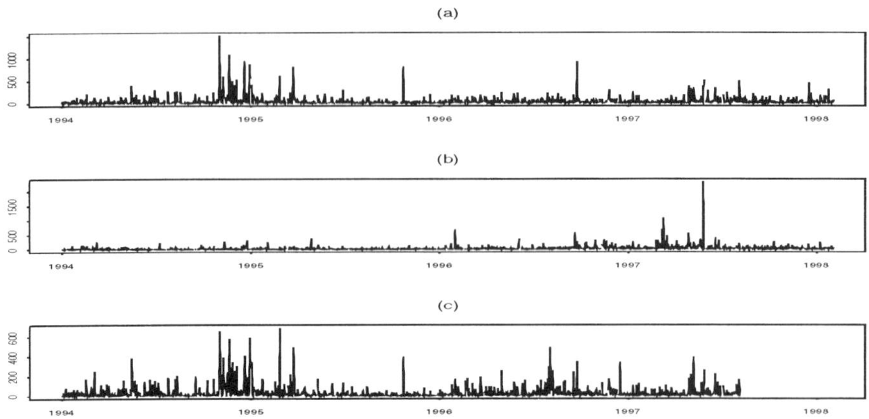

Figure 2: Square Exchange Rates: (a) Swiss Franc, (b) Australian Dollar. (c) Netherlands Guilder.

where $\Psi_0(k+1, k+1)$ is the $k+1$ diagonal element of Ψ_0 and ψ_i are the coefficients of the MA(∞) expansion of the ARFIMA model. For large k, $A_k \sim Ck^{2d-1}$, hence A_k converges to zero hyperbolically, i.e., as $k^{-\alpha}$, $\alpha > 0$, as $k \to \infty$.

On the other hand, $B_k = GF^{t_0}\Omega_0 F^{t_0\prime}G' = \omega_0(k+1, k+1)$ where $\omega_0(k+1, k+1)$ is the $k+1$ diagonal element of Ω_0. But,

$$\begin{aligned} \omega_0(k+1, k+1) &\leq \omega_0(k+t_0+1, k+t_0+1) + \sum_{i=0}^{t_0} \psi_{i+k+1}^2 \qquad (28) \\ &= \sum_{i=0}^{\infty} \psi_{i+t_0+k+1}^2 + \sum_{i=0}^{t_0-1} \psi_{i+k+1}^2 = \sum_{i=k+1}^{\infty} \psi_i^2. \end{aligned}$$

Thus, $B_k \sim Ck^{2d-1}$, for large k, and therefore B_k converges to zero hyperbolically, i.e., as $k^{-\alpha}$, $\alpha > 0$, as k goes to infinity. Part (b) follows directly form part (a). □

An application of the state space techniques to the study of financial time series with missing values is presented in the next section.

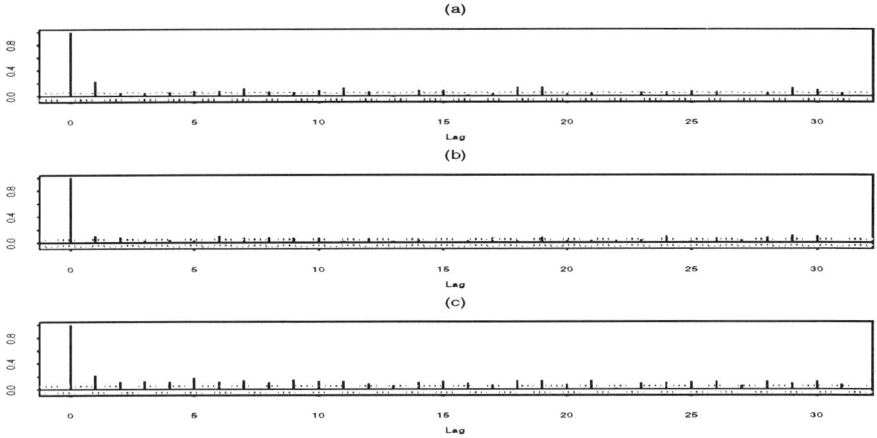

Figure 3: Square Exchange Rates: Autocorrelation Function (a) Swiss Franc, (b) Australian Dollar. (c) Netherlands Guilder.

4 Application

Evidence of long memory behavior in foreign exchange rates has been reported by many authors, see for example Cheung [4], Lobato [10] and Lobato and Robinson [11]. In this section we analyze the exchange rates of the Swiss Franc, Australian Dollar and Netherlands Guilder relative to the US Dollar. These data are available from Financial Data and Resources Locator, (www.ntu.edu.sg/library). Figure 2 displays the square of the first difference log exchange rates for these three countries. There are 1495 daily observations, from January 1994 to January 1998.

The sample autocorrelation functions (ACF) of these time series are shown in Figure 3. The coefficients of these autocorrelation functions decay slowly and some of them are significant even after 30–day lags.

According to Beran [1], log–var plots are useful to explore possible long range dependency in time series. Let $Var(\bar{x}_k)$ be the variance of the mean of

Table 1: Foreign Exchange Data: Estimates of the ARFIMA model .

Series	Missing Obs.	$\hat{d}$	$\hat{\sigma}_d$	t	σ_ϵ
Swiss Franc	107	0.1738	0.0208	8.3480	97.73
Australian Dollar	107	0.1100	0.0187	5.8808	91.22
Netherlands Guilder	277	0.1688	0.0208	8.1296	66.45

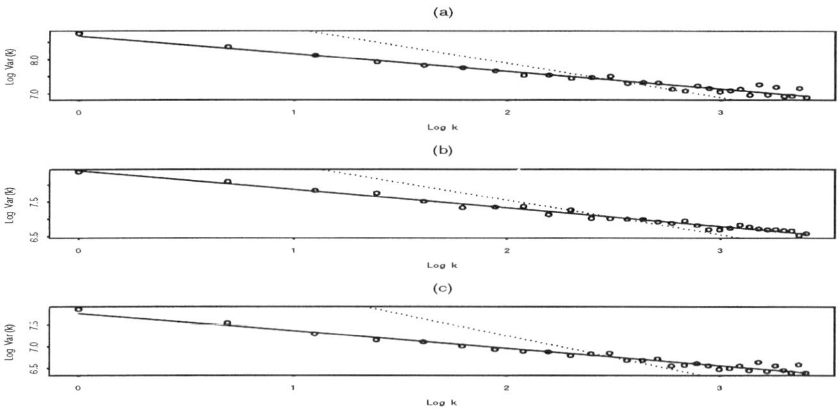

Figure 4: Square Exchange Rates: Log Var Plots ($\log(\gamma_k)$ versus $\log(k)$) (a) Swiss Franc, (b) Australian Dollar. (c) Netherlands Guilder.

k consecutive observations. If the process has long memory, then $Var(\bar{x}_k) \sim Ck^{2d-1}$. Thus, $\log[Var(\bar{x}_k)] \sim C_1 + (2d-1)\log(k)$. The log–var plots shown in Figure 4 indicate some long memory features in the ACF. For the three series, the slope of a fitted straight line of $\log[Var(\bar{x}_k)]$ versus $\log(k)$ (heavy line) does not equal minus one (dotted line), as expected for a short memory process.

To account for the long range dependency behavior observed in the sample autocorrelation functions and the log–var plots, we fitted an ARFIMA(0,d,0) model. The maximum likelihood estimates are presented in Table 1. As shown in the second column of this table, there are several missing values in the three series. The t-statistics displayed in the fifth column indicate that the values of the long memory parameter d (third column) are highly significant.

The prediction standard errors are shown in figure 5. It can be observed that right after the beginning of a data gap, the standard deviations of the one step forecasting error increases and then they drop to the σ_ϵ level. The Swiss Franc and Australian Dollar time series display single missing observations or very short gaps. However, the Netherlands Guilder series presents both isolated missing values and long data gaps.

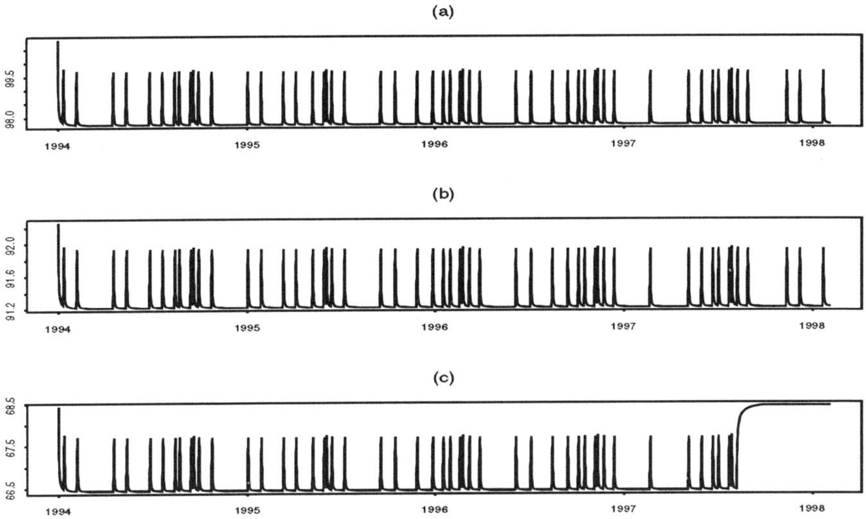

Figure 5: Square Exchange Rates Prediction Standard Deviations: (a) Swiss Franc, (b) Australian Dollar. (c) Netherlands Guilder.

5 Conclusions

In this paper, state space techniques are applied to the analysis of long memory time series with missing values. As shown by the study of the exchange rates, these procedures allows for easy handling of data gaps through appropriate Kalman filter recursions.

Acknowledgments

I would like to thank the organizers and the participants of the Hong Kong International Workshop on Statistics in Finance for useful comments on this paper. This work was partially supported by a Grant 1980859 from Fondecyt.

References

1. J. Beran, *Statistics for Long-Memory Processes*, New York: Chapman & Hall (1994).
2. P. Brockwell and R. Davis, *Time Series: Theory and Methods*, (Springer, New York, 1991).
3. N. H. Chan and W. Palma, *Ann. Statist.* **26**, 719 (1998)
4. Y. W. Cheung, *J. Bus. & Econ. Stat.* **11**, 93 (1993).
5. R. Dahlhaus, *Ann. Statist.* **17**, 1749 (1989)
6. R. H. Jones, *Technometrics* **22**, 389 (1980).

7. G. Kitagawa, *In Time Series Analysis of Irregularly Observed Data*, ed. E. Parzen, (Springer, New York, 1986).
8. R. Kohn and C. F. Ansley, *J. Amer. Statist. Assoc.* **81**, 751 (1986).
9. W. K. Li and A. I. McLeod, *Biometrika* **73**, 217 (1986)
10. N. Lobato, *J. Econometrics* **90**, 129 (1999)
11. N. Lobato and P. Robinson, *Rev. Econ. Stud.* **65**, 475 (1998)
12. W. Palma and N. H. Chan, *Journal of Forecasting* **62**, 183 (1997).
13. B. K. Ray and R. S. Tsay, *Biometrika* **84**, 791 (1997).
14. P. M. Robinson, *In Advances in Econometrics*, ed. C.A. Sims (Cambridge University Press, Cambridge, 1994).
15. F. Sowell, *J.Econometrics* **53**, 165 (1992).
16. G. C. Tiao and R. S. Tsay, *J. of Forecasting* **13**, 109 (1994).
17. H. Tong, *In Nonlinear Dynamics and Time Series*, ed. C. D. Cutler and D. T. Kaplan, (American Mathematical Society, Rhode Island, 1997).

SECOND ORDER TAIL EFFECTS

CASPER G. DE VRIES
Tinbergen Institute Rotterdam,
Erasmus Universiteit Rotterdam and NIAS
E-mail: cdevries@few.eur.nl

Semi-parametric extremal analysis can be a useful tool to calculate the Value-at-Risk (VaR) for loss probabilities which are at and below the inverse of the sample size. We first review the standard estimation procedures and VaR implications on the basis of the first order expansion to the tail probabilities of heavy tail distributed random variables. Subsequently we present some new results that are based on using a second order expansion of the tail risk. In particular, we discuss the issue of efficiency in estimation using high or low frequency data; and we investigate the relation between the VaR over a short and a long investment horizon.

1 Introduction

Financial asset data sets nowadays cover millions of high frequency price quotes. These data sets are well suited for studying the market risk on very large losses. Regulators of the financial industry currently require that commercial banks be able to report, on a daily basis, a loss estimate over a ten-day trading horizon for their entire trading portfolio given a certain preassigned low risk level. The loss estimate is called the Value-at-Risk (VaR). For internal risk management purposes the larger investment banks also back out a VaR estimate for a one-day trading horizon. Non-financial corporations nowadays do include long horizon VaR forecasts in their yearly statements. Out of convenience the continuously compounded asset returns are often presumed to be normally distributed, see J.P.Morgan (1995), Jorion (1997), and Dowd (1998). As it happens, however, asset returns are heavy tailed distributed. If we work from this assumption, the VaR can be well estimated by employing extreme value techniques, see e.g. Dacorogna et al. (1995), Longin (1997), Danielsson and De Vries (1997, 1998) and Dowd (1998). The approach is a go between the traditional finance based normal approach and the historical simulation based non-parametric approach.

In the paper we first briefly review the motivation behind the by now standard estimation procedures by means of a first order expansion to the tail probabilities of heavy tail distributed random variables. We discuss how the first order approach implies a particular relationship between the VaR over short and longer investment horizons. Subsequently we present some new results that are based on using a second order expansion of the tail risk. In particular we discuss the issue of efficiency in estimation using high and low frequency data; and we investigate the relation between the VaR over a short and a long investment horizon.

2 The First Order Approach to Heavy Tails and VaR

Suppose that the returns are i.i.d. and have tails which vary regularly at infinity. In that case

$$F(-x) = ax^{-\alpha}[1 + o(1)] \quad \text{as } x \to \infty, \quad \text{and } a > 0. \tag{1}$$

These distributions are said to exhibit heavy tails since the m-th moment $E[X^m]$ is unbounded when $\alpha < m$, whereas in case of e.g. the normal d.f. for any finite m the $E[X^m]$ is bounded. Given parameter estimates for the scale coefficient a and tail index α, the VaR x can be calculated upon inverting $ax^{-\alpha}$ for a given small risk level p: $x_p \approx (a/p)^{1/\alpha}$. We first discuss how the parameters can be estimated, and subsequently discuss the VaR application in more detail.

2.1 *Estimation*

The standard estimation procedures can be motivated as follows. Suppose the Pareto law $G(-x) = ax^{-\alpha}$ holds exact below a certain threshold $-s$, where $s > 0$. The conditional distribution reads $G_{X|X\le -s}(-x) = (x/s)^{-\alpha}$. One can go from this to the associated conditional density with tail index $\alpha + 1$: $g_{X|X\le -s}(-x) = \alpha(x/s)^{-\alpha-1}(1/s)$. Take logarithms to get

$$\log g_{X|X\le -s}(-x) = \log\alpha - (\alpha+1)\log\frac{x}{s} - \log s.$$

Substitute in this expression the random variable $-X_i$ for the x, whenever $X_i < -s$. Differentiate with respect to α, sum the result over the observations X_i which fall below $-s$, and equate to 0 in order to obtain the Maximum Likelihood estimator of the tail index:

$$\widehat{\frac{1}{\alpha}} = \frac{1}{M}\sum_{i=1}^{M}\log\frac{-X_i}{s}, \quad X_i < -s, \tag{2}$$

and where M is the random number of extreme observations X_i that fall below the threshold $-s$. For a large enough s, the conditional Pareto density $g_{X|X\le -s}(-x)$ may also be a good approximation to the true conditional density $f_{X|X\le -s}(-x)$, when the conditional distribution is not exactly Pareto but rather satisfies (1). The estimator (2) applied to the extreme observations from a heavy tailed distribution that adheres to (1) is known as the Hill (1975) estimator. We note that the estimator (2) is conditional on the appropriate choice of the threshold s; but how this choice has to be made cannot be discussed without going into the second order expansion. The assumption of independence is also crucial; although the estimator can be shown to be consistent for important classes of stochastic processes.

Likewise we can motivate the estimator for the extreme quantiles or VaR. Let x_p and x_t be two extreme quantiles with associated probabilities p and t respectively, that adhere to the law $G_{X|X\leq -s}(-x) = (x/s)^{-\alpha}$. Then $t/p = (x_t/x_p)^{-\alpha}$, and hence $x_p = x_t\,(t/p)^{1/\alpha}$. Suppose $p < 1/n < t$, where n is the sample size; moreover let t be such that M, $M < n$, is the closest integer equal to nt. Then we can estimate the VaR x_p by

$$\widehat{x_p} = x_t \left(\frac{M/n}{p}\right)^{\widehat{1/\alpha}}. \tag{3}$$

Since the statistical properties of $\widehat{x_p}$ are dominated by the properties of the exponent $\widehat{1/\alpha}$, we can limit the discussion towards discussing the properties of the tail index estimator.

2.2 Value at Risk at Different Horizons

Suppose a bank has estimated its one-day VaR from past daily return observations. It also has to calculate the VaR for a ten-day investment horizon to fulfill its regulatory requirements. The industry often works from the assumption of normality and calculates the ten-day VaR by sizing up the one-day estimate with a factor $\sqrt{10}$, since this is the well known convolution rule for summing i.i.d. normal random variables. The square-root procedure reduces the burden of estimation on risk managers. If the observations are heavy tailed distributed, this simple convolution rule no longer applies. Nevertheless for the tail risk, aggregation is still simple under the i.i.d. assumption.

Let the returns X_i have a distribution as in (1). For the sum $\Sigma_i^k X_i$ (holding k fixed), we have by Feller's theorem (1971, VIII.8)

$$P\left\{\Sigma_i^k X_i < -x\right\} = kax^{-\alpha}[1 + o(1)], \quad \textit{as } x \to \infty, \tag{4}$$

and where the scale factor 'a' is as in (1). We pointed out that banks for internal purposes often calculate the VaR over a one day investment horizon, but that regulators require a longer horizon. Corporations for their yearly reports need an even longer horizon, see the recently launched CorporateMetrics

(1999) product by the RiskMetrics group. The question therefore is how to go from the high frequency estimate to the low frequency estimate without having to reestimate the parameters on a reduced sample size, and thus possibly losing efficiency. In Dacorogna et al. (1995, 1998) the following rule was presented:

Proposition 1 (The α-root rule) *Suppose X has finite variance, so that $\alpha > 2$. At a constant risk level p, increasing the time horizon k increases the VaR for the normal model percentagewise by more, i.e. by $\sqrt{k}$, than for the fat tailed model, where the increase is a factor $k^{1/\alpha}$.*

Proof. *Rescale x on the left hand side in (4) by $k^{1/\alpha}$, this gives $ax^{-\alpha}$ on the right hand side and hence equals the first order term in (1).* ■

The opposite holds if the distribution is so heavy tailed that the second moment is unbounded, i.e. if $\alpha < 2$ then $k^{1/\alpha} > \sqrt{k}$. In the related economics literature on diversification it has been noted that the effect of diversification is less pronounced in comparison with the normal distribution, if the returns are sum-stable distributed with $\alpha < 2$, see Fama and Miller (1972, p. 270). They note that for $\alpha < 1$ diversification actually increases the dispersion. We are not aware of a discussion in the finance literature of the case $\alpha > 2$ but finite, for either the issue of diversification nor for the issue of tail risk (VaR) aggregation over time

3 The Second Order Approach to Heavy Tails

Throughout this section we assume that the following second order expansion applies:

$$F(-x) = ax^{-\alpha}[1 + bx^{-\beta} + o(1)], \quad \text{as } x \to \infty, \quad \text{and } a > 0. \tag{5}$$

Freely floating foreign exchange rate returns are often more or less symmetrically distributed about a zero mean. Therefore, in what follows we will often assume that the lower and upper side tails are similar up to and including the second order term

$$\begin{aligned} P\{X \leq -x\} &= ax^{-\alpha}(1 + bx^{-\beta} + o(x^{-\beta})), \\ P\{X > x\} &= ax^{-\alpha}(1 + bx^{-\beta} + o(x^{-\beta})). \end{aligned} \tag{6}$$

The differences may come from the o-terms. Note that the second order term is assumed to be of the same type as the first order term. Some motivation for this choice can be found in the following observations. If the second order term were of the form $\log x$, some of the results below would not apply due to the slower rate of convergence; for other functional forms like $\exp(-x)$ convergence is so rapid that the second order term plays no role of importance. The expansion (5) applies for symmetric heavy tailed distributions like the Student-t, which is often used to model the unconditional distribution of asset returns, and it applies to the stationary distribution of the ARCH(1) process, which is used for modelling the conditional asset returns.

3.1 statistical properties

On basis of the expansion (5) one can derive the first two moments of the Hill estimator (2) by elementary calculus. The conditional k−th order log empirical moment from a sample $X_1, ..., X_n$ of n i.i.d. draws from $F(x)$ is defined as follows:

$$u_k(s_n) \equiv \frac{1}{M} \sum_{i=1}^{M} \chi_{(X_i < -s_n)} (\log \frac{-X_i}{s_n})^k, \tag{7}$$

where s_n is a threshold that depends on n, M is the random number of left tail excesses, and where $\chi_{(.)}$ is the indicator function. Note that $u_k(s_n)$ is a function of the highest realizations only. We will sometimes suppress the reference to n in s_n when this does not create confusion. The theoretical properties of the Hill estimator $u_1(s_n)$ are well documented by e.g. Hall (1982) and Goldie and Smith (1987).

The properties of the Hill estimator derive from the following Lemma

Lemma 2 *Given the model (5), for $k \geq 1$, and as n, $s_n \to \infty$, while $s_n/n \to 0$,*

$$\mathrm{E}[u_k(s)] = \Gamma(k+1) \left(\frac{1}{\alpha^k} + \frac{bs^{-\beta}}{(\alpha+\beta)^k} \right) + o\left(s^{-\beta}\right). \tag{8}$$

Proof. From calculus after two transformations of variables we have the following result:

$$\begin{aligned} \alpha \int_s^\infty (\log \frac{x}{s})^k x^{-\alpha-1} dx &= \alpha s^{-\alpha} \int_1^\infty (\log y)^k y^{-\alpha-1} dy \\ &= \alpha s^{-\alpha} \int_0^\infty t^k \left(e^t\right)^{-\alpha-1} e^t dt \\ &= \alpha^{-k} s^{-\alpha} \int_0^\infty x^k e^{-x} dx \\ &= \frac{\Gamma(k+1)}{\alpha^k} s^{-\alpha}. \end{aligned}$$

Hence, the conditional expectation in (8) follows from the assumption (5) and the calculus result

$$\mathrm{E}\left[u_k(s)\right] = \frac{1}{1-F(s)} \int_s^{\infty} \left(\log \frac{x}{s}\right)^k f(x)\, dx$$
$$= \frac{\Gamma(k+1)}{1+bs^{-\beta}} \left[\frac{1}{\alpha^k} + \frac{bs^{-\beta}}{(\alpha+\beta)^k}\right] + o\left(s^{-\beta}\right).$$

■

It immediately follows that for $k = 1$:

Corollary 3 *The asymptotic bias of the Hill estimator $u_1(s_n)$ from (2) is*

$$\mathrm{E}\left[u_1(s_n) - \frac{1}{\alpha}\right] = -\frac{b\beta}{\alpha(\alpha+\beta)} s_n^{-\beta} + o\left(s_n^{-\beta}\right). \tag{9}$$

After some manipulation and application of the Lemma (2) for $k = 1, 2$, one obtains the asymptotic variance of the Hill estimator.

Corollary 4 *For the threshold $s_n \to \infty$, but $s_n^{\alpha}/n \to 0$,*

$$\mathrm{Var}\left[u_1(s_n) - \frac{1}{\alpha}\right] = \frac{s_n^{\alpha}}{an} \frac{1}{\alpha^2} + o\left(\frac{s_n^{\alpha}}{n}\right). \tag{10}$$

These two results can be readily combined to obtain the asymptotic mean squared error (AMSE) of $u_1(s_n)$

$$\mathrm{AMSE}\left(u_1(s_n)\right) \approx \frac{1}{a\alpha^2} \frac{s_n^{\alpha}}{n} + \frac{b^2\beta^2}{\alpha^2(\alpha+\beta)^2} s_n^{-2\beta}. \tag{11}$$

From this expression it is easy to see that for $n \to \infty$, the rate by which $s_n \to \infty$ determines which of the two terms in (11) asymptotically dominates the other, or that they just balance.

Rewrite (11) in shorthand notation as AMSE $= An^{-1}s^{\alpha} + Ds^{-2\beta}$. From the first order condition $\alpha An^{-1}s^{\alpha-1} - 2\beta Ds^{-2\beta-1} = 0$, the unique AMSE minimizing threshold level $\bar{s}$ is found as

$$\bar{s}_n = \left(\frac{2\beta D}{\alpha A}\right)^{\frac{1}{2\beta+\alpha}} n^{\frac{1}{2\beta+\alpha}}.$$

To summarize, we have the following result:

Proposition 5 *As $n \to \infty$ the* AMSE *minimizing asymptotic threshold level $\overline{s}_n$ is*

$$\overline{s}_n(u_1) = \left[\frac{2ab^2\beta^3}{\alpha(\alpha+\beta)^2}\right]^{\frac{1}{(\alpha+2\beta)}} n^{\frac{1}{(\alpha+2\beta)}}. \tag{12}$$

And the associated asymptotically minimal MSE *of $u_1(s_n)$ is*

$$\overline{AMSE}\left[u_1(\overline{s}_n)\right] = \frac{1}{a\alpha}\left[\frac{1}{\alpha}+\frac{1}{2\beta}\right]\left[\frac{2ab^2\beta^3}{\alpha(\alpha+\beta)^2}\right]^{\frac{\alpha}{2\beta+\alpha}} n^{-\frac{2\beta}{2\beta+\alpha}} + o\left(n^{-\frac{2\beta}{2\beta+\alpha}}\right). \tag{13}$$

From (11-13) it is straightforward to show that if s_n tends to infinity at a rate below $n^{1/(2\beta+\alpha)}$, the bias part in the MSE dominates, while conversely the variance part dominates if s_n tends to infinity more rapidly than $n^{1/(2\beta+\alpha)}$. It is also easy to see that the number of excedances M is such that

$$n^{\frac{-2\beta}{\alpha+2\beta}} M(u_1(\overline{s}_n)) \to a\left[\frac{2ab^2\beta^3}{\alpha(\alpha+\beta)^2}\right]^{-\frac{\alpha}{\alpha+2\beta}} \quad \text{in } p \text{ as } n \to \infty. \tag{14}$$

Further asymptotic properties of the Hill estimator, like asymptotic normality given that $\overline{s}_n$ is used in (2), are shown in e.g. Goldie and Smith (1987). Danielsson et al. (1997) discuss how a bootstrap of the AMSE can be used to back out the optimal threshold $\overline{s}_n$ in practice, such that the Hill estimator retains its asymptotic normality property. In this bootstrap procedure the empirical minimum of the bootstrapped MSE is used to estimate $\overline{s}_n$ consistently, and the procedure guarantees that the rate conditions assumed in the above results are automatically satisfied. By doing so one balances the two vices of bias squared and variance such that these disappear at the same rate. For dependent data it is sometimes known how the variance is affected, see e.g. the recent work by Drees (1999) and Starica (1999) for the ARCH(1) process, but other aspects, like the choice of the threshold s_n, are still open issues.

3.2 Time Aggregation and Efficiency

The log-returns are time additive, i.e. the two week return is the sum of the one week returns. Nowadays financial data sets can be obtained at even the finest time grid around, which is the trading time scale. The question is which data should be used for estimation purposes. In particular we ask ourselves the following question, if one needs results for a long investment horizon, should one nevertheless use the high frequency data for estimation, and then use a rule like the α-root rule to extrapolate to the low frequency level? We give an answer in terms of the asymptotic mean squared error efficiency.

Assume that $\alpha > 2$, because this is the relevant case for most financial data. In that case both the mean and the variance are bounded. We first obtain a general lemma on second order convolution behavior. This result is needed because, as was shown above, the $AMSE$ of the tail index estimator is a function of the first and second order parameters. The existing literature only gives a result on second order convolution behavior for positive random variables, see Geluk, De Haan, Resnick and Starica (1997). But since the log-asset returns can be positive and negative, we need to analyze this case afresh. To restrict the number of different combinations that will arise, we assume that the tails are similar. We find that because the distribution of asset returns is two-sided, a new factor depending on $E[X^2]$ enters.

Lemma 6 (Second order convolution) *Suppose that the tails are second order similar, i.e. as* $x \to \infty$

$$\begin{aligned} P\{X \leq -x\} &= ax^{-\alpha}(1 + bx^{-\beta} + o(x^{-\beta})), \\ P\{X > x\} &= ax^{-\alpha}(1 + bx^{-\beta} + o(x^{-\beta})), \end{aligned} \tag{15}$$

and $a > 0, b \neq 0$. *Moreover, assume that* $\alpha > 2$ *and* $\beta > 0$ *so that* $E[X]$ *and* $E[X^2]$ *are bounded. Suppose* X_1 *and* X_2 *are i.i.d. and satisfy (15). Then for the 2-convolution*

$$\begin{aligned} P\{X_1 + X_2 > s\} &= P\{X_1 + X_2 \leq -s\} \\ &= 2as^{-\alpha}(1 + bs^{-\beta} + \alpha E[X]s^{-1} + \frac{\alpha(\alpha+1)}{2}E[X^2]s^{-2}) \\ &\quad + o(s^{-\alpha-2}) + o(s^{-\alpha-\beta}) \end{aligned} \tag{16}$$

as $s \to \infty$.

The Lemma (6) was obtained in Dacorogna et al. (1998) by elaborate calculus arguments. We develop some intuition for the result by a novel argument. The probability $P\{X_1 + X_2 > s\}$ can be split into just two parts:

$$\begin{aligned} &P\{X_1 + X_2 > s\} \approx P\{X_1 + X_2 > s, X_2 \leq \frac{s}{2}\} + \\ &P\{X_1 + X_2 > s, X_1 \leq \frac{s}{2}\} \end{aligned} \tag{17}$$

The remaining other part $P\{X_1 > \frac{s}{2}, X_2 > \frac{s}{2}\} = P\{X_1 > \frac{s}{2}\}^2 = O(s^{-2\alpha})$ is of smaller order and can be ignored since it is assumed that $\alpha > 2$.

To determine $P\{X_1 + X_2 > s, X_2 \leq \frac{s}{2}\}$, we first compute the conditional probability $P\{X_1 + X_2 > s \mid X_2 = c\} = P\{X_1 + c > s\}$, say. This conditional probability is obtained from the marginal by translation. Consider the law $P\{X > x\} = ax^{-\alpha}(1 + bx^{-\beta} + o(x^{-\beta}))$ as $x \to \infty$, and suppose we shift X by adding the constant c. This changes the probability into $P\{X + c > x\} = a(x-c)^{-\alpha}(1 + b(x-c)^{-\beta} + o(x^{-\beta}))$. Use the Taylor expansion to write, assuming that $x > c$,[1]

$$\begin{aligned}(x-c)^{-\gamma} &= x^{-\gamma}(1 - \frac{c}{x})^{-\gamma} \\ &= x^{-\gamma}\{1 + \gamma\frac{c}{x} + \frac{\gamma(\gamma+1)}{2}(\frac{c}{x})^2 + O((\frac{c}{x})^3)\}.\end{aligned}$$

Use this twice to rewrite $P\{X + c > x\}$ as:

$$\begin{aligned}P\{X + c > x\} = ax^{-\alpha}[1 + \alpha c x^{-1} + \frac{\alpha(\alpha+1)}{2}c^2x^{-2} + bx^{-\beta} \\ +o(x^{-\beta}) + o(x^{-2})].\end{aligned} \tag{18}$$

The following conditional probability can be split into three parts

$$P\{X_1 + X_2 > s, -\frac{s}{2} \leq X_2 \leq \frac{s}{2}\} = \int_{-\infty}^{\infty} P\{X + c > s\}dF(c)$$

$$- \int_{-\infty}^{-s/2} P\{X + c > s\}dF(c) - \int_{s/2}^{\infty} P\{X + c > s\}dF(c).$$

In all three integrals substitute the right hand side form of (18) for $P\{X + c > s\}$. The second and third integral are of small order $O(s^{-2\alpha})$. For example, since for $s \to \infty$

$$\int_{s/2}^{\infty} P\{X + c > s\}dF(c) =$$

$$\int_{s/2}^{\infty} as^{-\alpha}(1 + o(1))\{\alpha x^{-\alpha-1}(1 + o(1))\}dx =$$

$$O(s^{-2\alpha}).$$

The first probability can be found by using the translation result

$$\int_{-\infty}^{\infty} P\{X_1 + c > s\}dF(c) = E_c[P\{X_1 + c > s\}]$$

[1] See also Dacorogna et al.(1995) where this expansion is used to show that the Hill estimator is not location invariant.

$$= E_c[as^{-\alpha}\{1 + \alpha cs^{-1} + \frac{\alpha(\alpha+1)}{2}c^2s^{-2} + bs^{-\beta} + o(s^{-\beta}) + o(s^{-2})\}]$$

$$= as^{-\alpha}\{1 + bs^{-\beta} + \alpha E[X_2]s^{-1} + \frac{\alpha(\alpha+1)}{2}E[X_2^2]s^{-2} + o(s^{-\beta}) + o(s^{-2})\}.$$

The last expression gives $P\{X_1 + X_2 > s, -\frac{s}{2} \leq X_2 \leq \frac{s}{2}\}$, but we need $P\{X_1 + X_2 > s, X_2 \leq \frac{s}{2}\}$, see (17). However, as before, the probability $P\{X_1 + X_2 > s, X_2 \leq -\frac{s}{2}\}$ is of small order and can be ignored. By symmetry the same result is obtained for $P\{X_1 + X_2 > s, X_1 \leq \frac{s}{2}\}$. Putting these two probabilities together yields the claim.

From this second order convolution result we can infer how the $AMSE$ will be affected by the choice of the return frequency in the estimation, see Dacorogna et al. (1995,1998):

Proposition 7 *Suppose the X_i are i.i.d. with a distribution $F(x)$ that is symmetric around zero, $E[X] = 0$, and varies regularly at infinity as in (5) with $\alpha > 2$. Then a w-convolution affects the leading term in the $\overline{AMSE}\left[u_1\left(\overline{s}_n\right)\right]$ from (13) as follows:*

(i) $\beta < 2$. There is no effect;

(ii) $\beta = 2$. The $AMSE$ changes by a factor

$$\left[\left(1 + \frac{1}{2}\alpha(\alpha+1)(w-1)E[X^2]/b\right)^2\right]^{\alpha/(2\beta+\alpha)};$$

(iii) $\beta > 2$. The $AMSE$ changes by a factor

$$c\left[\frac{1}{2}\alpha(\alpha+1)(w-1)E[X^2]\right]^{\frac{2\alpha}{4+\alpha}}\left(\frac{1}{b^2}\right)^{\alpha/(2\beta+\alpha)},$$

and where

$$c = \frac{4+\alpha}{2\beta+\alpha}\left(\frac{2}{\alpha+2}\right)^{\frac{2\alpha}{4+\alpha}}\left(\frac{\alpha+\beta}{\beta}\right)^{\frac{2\alpha}{2\beta+\alpha}}\left(\frac{\alpha}{4an}\right)^{\frac{4}{4+\alpha}}\left(\frac{2\beta an}{\alpha}\right)^{\frac{2\beta}{2\beta+\alpha}}.$$

The upshot of Proposition 7 is that either time aggregation has no effect, i.e. when $\beta < 2$, or that the $AMSE$ deteriorates, possibly only after the first few convolutions when $b < 0$ and $\beta = 2$. If $\beta > 2$ the $AMSE$ always deteriorates after the first convolution. While it can thus not be ruled out that higher frequencies deteriorate the $AMSE$ properties of $\widehat{\alpha}$ for the first few convolutions, the majority of the cases goes into the other direction. For this reason it may be advisable to use the highest frequency data available for estimation, and subsequently to extrapolate to obtain the lower frequency result by means of a rule like the α-root rule from Proposition 1.

3.3 Second Order VaR

Suppose one follows the advice from the previous subsection and estimates the low frequency VaR from the high frequency VaR. By doing this one exploits the efficiency that the high frequency data deliver. On the negative side however, one may loose from the fact that the α-root rule from Proposition 1 is based on a first order approximation We investigate the possible loss in precision that may arise from neglecting the second order terms. Assume the mean is $E[X] = 0$. Consider the convolution result (16), but inflate the VaR s by a factor $2^{1/\alpha}$. This gives

$$P\{X_1 + X_2 \le -2^{1/\alpha}s\} = \\ as^{-\alpha}\{1 + b2^{-\beta/\alpha}s^{-\beta} + \frac{\alpha(\alpha+1)}{2}E[X^2]2^{-2/\alpha}s^{-2}\} \\ + o(s^{-\alpha-2}) + o(s^{-\alpha-\beta}).$$

Let $P\{X \le -s\} = as^{-\alpha}(1 + bs^{-\beta} + o(s^{-\beta})) = p$, say, and use this to rewrite the above

$$P\{X_1 + X_2 \le -2^{1/\alpha}s\} = \\ p + as^{-\alpha}\{-b(1 - 2^{-\beta/\alpha})s^{-\beta} + \frac{\alpha(\alpha+1)}{2}E[X^2]2^{-2/\alpha}s^{-2}\} + \\ o(s^{-\alpha-2}) + o(s^{-\alpha-\beta}).$$

If $b > 0$ and $\beta < 2$, then for sufficiently large s the α-root rule is overly conservative, since the second order term $-b(1 - 2^{-\beta/\alpha})s^{-\beta}$ is negative. If, however, $b < 0$, or if $\beta > 2$, then the second order term is positive, and the α-root rule is not prudent enough. To circumvent the bias in the low frequency VaR estimates that stems from the α-root rule, one could redo the quantile estimation on the low frequency data by means of (3), while retaining the tail index estimate from the high frequency data. Which procedure is better is an issue for further research.

4 Conclusion

The paper first reviews the standard estimation procedures and VaR implications on the basis of a first order expansion for the tail probabilities of heavy tail distributed random variables. Subsequently, it was argued why second order results are needed for determining the properties of the estimators.

We developed a new intuitive derivation of the second order convolution result. This second order convolution result is useful for the discussion of the efficiency in estimation. While for most cases using the high frequency data is mean-square efficient, we showed that there are some exceptions. The second order convolution result also enables one to determine the precision of the rule by which the VaR over a short investment horizon is related to the VaR over a long investment horizon.

Acknowledgments

Summary of presentation for the 'Workshop on Statistics in Finance', Hong Kong, July 1999. Some of this material was first presented at the conference on 'Extremes, Risk and Safety' in Gothenburg, August 1998. The paper is partially based on joint work with M. Dacorogna, J. Danielsson, J. Geluk, L. de Haan, U. Muller, L. Peng and O. Pictet. I am grateful to R.Brinkman for helpful discussion and to a referee for careful reading of the manuscript.

References

1. M.M. Dacorogna, U.A. Muller, O.V. Pictet and C.G. de Vries, *Extremal returns in extremely large data sets.* (Tinbergen Institute discussion paper, TI95-70, 1995).
2. M.M. Dacorogna, U.A. Muller, O.V. Pictet and C.G. de Vries, *Extremal forex returns in extremely large data sets.* (mimeo, submitted, 1998).
3. K. Dowd, *Beyond value at risk, the new science of risk management,* (Wiley, Chichester, 1998).
4. Jansen D.J. Danielsson and C.G. de Vries, The methods of moments ratio estimator for the tail shape parameter. *Communications in Statistics, Theory and Methods* **25**, 711–720 (1996).
5. J. Danielsson, L. de Haan, L. Peng and C.G. de Vries, *Using a bootstrap method to choose the sample fraction in tail index estimation.* (Tinbergen Institute discussion paper, TI97-016/4, 1997), forthcoming in *Journal of Multivariate Analysis.*
6. J. Danielsson and C.G. de Vries, Tail index and quantile estimation with very high frequency data. *Journal of Empirical Finance* **4**, 241–257 (1997).
7. J. Danielsson and C.G. de Vries, *Value-at-Risk and extreme returns.* (Tinbergen Institute discussion paper TI98-017/2, 1998).
8. A.L.M. Dekkers, J.H.J. Einmahl and L. de Haan, On the estimation of the extreme-value index and large quantile estimation. *Annals of Statistics* **17**, 1795–1832 (1989).
9. H. Drees, *Weighted approximations of tail processes under mixing conditions.* University of Cologne, mimeo. (1999).

10. E.F. Fama and M.H. Miller, *The theory of finance.* (Dryden Press, Hinsdale, 1972).
11. W. Feller, *An introduction to probability theory and its applications, volume II.* (John Wiley, New York, 2nd edition, 1971).
12. J. Geluk, L. de Haan, S. Resnick and C. Starica, Second order regular variation, convolution, and the central limit theorem. *Stochastic Processes and their Applications* **69**, 139–159 (1997).
13. C.M. Goldie and R.L. Smith. Slow variation with remainder: Theory and applications. *Quarterly Journal of Mathematics*, Oxford 2nd series, **38**, 45–71 (1987).
14. P. Hall, On some simple estimates of an exponent of regular variation. *Journal of the Royal Statistical Society*, Series B, **44**, 37–42 (1982).
15. B.M. Hill, A simple general approach to inference about the tail of a distribution. *Annals of Statistics* **3**, 1163–1173 (1975).
16. P. Jorion, *Value-at-Risk.* (Irvin: McGraw Hill, 1997)
17. Morgan Guarantee Trust Company, *RiskMetrics Technical Document.* New York: J.P.Morgan Bank (1995).
18. F.M. Longin, *From value-at-risk to stress testing:the extreme value approach.* (CERSSEC working paper 97-004, 1997).
19. L. Peng, *Second order condition and Extreme value estimation.* Ph.D. dissertation #178. (Tinbergen Institute, Erasmus University Rotterdam, 1997).
20. RiskMetrics Group, *CorporateMetrics Technical Document* (1999). www.riskmetrics.com.
21. C. Starica, *On the tail empirical process of solutions of stochastic difference equations.* (Chalmers University, mimeo. 1999).

Part III: Volatility

RECENT DEVELOPMENTS IN HETEROSKEDASTIC TIME SERIES

N. H. CHAN
Department of Statistics, Carnegie Mellon University
Pittsburgh, PA 15213-3890, USA
E-mail: chan@stat.cmu.edu

G. PETRIS
Department of Mathematical Sciences, University of Arkansas
Fayetteville, AR 72701, USA
E-mail: gpetris@comp.uark.edu

This article surveys some of the recent developments in the modeling of heteroskedastic financial time series. Both discrete-time and continuous-time frameworks for some commonly used models and their estimating methodologies are discussed. In particular, the recently popularized long-memory heteroskedastic models are reviewed. A simulation-based Bayesian approach for long-memory stochastic volatility models is proposed. The paper concludes with an illustration of the proposed method applying to a value-weighted index from the Center for Research in Security Prices.

1 Introduction

Empirical analysis of financial data has by now provided overwhelming evidence that stock returns cannot be satisfactorily modeled by linear ARMA models. This paper reviews current developments in extending the linear framework to model the heteroskedastic behavior of stock return data and suggests a new approach to model the long-memory behavior of the stock returns. It is organized as follows. A survey of recent developments in modeling the heteroskedasticity of a financial series is given in section 2. Section 3 discusses the long-memory phenomenon and some recent findings of modeling long-memory heteroskedastic series. The long-memory stochastic volatility model and its state space formulation, together with a description of the MCMC sampling scheme and an example are also given in section 3. Concluding remarks are given in section 4.

2 Heteroskedasticity

Due to the celebrated random walk hypothesis for an efficient market, a random walk model (or variants of it) has been one of the most commonly used tools to model equity returns for decades. Specifically, let P_t denote the price of a stock at the end of period t and let $y_t = (P_t - P_{t-1})/P_{t-1} \approx \log P_t - \log P_{t-1}$ denote the return at the end of period t. The random walk model simply states that the return series $\{y_t\}$ is like a white noise sequence $\{Z_t\}$, i.e., $\log P_t$ follows and ARIMA(0,1,0) model. However, ample evidences about the inadequacy of modeling $\{y_t\}$ as white noise have been documented in the literature, see for example, Campbell and Lo [9]. These evidences are usually referred as stylized facts which can be gathered as follows.

- Leptokurtosis. The return series usually exhibits a heavy-tailed phenomenon which cannot be represented by a Gaussian-like assumption.
- Heteroskedasticity. The clustering of variation of the return series suggests strong heteroskedastic behavior which is at odd with the constant variance assumption of $\{y_t\}$ when it is modeled as white noise in a random walk model.
- Persistence of volatility. The autocorrelation function of the square of the returns decays slowly, suggesting certain kind of long-memory behavior.
- Negative correlation among returns and volatilities. There is a certain amount of asymmetry between returns and risks.

Since the 80s new models have been proposed to account for these phenomena. Instead of being a white noise sequence, the return series $\{y_t\}$ is generalized as

$$y_t = \sigma_t Z_t, \tag{1}$$

where $\{Z_t\}$ is a white noise sequence, usually Gaussian, but the conditional variance σ_t varies over time. In the next two subsections, we review some recent developments in the modeling of the volatility process $\{\sigma_t\}$. For an early account of some of these developments, see Shephard [31].

2.1 Discrete-Time Models

In a discrete-time setting, autoregressive conditionally heteroskedastic (ARCH) models were first proposed by Engle [15] and then extended by Bollerslev [7] to the generalized ARCH (GARCH) model. In these models, the volatility σ_t is assumed to be a predictable process, i.e. a deterministic function of the past. For a GARCH model, σ_t takes the form

$$\sigma_t^2 = \alpha_0 + \sum_{i=1}^{p} \alpha_i y_{t-i}^2 + \sum_{j=1}^{q} \beta_j \sigma_{t-j}^2. \tag{2}$$

Estimation of the parameters $(\alpha_0, \cdots, \beta_q)$ for GARCH models is customarily done using quasi-maximum-likelihood (QML) procedures. Although a GARCH model has a natural interpretation in terms of (2), it is somewhat inflexible and specific constraints need to be imposed on the parameters to ensure that the model is well-defined. Extensions of GARCH models such as E-GARCH, T-GARCH have also been proposed to capture other market features, see Nelson [28]. However, many empirical studies indicate that these extensions only provide marginal improvements over the nonstationary integrated GARCH(1,1) model which seems to fit many financial return series reasonably well.

Stochastic volatility (SV) is an alternative class of models that accounts for volatility clustering. Here the instantaneous variance of the observed series is modeled as a non-observable, or latent, process. Let $\{y_t\}$ denote the return of an equity. A basic setup of a stochastic volatility model takes the form

$$\begin{cases} y_t &= \sigma_t \xi_t, \\ \sigma_t &= \exp(v_t/2), \end{cases} \tag{3}$$

where $\{\xi_t\}$ is usually assumed to be a sequence of independent standard Normal random variables and the log volatility sequence $\{v_t\}$ satisfies an ARMA relation

$$\phi(B)v_t = \theta(B)\eta_t. \tag{4}$$

Here, $\{\eta_t\}$ is Gaussian white noise with variance τ, $\phi(\cdot)$ and $\theta(\cdot)$ are polynomials of order p, q, respectively, with all their roots outside the unit circle and with no common root, and B is the backshift operator $By_t = y_{t-1}$. Conceptually, this represents an extension with respect to GARCH models, since the evolution of the volatility is not completely determined by the past observations, but it includes a stochastic component and allows for a more flexible mechanism. Unfortunately, since $\{\sigma_t\}$ is not observable, the method of QML cannot be directly applicable. By letting $x_t = \log y_t^2$, $u_t = \log Z_t^2$, and taking log and squaring (3), we have

$$x_t = v_t + u_t, \tag{5}$$

$$\phi(B)v_t = \theta(B)\eta_t. \tag{6}$$

In this expression, the log volatility sequence satisfies a linear state space model with state equation (6) and observation equation (5), while the original process $\{\sigma_t\}$ follows a non-linear state space model. To complicate matters further, the observation error $u_t = \log \xi_t^2$ in (5) is non-Gaussian. Consequently, direct applications of the Kalman filter method for linear Gaussian state space models seem unrealistic. Several estimation procedures have been developed for

SV models to circumvent some of these difficulties. Melino and Turnbull [26] use a generalized method of moments (GMM), which is straightforward to implement, but not efficient. Harvey, Ruiz and Shephard [23] propose a QML approach, based on approximating the observation error $\{u_t\}$ by a mixture of Gaussian random variables which renders (5) and (6) into a linear Gaussian state-space setup. A Bayesian approach is taken by Jacquier, Polson and Rossi [24]. Kim, Shephard and Chib [25] suggest a simulation-based exact maximum likelihood estimator while Sandmann and Koopman (1998) propose a Monte Carlo maximum likelihood procedure. Although each of these methods is reported to work well under certain conditions, it is difficult to assess their overall performances across different data sets. Alternatively, the SV model can be considered as a discrete-time realization of a continuous-time process as follows.

2.2 Continuous-Time Models

Since the seminal work of Nelson [27] which shows that a GARCH type model can be approximated by a diffusion process as the time spans between observations tend to zero, we have witnessed a surge in research activities in continuous-time models, see for example, the recent monograph edited by Rossi [30] (1996). In addition, the celebrated Black-Scholes formula and the ready availability of high frequency tick-by-tick data provide a natural platform for using continuous-time diffusion processes to model financial assets. Let S_t denote a stock price at the end of period t and let $W_{1,t}, W_{2,t}$ denote two standard Brownian motions. One popular form of continuous-time model which incorporates a stochastic volatility factor is:

$$\frac{dS_t}{S_t} = \mu dt + \sigma_t dW_{1,t}, \tag{7}$$

where the log variance process $v_t = \log \sigma_t^2$ satisfies a mean-reverting relation as

$$dv_t = (\alpha - \beta v_t)dt + \eta dW_{2,t}, \tag{8}$$

where α, β and η are parameters governing the volatility diffusion equation (8). It is sometimes useful to assume the correlation coefficient between the two Brownian motions to be negative so that the stylized fact of negative correlations among risks and returns can be taken into account. Due to the latency of the process $\{v_t\}$, closed-form expressions for the discrete-time transition density of (7) are generally unknown, making QML procedures infeasible.

A number of attempts have been proposed to deal with consistent continuous-time estimation. Aït-Sahalia [1] proposes a semiparametric method to estimate the diffusion parameter based on the Kolmogorov forward equation. A different approach is to treat the estimation problem as a missing value problem in diffusion as discussed in Pedersen (1995). Elerian, Chib and Shephard [14] make use of this idea and develop a Markov Chain Monte Carlo estimation procedure for partially observed diffusion models. Another notable approach is the method of moments, mainly the GMM discussed in Hansen [22]. A lucid summary about GMM and its relationship with MLE can be found in the appendix of Campbell, Lo and McKinley [10]. More recently, the GMM method has been extended to a powerful tool, known as the efficient methods of moments (EMM), by Gallant and Tauchen [18] that deals with inferences for continuous-time processes.

To illustrate the EMM idea, consider the conditional distribution of the return series. EMM first estimates this distribution semiparametrically via QML. This provides an auxiliary model, then the scores of this auxiliary model are used to obtain moment conditions as in GMM for estimating the underlying diffusion parameters. As an example, consider the system (7) and (8) as the structural model, i.e., the true data generating mechanism. In this context, let $y_t = \frac{dS_t}{S_t}$ denote the return process and let $f(y_t|Y_{t-1}, \xi)$ denote the conditional density of y_t given the history $Y_{t-1} = \{y_{t-1}, \cdots, y_1\}$ in an auxiliary model. Suppose there are T data points and the complete history is denoted by $Y_T = \{y_T, \cdots, y_1\}$. The EMM consists of two stages.

Stage I. First, the auxiliary parameter vector, ξ, of the auxiliary model is estimated via QML, i.e., we find the $\hat{\xi}_T$ which satisfies the first-order conditions

$$\frac{1}{T}\sum_{t=1}^{T} \frac{\partial}{\partial \xi} \log f(y_t|Y_{t-1}, \hat{\xi}_T) = 0. \tag{9}$$

Note that the left hand side of (9) is simply the average of the score function of the auxiliary model evaluated at $\hat{\xi}_T$ and thus provides an estimate of the expected value of the score function of the auxiliary function. This equation provides the analogous orthogonal condition used in GMM.

As far as the form of the conditional density f of the auxiliary model is concerned, Gallant and Tauchen [18] propose a semi-nonparametric (SNP) method where f has the form

$$f_K(y_t|Y_{t-1}, \xi) = \frac{P_K^2(z_t)}{\int_{-\infty}^{\infty} P_K^2(u)\phi(u)du} \frac{\phi(z_t)}{\sigma_t}, \tag{10}$$

where $z_t = y_t/\sigma_t$ (assuming the mean of y is zero), $\phi(\cdot)$ denotes the standard normal density, and $P_K(\cdot)$ denotes the Hermite polynomial

$$P_K(u) = \sum_{i=0}^{K_z} a_i u^i. \tag{11}$$

The constant K_z denotes the order of the polynomial expansion that controls the deviation to normality (leptokurtosis). The coefficients a_i of the polynomials can be functions of the history and be part of the auxiliary parameter ξ, see Gallant and Tauchen [18]. Notice that (10) is fairly flexible. Additional features of the data can be accommodated by either increasing the order K_z or replacing the normal density by other distributions in (10). Specific examples of various forms of (10) for GARCH type models are studied in Gallant, Hsieh and Tauchen [17] and Andersen, Chung and Sørensen [2]. Extension to cases with jump components in the return process is given in Andersen, Benzoni, and Lund [3].

Stage II. In the second stage, EMM inverts the score equation (9) to obtain a consistent estimate of the structural parameter vector $\psi = (\mu, \alpha, \beta, \eta, \rho)'$ of the structural model (7) and (8). The key idea in EMM lies in replacing the orthogonal condition (9) under the auxiliary model by the structural model and using it as a moment condition in GMM. Specifically, if the auxiliary model is flexible enough to capture the statistical behavior of the observed series, one would also expect

$$m(\psi, \hat{\xi}) = E_\psi[\frac{\partial}{\partial \xi} \log f(y_t | Y_{t-1}, \hat{\xi})] \tag{12}$$

to be small. Unfortunately, due to the lack of a closed form expression for the transition density of the solution of (7), the expected value (12) cannot be evaluated. Instead, EMM suggests using simulations to approximate (12) by Monte Carlo integration. For a given value of the structural parameter ψ, a simulated series $\hat{y}_n(\psi), n = 1, \cdots, N$ is generated from the structural model. This simulated series is then used to evaluate the sample moments at the fixed QMLE $\hat{\xi}_T$ as

$$m_N(\psi, \hat{\xi}_T) = \frac{1}{N} \sum_{n=1}^{N} \frac{\partial}{\partial \xi} \log f(\hat{y}_n(\psi) | \hat{Y}_{N-1}(\psi), \hat{\xi}_T). \tag{13}$$

The EEM estimator of ψ is the value $\hat{\psi}_T$ that minimizes the weighted version of (13)

$$\hat{\psi}_T = \text{ argmin }_\psi [m_N(\psi, \hat{\xi}_T)' \hat{I}_T^{-1} m_N(\psi, \hat{\xi}_T)], \tag{14}$$

where $\hat{I}_T$ denotes a consistent estimator of the asymptotic covariance matrix of the sample score vector.

Under suitable regularity conditions, Gallant and Tauchen [18] and Gallant and Tauchen [19] show that the EMM estimator is consistent, asymptotically normal and efficient. Although computationally intensive, EMM is a sufficiently general method that can be used to deal with both discrete-time and continuous-time model when latent variables are involved. Furthermore, as shown in Gallant, Hsieh and Tauchen [17] and Andersen, Chung and Sørensen [2], calibration of the auxiliary model through diagnostic tests can be achieved by means of goodness-of-fit tests derived from the asymptotic results. The EMM seems to offer a powerful tool for statistical inference for continuous-time diffusion models.

3 Long-Memory Models

Long-memory models have been receiving considerable attentions in the literature for the past two decades. Related references on long-memory models can be found in the monograph by Beran [6] or the survey article by Baillie [4]. Although long-memory behavior is usually understood in terms of specific autoregressive fractionally integrated moving average (ARFIMA) models, other descriptions are available. For example, Hall [21] discusses how to define and measure long-range dependence of a given set of data in terms of the convergence rate of the statistic of interest when compared with short-range dependent data.

In the financial domain, a number of attempts have been made to extend GARCH and SV models to capture the long-term dependence structure in the volatilities that are reported in empirical studies, see for example Ding, Granger and Engle [13]. Recently, Baillie, Bollerslev and Mikkelsen [5] introduce the Fractionally Integrated GARCH (FIGARCH) class of models and propose using QML for estimations. On another front, Breidt, Crato, and deLima [8] extend stochastic volatility to the Long-Memory Stochastic Volatility (LMSV) class of models. The estimation method they propose is based on the spectral approximation to the Gaussian likelihood by means of the Whittle likelihood. In what follows, we shall discuss the FIGARCH and FISV models in more detail.

3.1 Fractionally Integrated GARCH

In order to accommodate the long-memory behavior of the volatility process, one way to generalize the GARCH model is to introduce a fractionally integrated factor $(1-B)^d$, $d \in (-0.5, 0.5)$ in the GARCH model as first suggested by Robinson [29]. Specifically, Baillie *et al.* [5] formulate a FIGARCH model as

$$(1-B)^d\phi(B)y_t^2 = \omega + (1-\theta(B))\nu_t, \tag{15}$$

where $\nu_t = y_t^2 - \sigma_t^2$, $\phi(B), \theta(B), \omega$ are the corresponding ARMA representation of the GARCH process $\{y_t, \sigma_t\}$ defined in (2) in terms of $\{\nu_t\}$. Expressing (15) as an infinite ARCH representation, Baillie *et al.* [5] propose to estimate the parameters of this model by QML. In order to achieve convergence, their procedure requires a large truncation lag (1000 in their study) for the infinite series representation. Assessing this truncation effect in other situations may be difficult.

A different idea is to make use of the ARFIMA structure. Since (15) can be written as an ARFIMA model in terms of the martingale difference process $\{\nu_t\}$, one can estimate this model by means of available approximated MLE methods for ARFIMA models. Unfortunately, since the process $\{\nu_t\}$ is non-Gaussian and highly skewed, direct applications of approximated Gaussian procedures seem unrealistic. Furthermore, higher moment conditions are usually required for maximum likelihood procedures in these situations which impose more conditions on the parameter space of the underlying model.

In summary, although one can use FIGARCH models to capture the long-memory persistency of volatility, estimation and inferences for these models are extremely tricky. In addition, a FIGARCH model suffers from the same drawbacks discussed in section 2.1 that existed in a GARCH model.

3.2 Fractionally Integrated Stochastic Volatility

Breidt *et al.* [8] introduce the idea of a fractionally integrated stochastic volatility (FISV) model as follows:

$$\begin{cases} y_t &= \sigma_t \xi_t, \\ \sigma_t &= \sigma \exp(v_t/2), \end{cases} \tag{16}$$

where $\{\xi_t\}$ is a sequence of independent standard Normal random variables, σ is a positive constant and the sequence $\{v_t\}$ satisfies the ARFIMA relation

$$(1-B)^d\phi(B)v_t = \theta(B)\eta_t. \tag{17}$$

Here $d \in (-0.5, 0.5)$, $\{\eta_t\}$ is Gaussian white noise with variance τ, $\phi(\cdot)$ and $\theta(\cdot)$ are polynomials of order p, q, respectively, with all their roots outside the unit circle and with no common root. To estimate the parameters of the model, they suggest to maximize the spectral likelihood. Alternatively, one can approach the inference for FISV models as follows. Equation (17) implies that $\{v_t\}$ has an infinite moving average representation in terms of the white noise $\{\eta_t\}$. One can truncate the infinite moving average to a finite number of terms M, say, to obtain an approximate representation of $\{v_t\}$. As demonstrated in Chan and Palma [11], one obtains a better approximation by considering the corresponding truncation of the moving average representation of the first difference of $\{v_t\}$:

$$\begin{aligned} \Delta v_t &= (1-B)^{-d+1}\left(\phi(B)\right)^{-1}\theta(B)\eta_t \\ &= \sum_{j=0}^{M} \varphi_j \eta_{t-j}. \end{aligned} \tag{18}$$

The coefficients φ_j of the truncated moving average representation of $\{\Delta v_t\}$ are functions of d, $\underline{\phi} \triangleq (\phi_1, \ldots, \phi_p)$ and $\underline{\theta} \triangleq (\theta_1, \ldots, \theta_q)$. Another advantage of considering first differences is that we are able to extend the parameter space by allowing d to vary in $(-0.5, 1.5)$. In this way the model includes both long-memory stationary and nonstationary series at the same time. Let $x_t \triangleq \log y_t^2$ and $u_t \triangleq \log \xi_t^2$, (16) implies

$$\Delta x_t = \Delta v_t + \Delta u_t. \tag{19}$$

We have therefore the following approximate state space model:

$$\begin{cases} \Delta x_t &= \Delta v_t + \varepsilon_t \\ \Delta v_t &= \sum_{j=0}^{M} \varphi_j \eta_{t-j} \\ \varepsilon_t &= u_t - u_{t-1}. \end{cases} \tag{20}$$

Since $\{\Delta v_t\}$ and $\{\varepsilon_t\}$ are independent moving average processes, (20) can be conveniently represented in terms of a Dynamic Linear Model (DLM) (West and Harrison [33]) as follows:

$$\begin{cases} \Psi_{t+1} = \left[\begin{array}{c|cc} 0 & 0 & \\ \hline 0 & 0 & 0 \\ & \mathbf{I_M} & 0 \end{array}\right] \Psi_t + \begin{bmatrix} -u_t \\ \eta_{t+1} \\ 0 \\ \vdots \\ 0 \end{bmatrix} \\ \Delta x_t - \left[1 \; \varphi_0 \ldots \varphi_M\right] \Psi_t + u_t \end{cases} \tag{21}$$

The DLM is completely specified once a distribution for the state vector at time $t = 1$ is given. If we imagine that the dynamics of the system can be extended into the past, the components of Ψ_1 have the following interpretation:

$$\begin{aligned} \Psi_{1,1} &= -u_0, \\ \Psi_{1,1+j} &= \eta_{2-j}, \qquad j = 1, \ldots, M+1. \end{aligned} \tag{22}$$

Guided by this interpretation, we assign Ψ_1 a Normal distribution with mean $(1.27, 0, \ldots, 0)'$ and variance $\mathrm{diag}(\pi^2/2, \tau, \ldots, \tau)$. Note that -1.27 and $\pi^2/2$ are the mean and variance of the $\log \chi_1^2$ distribution. In order to work with a Gaussian DLM, we approximate the distribution of $\{u_t\}$, which is the $\log \chi_1^2$ distribution, to a convenient accuracy with a finite mixture of normal distributions. Denoting by $\mathcal{L}(X)$ the distribution of X, for any random element X, we can write

$$\mathcal{L}(u_t) \approx \sum_{j=1}^{N} \pi_j \mathcal{N}(m_j, \sigma_j^2), \tag{23}$$

where $\mathcal{N}(m, \sigma^2)$ denotes a Gaussian distribution with mean m and variance σ^2, and the π_j's are positive weights adding to one. This suggests that we add to the model a vector of T discrete independent latent variables

$$\underline{K} = (K_1, \ldots, K_T), \tag{24}$$

whose distribution is defined by

$$P(K_t = j) = \pi_j, \qquad j = 1, \ldots, N, \quad t = 1, \ldots, T, \tag{25}$$

and we set

$$P(u_t \leq x | \underline{K}) = \Phi\Big(\frac{x - m_j}{\sigma_j}\Big) \qquad \text{if } K_t = j, \tag{26}$$

$j = 1, \ldots, N$, $t = 1, \ldots, T$, where Φ is the cumulative distribution function of the standard normal distribution. Up to the approximation (23), the marginal distribution of the sequence (u_t) has not changed; on the other hand, conditional on $\underline{K}$, the DLM (21) is Gaussian.

Although it is possible to use informative priors, and we encourage to do so when prior information is available, we consider here a noninformative (uniform) prior on d, $\underline{\phi}$, $\underline{\theta}$ and an inverse gamma conjugate prior for τ. Specifically, the prior density is assumed to be

$$p(d, \underline{\phi}, \underline{\theta}, \tau) = \textit{constant} \cdot \tau^{-\alpha_\tau - 1} \exp\Big(-\frac{\beta_\tau}{\tau}\Big). \tag{27}$$

Let us denote by Λ the parameter of the model, including the latent variables that we have introduced, i.e.,

$$\Lambda \triangleq \big((\Psi_t), \underline{K}, \tau, d, \underline{\phi}, \underline{\theta}\big). \tag{28}$$

To analyze the posterior distribution we need to generate a sample from the distribution of Λ, conditional on the observed sequence $\{\Delta x_t : t = 1, \cdots, T\}$.

Loosely speaking, each of the six components of Λ (which may itself be multidimensional, (Ψ_t) for example) is sampled from its full conditional distribution, i.e., its conditional distribution given the data and the other parameters.

Since, given all the other parameters and latent variables, the model reduces to the DLM (21), sampling from the full conditional distribution of (Ψ_t) is equivalent to sampling from the posterior distribution of the state vectors at time $t = 1, \ldots, T$ in a completely specified Gaussian DLM. This can be done efficiently using the forward filtering, backward sampling approach of Frühwirth-Schnatter [16].

Sampling from $\underline{K}$ is straightforward, when one realizes that the component of $\underline{K}$ are, under the full conditional distribution, independent and have a finite support. Also straightforward is sampling τ: since we choose a conjugate prior, its full conditional distribution is again an inverse gamma.

The full conditional densities of the remaining parameters do not have an analytic form which can be recognized as corresponding to any known and well-studied distribution. Therefore to draw from these distributions we use Metropolis-Hastings algorithm (see Tierney [32]). For each one-dimensional full conditional, the proposal distribution we use is based on a linear approximation of the logarithm of the target density, as described in Gilks and Wild [20]. More details on the simulation method can be found in Chan and Petris [12].

3.3 An Application

We apply the model and estimation technique described in the previous section to a financial time series. The data consists in the daily returns for the value-weighted market index from the Center for Research in Security Prices from July 1962 to July 1989. Following a common practice, the correlation in the return data due to the day of the week and month of the year was removed using standard filters, details can be found in Breidt *et al.* [8]. The series of these returns, together with the series of log squared returns, are plotted in Figure 1. There seems to be an increasing trend in the second series (log squared returns), suggesting strong persistence or even nonstationarity. Fitting a straight line to the log returns by ordinary least squares gives a t-value of 15 for the slope parameter. Since the standard normality assumptions clearly do not hold here, it is difficult to interpret this number, for example attaching a p-value to it. However, we are inclined to judge it as a large number, though informally, which prompts the use of a model that allows for a nonstationary behavior.

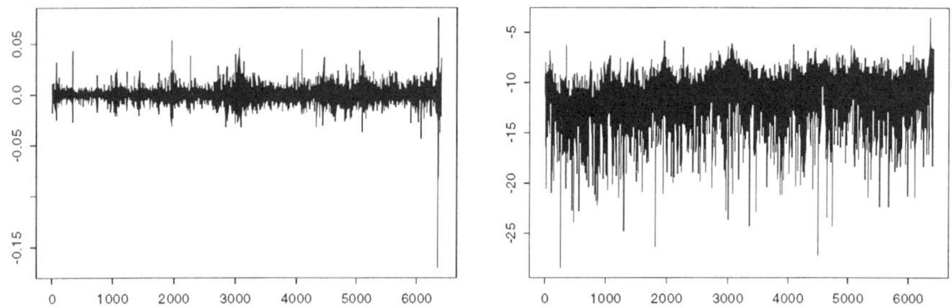

Figure 1: Daily returns (left) and log squared returns (right).

Table 1: Posterior summaries

	0.05	0.25	Mean	0.75	0.95
d	0.555	0.642	0.675	0.717	0.722
ϕ_1	0.589	0.590	0.595	0.596	0.602
τ	0.000129	0.000229	0.002655	0.005216	0.077035

We model the return data as

$$\begin{aligned} y_t &= \sigma_t \xi_t, \\ \sigma_t &= \sigma \exp(v_t/2), \\ (1-B)^d(1-\phi_1 B)v_t &= \eta_t, \end{aligned} \tag{29}$$

using the prior described in the previous section. Posterior summaries (the mean and four quantiles) for selected parameters resulting from the MCMC simulation are reported in Table I. The posterior distribution of d confirms our feeling about the nonstationarity of the volatility process.

One advantage of the Bayesian approach is that a full posterior distribution is available, so that inference on events or quantities depending on the parameters is conceptually straightforward. For example, one issue with this kind of data is whether the process driving the volatility is stationary or not. This formally corresponds to test the hypothesis that d is less than 0.5. For this data set, the posterior probability that the process is nonstationary ($d \geq 0.5$), evaluated from the Monte Carlo sample, turns out to be 99%. Note that the prior we use is noninformative with respect to this issue, in the sense that $P(-0.5 < d < 0.5) = P(0.5 \leq d < 1.5) = 1/2$.

Breidt *et al.* [8] find estimates $\hat{\phi}_1 = 0.932$ and $\hat{d} = 0.444$. It should be pointed out, in order to explain the discrepancy between those estimates and the corresponding posterior means reported in Table 1, that we are using a different model. In fact, even if the set of equations describing the observation process and the evolution of the volatility is the same, and is expressed in terms of the same parameters, our parameter space is different. While we allow the long-memory parameter d to vary in $(-0.5, 1.5)$, Breidt *et al.*[8] constrain this parameter to the stationarity region $(-0.5, 0.5)$. Notice that the two polynomials $(1-B)^{0.675}(1-0.590B)$ and $(1-B)^{0.444}(1-0.932B)$ are almost identical in the sense that their power tranfer functions are very close to each other at all frequencies except near zero, where their ratio goes to infinity.

4 Concluding Remarks

Returns on stocks or indexes typically show a nonlinear behavior. Several models have been proposed to describe this kind of data, usually assuming the unobservable volatility of the returns to follow either a stationary process (e.g., GARCH, SV), or a nonstationary one (e.g., IGARCH). The present paper, after reviewing the most popular models and estimation techniques for financial return data, introduces a model that encompasses stationarity and nonstationarity, as well as long-range dependence, a feature frequently observed in daily financial time series. The Bayesian approach taken here allows one, by combining a (typically noninformative) prior with the evidence provided by the data through the likelihood function, to obtain a readily interpretable posterior probability of the volatility process being stationary. In the example considered in Section 3.3, the evidence against stationarity is fairly strong. This is in accord with the recent findings that when daily returns are analyzed, one often ends up with an IGARCH(1,1) model or an SV model with parameters close to the boundary of the stationarity region.

A stylized fact about daily returns that has not been considered here is the excess kurtosis in the returns. This can be easily accommodated in our model by taking the distribution of ξ_t in equation (16) to be a Student's t with fixed degrees of freedom. Then in the mixture of normals (23), the weights π_j's, means m_j's, and variances σ_j^2's have to be revised so that the mixture approximate the corresponding moments of the $\log(t^2)$ distribution. Note that this does not make the analysis or the simulation scheme more involved. In a more general framework, one should be able to estimate the extent to which the ξ_t's are leptokurtic. One possibility is to consider for ξ_t a Student's t distribution with unknown degrees of freedom ν for a finite number of possible values of ν. This would only add one extra discrete distribution to sample in the MCMC step.

Several other topics remain open for future research, including the important problem of forecasting the volatility for risk management. The state-space formulation, together with the simulation approach, is perfectly suited to generate future paths of the volatility from the appropriate predictive distribution. Generating a stretch of future volatilities for a fixed value of the parameters ϕ_j in equation (18), determined by the current state of the chain, is as easy as generating from a moving average process with known parameters. From these future volatility scenarios, one can compute means, probability intervals, standard deviations. The added value brought in by the simulation approach is that one can also look at typical future behaviors of the volatility, in addition to pointwise summaries such as means and histograms. Clearly, as is well known, one needs to check the predictive properties of the model on past data, even if this is not a guarantee of future performances, before using these predictions.

Acknowledgments

We would like to thank Dr. Jay Breidt for kindly providing the data set analyzed in section 3.3 and a referee for helpful comments. Research supported in part by an Earmarked Grant No. HKUST6082/98T from the Research Grant Council of Hong Kong and by a National Science Foundation Group Infrastructure Grant to the Department of Statistics at Carnegie Mellon University.

References

1. Aït-Sahalia, Y. (1996). Nonparametric pricing of interest rate derivative securities. *Econometrica* **64**, 527-560.
2. Andersen, T.G., Chung, H.J. and Søresen, B.E. (1999). Efficient method of moments estimation of a stochastic volatility model: A Monte Carlo study. *Journal of Econometrics* **91**, 61-87.
3. Andersen, T.G., Benzoni, L. and Lund, J. (1999). Estimating jump-diffusions for equity returns. Technical Report, Finance Department, Northwestern University, Evanston, IL 60208, U.S.A.
4. Baillie, R.T. (1996). Long-memory processes and fractional integration in econometrics. *J. Econometrics* **73**, 5-59.
5. Baillie, R.T. and Bollerslev, T. and Mikkelsen, H.O. (1996). Fractionally integrated generalized autoregressive conditional heteroskedasticity. *Journal of Econometrics* **74**, 3-30.
6. Beran, J. (1994). *Statistics for Long-Memory Processes.* Chapman and Hall, New York.

7. Bollerslev, T. (1986). Generalized autoregressive conditional heteroskedasticity. *Journal of Econometrics* **31**, 307-327.
8. Breidt, F.J. and Crato, N. and de Lima, P. (1998). The detection and estimation of long-memory in stochastic volatility. *Journal of Econometrics* **83**, 325-348.
9. Campbell, J.Y. and Lo, A.W. (1999). *A Non-Random Walk Down Wall Street.* Princeton University Press, New Jersey.
10. Campbell, J.Y., Lo, A.W. and MacKinlay, A.C. (1997). *The Econometrics of Financial Markets.* Princeton University Press, New Jersey.
11. Chan, N.H. and Palma, W. (1998). State space modeling of long-memory processes. *Annals of Statistics* **26**, 719-740.
12. Chan, N.H. and Petris, G. (1999). Bayesian analysis of long-memory stochastic volatility models. *Technical report.* Department of Statistics, Carnegie Mellon University, Pittsburgh.
13. Ding, Z. and Granger, C. and Engle, R.F. (1993). A long-memory property of stock market returns and a new model. *Journal of Empirical Finance* **1**, 83-106.
14. Elerian, O., Chib, S. and Shephard, N. (1999). Likelihood inference for discretely observed non-linear diffusions. Technical Report, Nuffield College, Oxford University, Oxford, OX1 1NF, U.K.
15. Engle, R. (1982). Autoregressive conditional heteroskedasticity with estimates of the variance of UK inflation. *Econometrica* **50**, 987-1008.
16. Frühwirth-Schnatter, S. (1994). Data augmentation and dynamic linear models. *Journal of Time Series Analysis* **15**, 183-202.
17. Gallant, A.R., Hsieh, D. and Tauchen, G. (1997). Estimation of stochastic volatility models with diagnostics. *Journal of Econometrics* **81**, 159-192.
18. Gallant, A.R. and Tauchen, G. (1996). Which moments to match? *Econometric Theory* **12**, 657-681.
19. Gallant, A.R. and Tauchen, G. (1999). The relative efficiency of method of moments estimators. *Journal of Econometrics* **92**, 149-172.
20. Gilks, W. R. and Wild, P. (1992). Adaptive rejection sampling for Gibbs sampling. *Applied Statistics* **41**, 337-348.
21. Hall, P. (1997). Defining and measuring long-range dependence. In Cutler, C. and Kaplan, D.T. (Eds.) *Nonlinear Dynamics and Time Series: Building a Bridge between the Natural and Statistical Sciences.* American Mathematical Society, Rhode Island.
22. Hansen, L. (1982). Large sample properties of generalized method of moments estimators. *Econometrica* **50**, 1029-1054.
23. Harvey, A. and Ruiz, E. and Shephard, N. (1994). Multivariate stochastic

variance models. *Review of Economic Studies* **61**, 247-264.

24. Jacquier, E. and Polson, N. and Rossi, P. (1994). Bayesian analysis of stochastic volatility models. *Journal of Business and Economic Statistics* **12**, 371-389.
25. Kim, S. and Shephard, N. and Chib, S. (1998). Stochastic volatility: likelihood inference and comparison with ARCH models. *Review of Economic Studies* **65**, 361-393.
26. Melino, A. and Turnbull, S. (1990). Pricing foreign currency options with stochastic volatility. *Journal of Econometrics* **45**, 239-265.
27. Nelson, D. (1990). ARCH models as diffusion approximations. *Journal of Econometrics* **45**, 7-38.
28. Nelson, D. (1991). Conditional heteroskedasticity in asset return: a new approach. *Econometrica* **59**, 347-370.
29. Robinson, P. (1991). Testing for strong serial correlation and dynamics conditional heteroskedasticity in multiple regression. *Journal of Econometrics* **47**, 67-84.
30. Rossi, P. (1996). *Modeling Stock Market Volatility: Bridging the Gap to Continuous Time.* Academic Press, California.
31. Shephard, N. (1996). Statistical aspects of ARCH and stochastic volatility. In: Cox, D.R., Hinkley, D.V. and Barndorff-Nielsen, O.E. (Eds.) *Time Series Models: In econometrics, finance and other fields.* Chapman and Hall, New York.
32. Tierney, L. (1994). Markov chains for exploring posterior distributions. *The Annals of Statistics* **22**, 1701–1762.
33. West, M. and Harrison, J. (1997). *Bayesian Forecasting and Dynamic Models, 2nd Ed.* Springer-Verlag, New York.

BAYESIAN ESTIMATION OF STOCHASTIC VOLATILITY MODEL VIA SCALE MIXTURES DISTRIBUTIONS

S.T.B. CHOY and C.M. CHAN

Department of Statistics and Actuarial Science
The University of Hong Kong
Pokfulam Road, Hong Kong
E-mail: bchoy@hkustasc.hku.hk

This paper considers statistical inference of stochastic volatility (SV) models. The usual choice of normal and Student-t distributions for asset returns is replaced by the exponential-power (EP) distribution which can be light- and heavy-tailed than the normal distribution. This modification provides a wider choice of distributions for the SV models and simplifies the Markov chain Monte Carlo procedures for carrying out statistical analysis via uniform scale mixtures.

1 Introduction

Theoretically, stochastic volatility (SV) models are an alternative version of the autoregressive conditional heteroscedasticity (ARCH) models developed by Engle (1982), which are commonly used to model asset returns. For a review of ARCH models, see Bollerslev *et al.* (1992). The conditional variance of the ARCH models is assumed to be a function of the previous observations and past variances, since in real situations, the variance of the asset returns varies over time. Instead, the conditional variance is modelled with a stochastic process in the SV models and hence the estimation procedure of the SV models is noticeably harder than the ARCH family of models. Recently, a number of literatures attempt to produce efficient estimation procedures for the SV models. See, for example, Kim *et al.* (1998). In econometrics context, Jacquier *et al.* (1994) adopt the Bayesian approach to study the SV models while Harvey *et al.* (1994) extend the SV models to the multivariate case. In pricing options, Hull and White (1987) generalize the well-known Black-Scholes option pricing formula to allow for stochastic volatility.

Let r_t be the asset value of an equity or a portfolio of financial instruments at time $t = 0, 1, 2, \ldots, n$. The mean adjusted asset return y_t at time t is defined as

$$y_t = \ln\left(\frac{r_t}{r_{t-1}}\right) - \frac{1}{n}\sum_{i=1}^{n}\ln\left(\frac{r_i}{r_{i-1}}\right), \qquad t = 1, 2, \ldots, n$$

The simplest SV model for the returns y_t and log-volatilities h_t is specified by

$$y_t = \beta \exp(h_t/2)\,\varepsilon_t, \qquad t = 1, 2, \ldots, n$$

$$h_t = \begin{cases} \sigma\eta_1 / \sqrt{(1-\phi^2)} & t = 1 \\ \phi h_{t-1} + \sigma\,\eta_t & t > 1. \end{cases}$$

where ε_t and η_t are independent standard Gaussian processes. Here, β is a constant factor that represents the modal instantaneous volatility which is usually set to one in many literatures. σ is the variance of the log-volatility and ϕ is the persistence of the volatility which takes a value within the interval (-1,1) to satisfy the stationarity condition.

This SV model can be easily implemented using either likelihood or Bayesian approaches. However, in many situations, normality assumption for the distribution of asset returns may be inappropriate. Many financial practitioners and statisticians may use heavy-tailed distributions such as the Student-t and symmetric stable distributions for modeling asset returns. However, this extension increases the computational effort substantially. By representing the Student-t distribution as a scale mixtures of normals, (see Andrews and Mallows 1974), Jacquier *et al.* (1994) analyze the modified SV models using Markov chain Monte Carlo methods (see Gelfand and Smith, 1990, Smith and Roberts, 1993 and Tierney, 1994). In fact, the use of scale mixtures densities makes Bayesian computational easier to perform.

This paper aims to use the EP family of distributions which generalizes the normal distribution to a class of symmetric distributions of platykuric and leptokurtic shapes. The key of implementation of the EP distribution is to express the EP density into a scale mixtures of uniform form and we shall show that the required Bayesian computation can be simplified. In Section 2, we introduce the uniform scale mixtures form for the EP density and consider a Bayesian SV model with EP sampling distribution via this mixture representation. A full Bayesian analysis using Gibbs sampling approach is carried out in Section 3. In Section 4, an empirical application on daily closing prices of exchange rate is presented. We shall demonstrate the effects of parameter estimations on different choices of EP sampling distribution. In Section 5, we discuss the extension to allow the model volatility β and the kurtosis parameter α of the EP distribution to be random. Sampling techniques of random variates from these two extra full conditional densities in the Gibbs sampler is presented. In addition, for robustification purpose, we attempt to model the log-volatilities using the class of scale mixtures of normal distributions and take Student-t as a special case to obtain the system of full conditional densities. Then, we consider a EP-EP SV model, allowing both sampling distribution and distribution of log-volatilities h_t come from the EP family with different kurtosis parameters perhaps. We shall show that all full conditional distributions are of standard forms under this case. Finally, a concluding remark is presented in Section 6.

2 The Exponential-power SV Models

2.1 The EP distribution

The EP family of distributions provide both heavy- and light-tailed than the normal shape. Let θ be the mean, σ be the scale parameter and $\alpha \in (0, 2]$ be the kurtosis parameter that controls the thickness of the tails. The EP distribution is denoted by $EP(\theta, \sigma, \alpha)$ with a density function given by

$$f(x|\theta, \sigma, \alpha) \quad \propto \quad \exp\left(-\frac{1}{2}\left|\frac{x-\theta}{\sigma}\right|^{2/\alpha}\right). \tag{2.1}$$

The mean and variance are

$$E[X] = \theta \qquad \text{and} \qquad V[X] = \frac{2^{\alpha}\Gamma(3\alpha/2)}{\Gamma(\alpha/2)},$$

respectively. The EP distribution has been studied thoroughly by Box and Tiao (1973) and Choy and Walker (1998) for statistical modelling and Bayesian robustness. Choy and Smith (1997) adopt the normal scale mixtures property of the EP density for Bayesian inference using Markov chain Monte Carlo methods with α restricted to the range $1 < \alpha \leq 2$. Recently, Walker and Gutierrez-Pena (1999) discover the following uniform scale mixtures representation for the EP density:-

$$f(x|\theta, \sigma, \alpha) = \int_0^{\infty} U(x|\theta - \sigma u^{\alpha/2}, \theta + \sigma u^{\alpha/2}) Ga(u|1 + \alpha/2, 1/2) du$$

where $U(x|a, b)$ is the uniform density function defined on the interval (a, b) and $Ga(x|c, d)$ is the gamma density function with mean c/d. This representation is valid for the entire range of α and also allows us to rewrite the EP distribution into the following hierarchical form

$$X|U = u \sim U\left(\theta - \sigma u^{\alpha/2}, \theta + \sigma u^{\alpha/2}\right) \qquad \text{and} \qquad U \sim Ga\left(1 + \frac{\alpha}{2}, \frac{1}{2}\right)$$

where U is always referred to as the mixing parameter of the scale mixture representation. Note that the normal and Laplace (or double exponential) distributions are special cases of the EP family with $\alpha = 1$ and $\alpha = 2$, respectively.

2.2 Bayesian EP-N SV models

Although the assumption of normality for the ε_t of time series data has been widely used, many financial data exhibit fat-tailed behavior. The Student-t and symmetric stable distributions are commonly chosen alternatives to the normal family for modelling these data. In addition, they are used for robustness purpose. In this paper, we consider the family of EP distributions as a generalization of the normal family to model financial data. This family provides both leoptkutic and platykurtic shapes of distributions that the normal, Student-t and stable families do not offer. From a practical point of veiw, we believe that the EP distribution may be appropriate to model certain types of data and it is worthwhile to develop efficient methods for statistical analysis. A Gibbs sampling approach using the uniform scale mixtures is discussed in Section 3.

Without loss of generality, we assume that β is fixed. The usual choice of the normal distribution for the white noise ε_t of the SV model is replaced by the EP distribution with known kurtosis parameter α, i.e.

$$y_t|h_t \sim EP\left(0, \beta e^{h_t/2}, \alpha\right) \qquad\qquad t = 1, 2, \ldots, n$$

which is expressed into the following hierarchical form:-

$$y_t|h_t, u_t \sim U\left(-\beta e^{h_t/2} u_t^{\alpha/2}, \beta e^{h_t/2} u_t^{\alpha/2}\right)$$

$$u_t \sim Ga\left(1 + \frac{\alpha}{2}, \frac{1}{2}\right).$$

Normality assumption is still valid for the conditional and marginal distributions of the log-volatility h_t in this section, i.e.

$$h_t|h_{t-1}, \phi, \sigma^2 \sim N\left(\phi h_{t-1}, \sigma^2\right)$$

and

$$h_t|\phi, \sigma^2 \sim N\left(0, \frac{\sigma^2}{1 - \phi^2}\right).$$

Here we shall refer this SV model with EP white noise and normal log-volatility to as the EP-N SV model. In order to complete a full Bayesian framework for this SV model, we assign the following priors to other model parameters :-

$$\sigma^2 \sim IG\left(a_\sigma, b_\sigma\right) \qquad \text{and} \qquad \frac{\phi+1}{2} \sim Be\left(a_\phi, b_\phi\right)$$

where $Be(a,b)$ is the beta distribution with mean $a/(a+b)$ and $a_\sigma, b_\sigma, a_\phi$ and b_ϕ are pre-specified. Here the prior distribution for ϕ is a shifted beta distribution with density

$$p(\phi) \propto (1+\phi)^{a_\phi - 1}\,(1-\phi)^{b_\phi - 1} \qquad |\phi| < 1.$$

3 Gibbs Sampler for the EP-N SV Models

To carry out statistical analysis for complicated Bayesian models, the simulation-based Gibbs sampling approach has become one of the standard methods to be used. The Gibbs sampler allows us to study posterior characteristics via a sequence of iteratively simulated values drawn from a system of full distributional distributions. The efficiency of the Gibbs sampler can be substantially increased if the required samples are drawn from distributions of some standard forms. Now the joint distribution of $\boldsymbol{y} = (y_1, y_2, \ldots, y_n)$, $\boldsymbol{h} = (h_1, h_2, \ldots, h_n)$, $\boldsymbol{u} = (u_1, u_2, \ldots, u_n)$, ϕ and σ^2 is given by

$$p(\boldsymbol{y}, \boldsymbol{h}, \boldsymbol{u}, \phi, \sigma^2) = \prod_{t=1}^{n} p(y_t|h_t, u_t)\, p(h_1|\phi, \sigma^2) \prod_{t=2}^{n} p(h_t|h_{t-1}, \phi, \sigma^2)\, p(\boldsymbol{u})\, p(\phi)\, p(\sigma^2).$$

Write $\boldsymbol{h}_{-t} = (h_1, \ldots, h_{t-1}, h_{t+1}, \ldots, h_n)$ and $\boldsymbol{u}_{-t} = (u_1, \ldots, u_{t-1}, u_{t+1}, \ldots, u_n)$. Then the Gibbs sampling scheme performs successive random variate generation from the following full conditional distributions.

1. Full conditional densities of h_t:-

The full conditional density of h_t is given by

$$p(h_t|\boldsymbol{y}, \boldsymbol{h}_{-t}, \boldsymbol{u}, \phi, \sigma^2) \propto p(y_t|h_t, u_t) p(h_t|h_{t-1}, u_t, \phi, \sigma^2) p(h_{t+1}|h_t, u_{t+1}, \phi, \sigma^2)$$

for $t = 1, 2, ..., n$. We can then show that these full conditional distributions are truncated normal of the form

$$h_t|\boldsymbol{y}, \boldsymbol{h}_{-t}, \boldsymbol{u}, \phi, \sigma^2 \sim \begin{cases} N\left(\phi h_{t+1} - \sigma^2/2,\, \sigma^2\right) & t = 1 \\ N\left(\frac{\phi(h_{t-1}+h_{t+1})-\sigma^2/2}{1+\phi^2},\, \frac{\sigma^2}{1+\phi^2}\right) & 2 \le t \le n-1 \\ N\left(\phi h_{t-1} - \sigma^2/2,\, \sigma^2\right) & t = n \end{cases}$$

subject to

$$h_t > \ln y_t^2 - \ln \beta^2 - \alpha \ln u_t \qquad t = 1, 2, \ldots, n.$$

The algorithm proposed by Robert (1995) is an efficient method for generating random variates from the truncated normal distribution.

2. Full conditional densities of u_t and σ^2:-
Representing the EP density into a uniform scale mixtures form, we can show that the full conditional distribution of the mixing parameter u_t is a truncated exponential distribution of the form

$$u_t | \boldsymbol{y}, \boldsymbol{h}, \boldsymbol{u}_{-t}, \phi, \sigma^2 \sim Exp(0.5) \qquad t = 1, 2, \ldots, n$$

subject to

$$u_t^\alpha > \frac{y_t^2}{\beta^2} e^{-h_t}.$$

Inversion method can be used to sample random variates from the truncated exponential distribution. For σ^2, using conjugate prior leads to an inverse gamma full conditional distribution and σ^2 is then straightforwardly sampled from

$$\sigma^2 | \boldsymbol{y}, \boldsymbol{h}, \boldsymbol{u}, \phi \sim IG\left(a_\sigma + \frac{n}{2}, b_\sigma + \frac{1}{2}\left((1 - \phi^2) h_1^2 + \sum_{t=2}^{n} (h_t - \phi h_{t-1})^2 \right)\right).$$

3. Full conditional density of ϕ:-
Obviously, the full conditional density of ϕ is given by

$$p(\phi | \boldsymbol{y}, \boldsymbol{h}, \boldsymbol{u}, \sigma^2) \propto p(h_1 | \phi, \sigma^2) \prod_{t=2}^{n} p(h_t | h_{t-1}, \phi, \sigma^2) p(\phi).$$

It can be easily verified that $\prod_{t=2}^{n} p(h_t | h_{t-1}, \phi, \sigma^2)$ is proportional to a normal density of ϕ with mean μ_ϕ and variance σ_ϕ^2 where

$$\mu_\phi = \frac{\sum_{t=1}^{n-1} h_t h_{t+1}}{\sum_{t=1}^{n-1} h_t^2} \qquad \text{and} \qquad \sigma_\phi^2 = \frac{\sigma^2}{\sum_{t=1}^{n-1} h_t^2}.$$

Kim *et al.* (1998) suggest using the Metropolis-Hastings algorithm to draw proposed samples of ϕ from the normal $N\left(\mu_\phi, \sigma_\phi^2\right)$ distribution. Alternatively, we can express the full conditional density by a product of a truncated normal and a shifted beta density functions, i.e.

$$p(\phi | \boldsymbol{y}, \boldsymbol{h}, \boldsymbol{u}, \sigma^2) \propto N\left(\phi \left| \frac{\sum_{t=1}^{n-1} h_t h_{t+1}}{\sum_{t=2}^{n-1} h_t^2}, \frac{\sigma^2}{\sum_{t=2}^{n-1} h_t^2} \right. \right) (1 + \phi)^{a_\phi - 1/2} (1 - \phi)^{b_\phi - 1/2}$$

for $|\phi| < 1$. Sampling random variates from this full conditional density can be easily done using the rejection sampling method with proposed samples drawn from the truncated normal distribution. Of course, the Metropolis-Hastings method can also be used as an alternative.

4 Example

For illustration purpose, we analyze the daily closing prices of US dollars to Sterling pounds exchange rates. The data set contains 1000 mean adjusted daily exchange rate returns collected from January 2, 1981 and the plotted are given in Fig.6. Without loss of generality, we set $\beta = 1$. Inverse gamma $IG(a_\sigma, b_\sigma)$ distribution with $a_\sigma = b_\sigma = 0.001$ is assigned to σ^2 to reflect non-informative prior knowledge about σ^2. For ϕ, we choose $a_\phi = 20$ and $b_\phi = 1$. Since the conditional densities of $\boldsymbol{h}$, $\boldsymbol{u}$ and σ^2 are all of standard forms, sampling ϕ values is the most computationally expensive part of the Gibbs sampling method.

In practice, it is well-known that simulated values from the Gibbs sampler are highly correlated for time series data and hence the Markov chain must be run sufficiently long. In this example, the Gibbs sampler cycled through the full conditional distributions of $\boldsymbol{h}$, $\boldsymbol{u}$, ϕ and σ^2 for 35000 iterations with the first 5000 iterations as the "burn-in" period. Ergodic average plots of h_1, u_1, (and similarly for other h_t's and u_t's), σ and ϕ are displayed in Fig.1 and we notice that there has no convergence problem. Simulated values after the "burn-in" period are picked at every 30 iterations to mimic a random sample of size 1000 from the target joint posterior distribution to provide posterior summaries. Autocorrelation functions of h_1, u_1, σ and ϕ are plotted in Fig.2 using these 1000 sampled values. The plots are quite satisfactory except for σ whose autocorrelation function dies out relatively slow and we therefore believe that the selected 1000 values from the Markov chain are quite independent.

For different choices of kurtosis α for the EP-N SV model, Bayes estimates of σ and ϕ are summarized in Table 1. For small values of α, the middle part of the EP sampling distribution are fatter than the normal distribution. The observations are therefore more volatile than under a normal sampling distribution. The estimate of σ is therefore larger but drops very quickly as α increases. Similarly, for small α, the persistence ϕ is small. See Fig.3 for a clear pictures. The estimates of the log-volatility h_t, $t = 1, \ldots, 1000$, are given in Fig.4 for $\alpha = 0.5$, 1.0, 1.5, 2.0. We note that smaller log-volatility associates with larger value of α. This is partially due to the fact that the variance of the EP component increases with α. In fact, after adjusting for

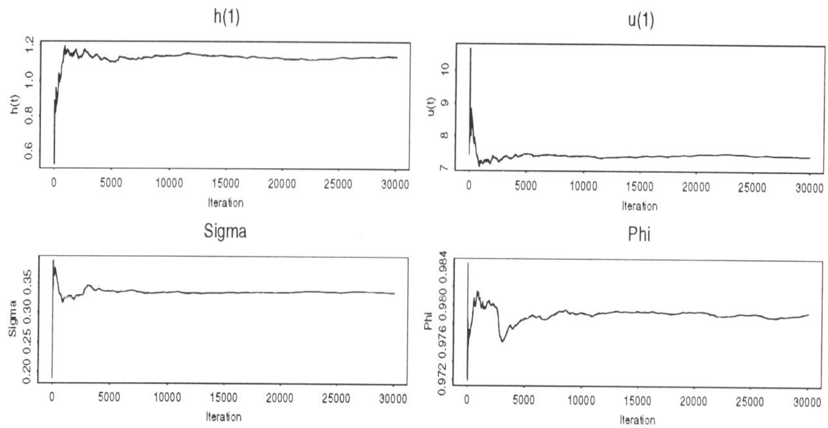

Figure 1: Ergodic averages plots of h_1, u_1, σ and ϕ from an EP-N SV model with $\alpha = 1.5$.

this effect by redefining the EP density to have a variance equal to σ^2, the pattern in Fig.4 is preserved. In addition, further simulation study shows that ϕ is quite insensitive to the choice of hyperparameters a_ϕ and b_ϕ. Fig.5 gives the histograms of h_t for N-N ($\alpha = 1$) and Laplace-N ($\alpha = 2.0$) SV models. The means of h_t is roughly equal to -1.0 for normal case and -3.0 for Laplace case. Therefore, if a Laplace distribution is assumed to model the exchange rate returns, the log-volatilities will be substantially reduced. The use of uniform scale mixtures for the EP distribution allows us to perform a global diagnosis of possible outliers as the scale mixtures of normal distributions do. Extreme values of y_t will associate with large u_t value. For $\alpha = 1.5$, for example, Fig.6 exhibits the posterior means of the u_t's and it verifies that the most volatile trading day corresponds to $y_t = 3.84, E[h_t|\boldsymbol{y}] = -0.5408$ and $E[u_t|\boldsymbol{y}] = 10.7349$. The histograms of u_t, for α =1.0 and $\alpha = 2.0$ are given in Fig.5.

On the other hand, whether the EP distribution is a better choice than the normal distribution in the SV model is worth considering. A model selection criterion based on the posterior predictive distribution is suggested by San Martini and Spezzaferri (1984). Let M_α be the EP-N SV model with kurtosis $\alpha = 0.25, 0.5, \ldots, 2.0$. Assuming that these models are equally likely and one of them is the true model, the model selection criterion is to choose the model

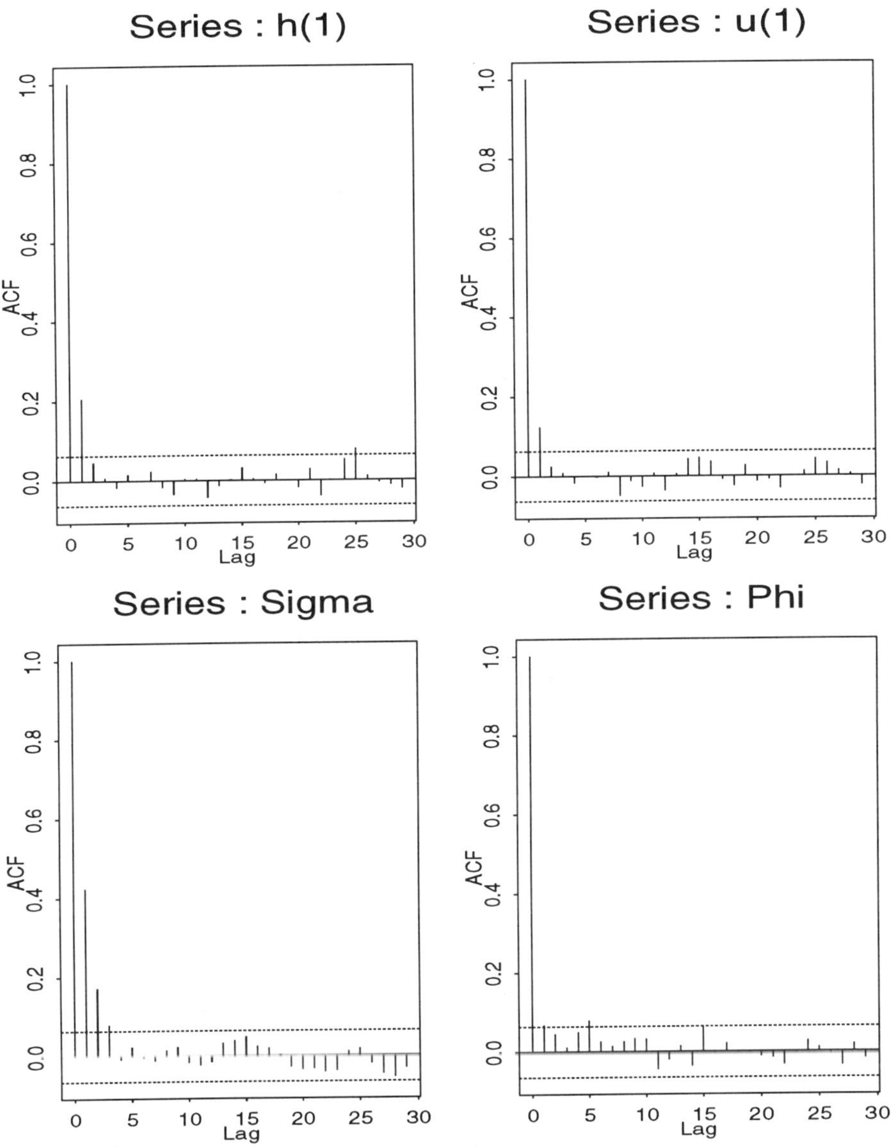

Figure 2: Autocorrelation functions for h_1, u_1, σ and ϕ series from a EP-N SV model with $\alpha = 1.5$.

α	σ	ϕ
0.25	0.7348 (0.0698)	0.6082 (0.0784)
0.50	0.5371 (0.0635)	0.8132 (0.0447)
0.75	0.4218 (0.0536)	0.9249 (0.0231)
1.00	0.3748 (0.0492)	0.9664 (0.0132)
1.25	0.3511 (0.0477)	0.9819 (0.0082)
1.50	0.3463 (0.0452)	0.9884 (0.0057)
1.75	0.3564 (0.0436)	0.9913 (0.0044)
2.00	0.3609 (0.0441)	0.9933 (0.0035)

Table 1: Bayes estimates (with standard error in parentheses) of σ and ϕ for various values of kurtosis parameter α.

with the largest posterior expected utility $U(\alpha)$, defined by

$$U(\alpha) = \frac{1}{n}\sum_{t=1}^{n} \ln p(y_t|M_\alpha)$$

which can be computed using Gibbs sampling outputs. In other words, $nU(\alpha)$ is the predictive log-likelihood function of model M_α. The expected utilities are given in Table 2 and the results are in favour of the Lapace-N SV model. For further comparisons, we consider the Student-N SV model with different degrees of freedom ν and the expected utilities can be found in Table 3. The best model corresponds to the Cauchy case which is less competitive than the Lapace-N model in this simulation study.

α	0.25	0.50	0.75	1.00	1.25	1.50	1.75	2.00
$U(\alpha)$	-0.664	-0.663	-0.662	-0.574	-0.537	-0.506	-0.487	-0.472

Table 2: Expected utilities for the EP-N SV models with different kurtosis parameter α.

ν	1	3	5	10	15	20
$U(\nu)$	-0.488	-0.516	-0.529	-0.550	-0.558	-0.562

Table 3: Expected utilities for the Student-N SV models with different degrees of freedom ν.

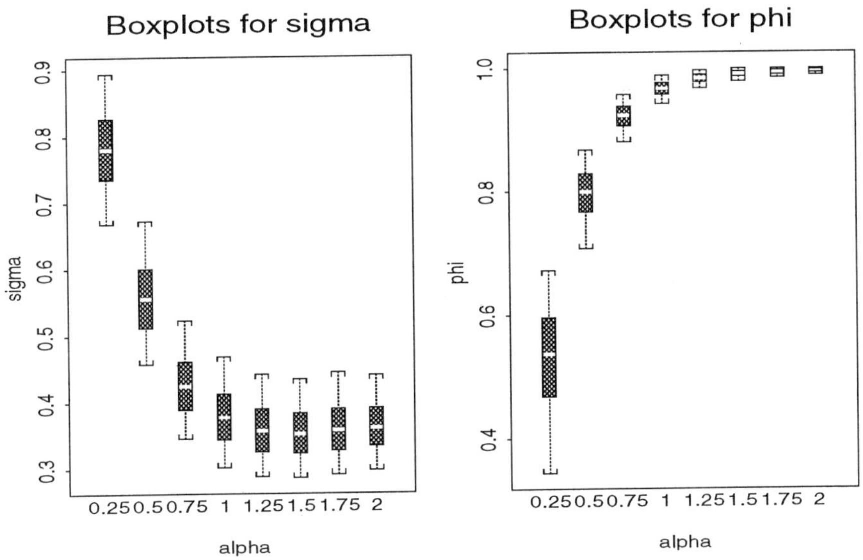

Figure 3: Boxplots of σ and ϕ for various values of α of the EP-N SV model.

5 Extension

5.1 β and α are random

The SV models can be made more realistic by assuming that the modal volatility β is a random quantity. In addition, the asset returns can be modeled by a general EP shape with unknown kurtosis α. For conjugacy, an inverse gamma $IG(a_\beta, b_\beta)$ prior distribution can be assigned to β and a suitable choice of prior distribution for α is a shifted beta distribution with parameters a_α and b_α since $\alpha \in (0, 2]$. Assuming randomness for β and α, the Gibbs sampler will cycle through two extra full conditional distributions - the full conditionals of β and α. By conjugacy, it can be easily shown that the full conditional distribution of β is a right-truncated inverse gamma distribution of the form

$$\beta | \boldsymbol{h}, \boldsymbol{u}, \phi, \sigma^2, \alpha \sim IG(a_\beta + n, b_\beta)$$

subject to

$$\beta \; > \; \sup\, |y_t| e^{-h_t/2}\, u_t^{-\alpha/2} \qquad t = 1, 2, \ldots, n.$$

Simulation from the truncated inverse gamma distribution can be done by modifying the algorithm proposed by Philippe (1997). For α, the full condi-

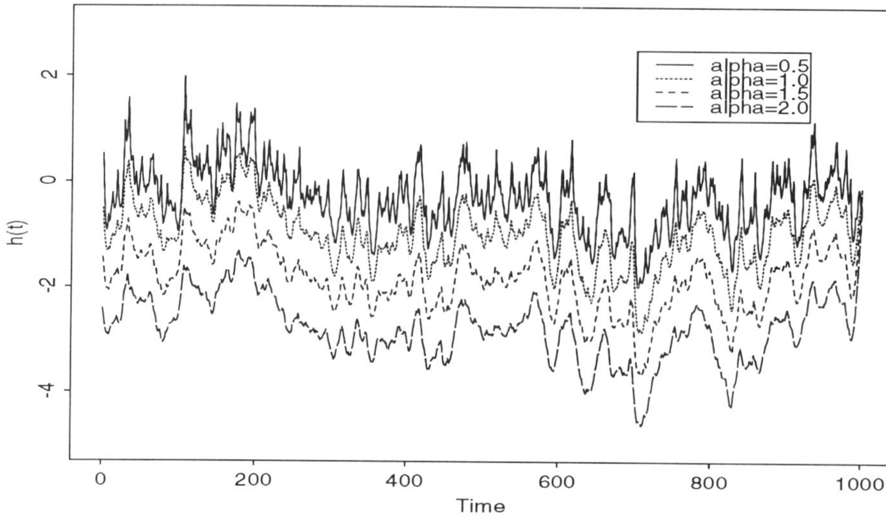

Figure 4: Posterior means of h_t, $t = 1, \ldots, 1000$ for $\alpha = 0.5, 1.0, 1.5$ and 2.0, respectively.

tional is of the form

$$p(\alpha|\boldsymbol{h}, \boldsymbol{u}, \phi, \sigma^2, \beta) \propto p(\boldsymbol{y}|\boldsymbol{h}, \boldsymbol{u}, \alpha, \beta)p(\boldsymbol{u}|\alpha)p(\alpha).$$

After some algebra, we get

$$p(\alpha|\boldsymbol{h}, \boldsymbol{u}, \phi, \sigma^2, \beta) \propto \left(2^{\alpha/2}\,\Gamma(1 + \frac{\alpha}{2})\right)^{-n} \alpha^{a_\alpha - 1}(2 - \alpha)^{b_\alpha - 1} I_\alpha(s, 2)$$

where

$$s = \sup\left(0, \left\{\frac{1}{\ln u_t}(\ln y_t^2 - \ln \beta^2 - h_t), 1 \le t \le n\right\}\right)$$

and

$$I_\alpha(s_1, s_2) = \begin{cases} 1 & s_1 \le \alpha \le s_2 \\ 0 & \text{otherwise.} \end{cases}$$

To simulate random variates from this conditional density, we adopt the Metropolis Hastings algorithm to draw proposed samples from either the uniform $U(s, 2)$ or the shifted beta $Be(a_\alpha, b_\alpha)$ distribution. Other methods including the ratio-of-uniforms (see Wakefield *et al.*, 1991) can also be used. In particular, if a

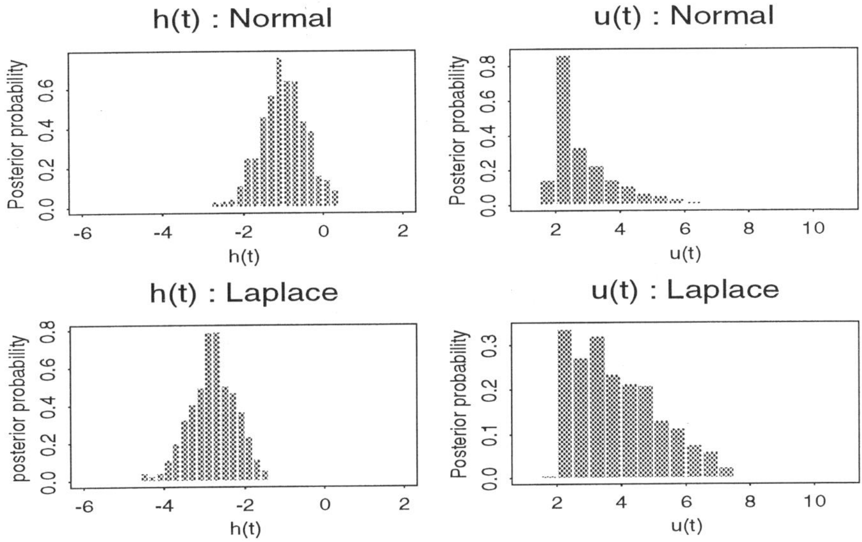

Figure 5: Histrograms of h_t and u_t for N-N SV model and Laplace-N SV model.

uniform prior is assumed for α, i.e. $a_\alpha = b_\alpha = 1$, the full conditional density becomes

$$p(\alpha|\boldsymbol{h}, \boldsymbol{u}, \phi, \sigma^2, \beta) \propto \left(2^{\alpha/2}\,\Gamma(1 + \frac{\alpha}{2})\right)^{-n} I_\alpha(s, 2)$$

from which random variates can be easily obtained using rejection sampling method with proposed samples drawn from the uniform $U(s, 2)$ distribution.

5.2 Fat-tailed Scale Mixtures of Normal Distribution for Log-volatility

For robustification purpose, fat-tailed distribution can be introduced to model the log-volatility h_t. This class of distributions includes the Student-t, symmetric stable, exponential-power and logistic distributions which are members of the class of scale mixtures of normal family. Taking the Student-t distribution with known degrees of freedom ν as an example to model the log-volatility h_t, the marginal distribution of h_t can be replaced by

$$h_t|\phi, \sigma^2, \lambda_t \sim N\left(0, \frac{\sigma^2}{\lambda_t(1-\phi^2)}\right) \quad \text{and} \quad \lambda_t \sim Ga\left(\frac{\nu}{2}, \frac{\nu}{2}\right)$$

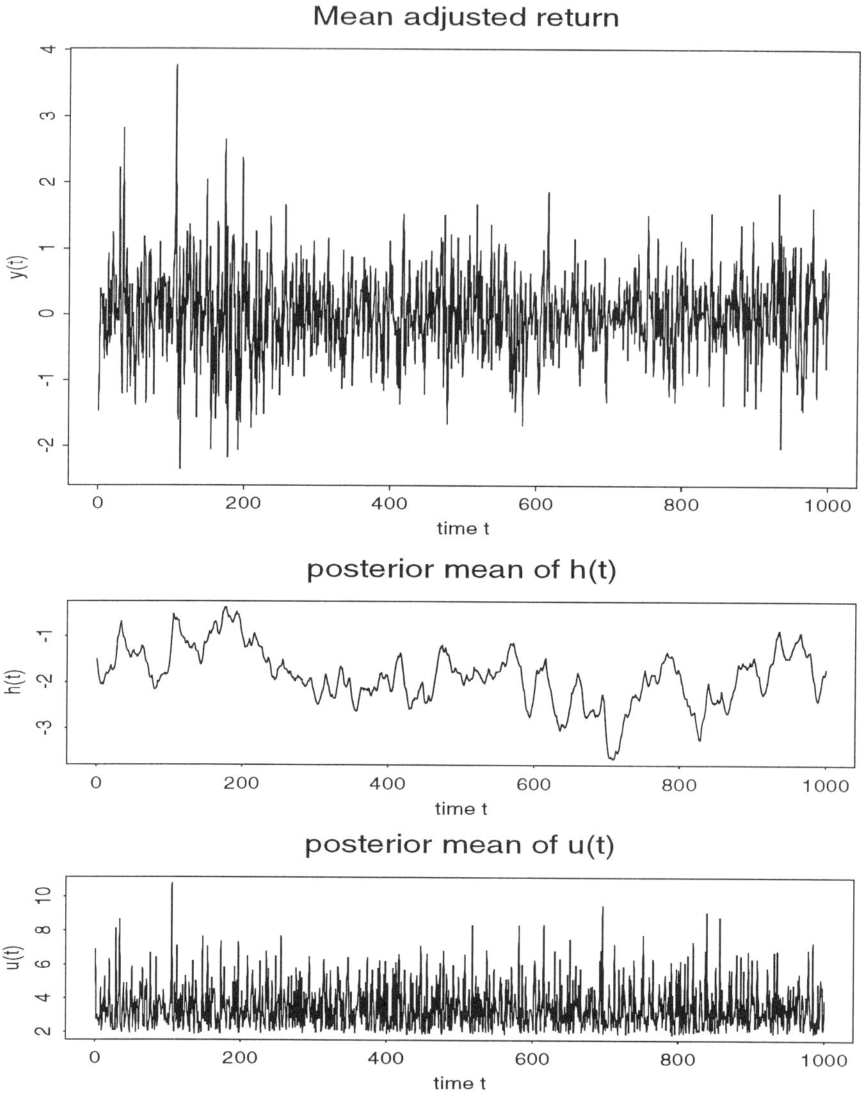

Figure 6: Time series plots of the mean adjusted exchange rate returns, posterior means of h_t and posterior means of u_t of the EP-N model with $\alpha = 1.5$.

where λ_t is the second stage mixing parameter and the conditional distribution of h_t becomes

$$h_t|h_{t-1},\phi,\sigma^2,\lambda_t \sim N\left(\phi h_{t-1},\frac{\sigma^2}{\lambda_t}\right) \quad \text{and} \quad \lambda_t \sim Ga\left(\frac{\nu}{2},\frac{\nu}{2}\right).$$

The use of normal scale mixtures form for some well-known distributions can facilitate a more efficient Gibbs samplers for Bayesian analysis. See Pitt and Walker (1998) for applications in SV models, Choy and Smith (1997) and Fernandez and Steel (1998) for general applications. Let $\boldsymbol{\lambda} = (\lambda_1, \lambda_2, \ldots, \lambda_n)$, the full conditional distribution for h_t becomes

$$h_t|\boldsymbol{y},\boldsymbol{h}_{-t},\boldsymbol{u},\boldsymbol{\lambda},\phi,\sigma^2 \sim \begin{cases} N\left(\frac{\lambda_t\phi h_{t+1}-\sigma^2/2}{1-(1-\lambda_t)\phi^2},\frac{\sigma^2}{1-(1-\lambda_t)\phi^2}\right) & t=1 \\ N\left(\frac{\phi(\lambda_t h_{t-1}+\lambda_{t+1}h_{t+1})-\sigma^2/2}{\lambda_t+\lambda_{t+1}\phi^2},\frac{\sigma^2}{\lambda_t+\lambda_{t+1}\phi^2}\right) & 2\le t\le n-1 \\ N\left(\phi h_{n-1}-\frac{\sigma^2}{2\lambda_t},\frac{\sigma^2}{\lambda_t}\right) & t=n \end{cases}$$

subject to

$$h_t > \ln y_t^2 - \ln\beta^2 - \alpha\ln u_t \qquad t = 1,2,\ldots,n.$$

Denote $\boldsymbol{\lambda}_{-i} = (\lambda_1,\ldots,\lambda_{i-1},\lambda_{i+1},\ldots,\lambda_n)$. We have

$$\lambda_t|\boldsymbol{y},\boldsymbol{h},\boldsymbol{u},\boldsymbol{\lambda}_{-1},\phi,\sigma^2 \sim \begin{cases} G\left(\frac{\nu+1}{2},\frac{\nu}{2}+\frac{(1-\phi^2)h_t^2}{2\sigma^2}\right) & t=1 \\ G\left(\frac{\nu+1}{2},\frac{\nu}{2}+\frac{(h_t-\phi h_{t-1})^2}{2\sigma^2}\right) & 2\le t\le n \end{cases}$$

By inspecting the posterior means or medians of the λ_t's, trading days with excessive volatilities which associate with small values of these statistics can be identified. More importantly, the use of fat-tailed distribution allows the SV models to accommodate these extremes and provides an automatic mechanism to downweight the effects of the extremes in statistical inference.

5.3 EP Distribution for Log-volatility

We have seen that the use of uniform scale mixtures for the EP density function can simplify the computational effort for the Gibbs sampler in Bayesian inference. Here we further extend to use the EP distribution with known kurtosis parameter γ for the log-volatility. That is, we are considering an EP-EP SV model specified by

$$y_t|h_t,u_t \sim U\left(-\beta e^{h_t/2}u_t^{\alpha/2},\ \beta e^{h_t/2}u_t^{\alpha/2}\right) \qquad 1\le t\le n$$

$$h_1|\lambda_1, \phi, \sigma^2 \sim U\left(-\frac{\sigma\lambda_1^{\gamma/2}}{\sqrt{1-\phi^2}}, \frac{\sigma\lambda_1^{\gamma/2}}{\sqrt{1-\phi^2}}\right)$$

$$h_t|h_{t-1}, \lambda_t, \phi, \sigma^2 \sim U\left(\phi h_{t-1} - \sigma\lambda_t^{\gamma/2}, \phi h_{t-1} + \sigma\lambda_t^{\gamma/2}\right) \qquad 2 \le t \le n$$

$$u_t \sim G\left(1+\frac{\alpha}{2}, \frac{1}{2}\right)$$

$$\lambda_t \sim G\left(1+\frac{\gamma}{2}, \frac{1}{2}\right)$$

$$\sigma^2 \sim IG(a_\sigma, b_\sigma)$$

$$\frac{\phi+1}{2} \sim Be(a_\phi, b_\phi)$$

where λ_t's are the second stage mixing parameters. Write $\boldsymbol{\lambda} = (\lambda_1, \lambda_2, \ldots, \lambda_n)$. The full conditional density of h_t is given by

$$p(h_t|\boldsymbol{y}, \boldsymbol{h}_{-t}, \boldsymbol{u}, \boldsymbol{\lambda}, \phi, \sigma^2) \propto \exp\left(-\frac{h_t}{2}\right) \qquad a_t \le h_t \le b_t$$

where

$$a_1 = \sup\left(\ln y_1^2 - \ln\beta^2 - \alpha\ln u_1, \frac{\sigma}{\sqrt{1-\phi^2}}\lambda_1^{\gamma/2}, \frac{1}{\phi}\left(h_2 - \sigma\lambda_2^{\gamma/2}\right)\right)$$

$$b_1 = \inf\left(\frac{\sigma}{\sqrt{1-\phi^2}}\lambda_1^{\gamma/2}, \frac{1}{\phi}\left(h_2 + \sigma\lambda_2^{\gamma/2}\right)\right)$$

$$a_n = \sup\left(\ln y_n^2 - \ln\beta^2 - \alpha\ln u_n, \phi h_{n-1} - \sigma\lambda_n^{\gamma/2}\right)$$

$$b_n = \phi h_{n-1} + \sigma\lambda_n^{\gamma/2}$$

and for $t = 2, 3, \ldots, n-1$,

$$a_t = \sup\left(\ln y_t^2 - \ln\beta^2 - \alpha\ln u_t, \phi h_{t-1} - \sigma\lambda_t^{\gamma/2}, \frac{1}{\phi}\left(h_{t+1} - \sigma\lambda_{t+1}^{\gamma/2}\right)\right)$$

$$b_t = \inf\left(\phi h_{t-1} + \sigma\lambda_t^{\gamma/2}, \frac{1}{\phi}\left(h_{t+1} + \sigma\lambda_{t+1}^{\gamma/2}\right)\right).$$

Although the full conditional density of h_t is proportional to $\exp(-h_t/2)$, there is no guarantee that all a_t's are positive numbers and therefore we cannot regard the full conditional distribution of h_t as a truncated exponential Exp(0.5) distribution. But if we define $h_t' = h_t - a_t$, then the full conditional distribution of h_t' is a left-truncated exponential Exp(0.5) distribution, i.e.

$$h_t'|\boldsymbol{h}_{-t}, \boldsymbol{u}, \boldsymbol{\lambda}, \phi, \sigma^2 \sim Exp(0.5) \qquad h_t' < b_t - a_t.$$

We can sample h_t' from the truncated exponential Exp(0.5) distribution using inversion method and hence return a sampled value of h_t. For the mixing parameters u_t's and λ_t's, the full conditionals are truncated exponential distributions of the form

$$u_t|\boldsymbol{y}, \boldsymbol{h}, \boldsymbol{u}_{-t}, \boldsymbol{\lambda}, \phi, \sigma^2 \sim Exp(0.5) \qquad u_t^{\alpha} > \frac{y_t^2}{\beta^2} e^{-h_t}$$

and

$$\lambda_t|\boldsymbol{y}, \boldsymbol{h}, \boldsymbol{u}, \boldsymbol{\lambda}_{-t}, \phi, \sigma^2 \sim Exp(0.5) \qquad \lambda_t > \lambda_{ot}$$

where

$$\lambda_{ot}^{\gamma} = \begin{cases} \frac{(1-\phi^2)h_1^2}{\sigma^2} & t = 1 \\ \frac{(h_t - \phi h_{t-1})^2}{\sigma^2} & 2 \le t \le n. \end{cases}$$

For σ^2, the full conditional distribution is a truncated inverse gamma distribution

$$\sigma^2|\boldsymbol{y}, \boldsymbol{h}, \boldsymbol{u}, \boldsymbol{\lambda}, \phi \sim IG\left(a_\sigma + n/2, b_\sigma\right)$$

subject to

$$\sigma^2 > \sup\left(\frac{(1-\phi^2)h_1^2}{\lambda_1^{\gamma}}, \left\{\frac{(h_t - \phi h_{t-1})^2}{\lambda_t^{\gamma}}, 2 \le t \le n\right\}\right)$$

Finally, for the persistence parameter ϕ, it can be shown that

$$\frac{1+\phi}{2}|\boldsymbol{y}, \boldsymbol{h}, \boldsymbol{u}, \boldsymbol{\lambda}, \sigma^2 \sim Be\left(a_\phi + 1/2, b_\phi + 1/2\right) \qquad \phi_l \le \phi \le \phi_u$$

where

$$\phi_l = \sup\left(-\left|1 - \frac{h_1^2}{\lambda_1^{\gamma}\sigma^2}\right|, \left\{\frac{h_t - \sigma\lambda^{\gamma/2}}{h_{t-1}}, 2 \le t \le n\right\}\right)$$

$$\phi_u = \inf\left(\left|1 - \frac{h_1^2}{\lambda_1^{\gamma}\sigma^2}\right|, \left\{\frac{h_t - \sigma\lambda^{\gamma/2}}{h_{t-1}}, 2 \le t \le n\right\}\right).$$

Since all conditional distributions are all of standard forms, there has no difficulty in performing random variate generation. The extension from normal family to EP family for the SV models will not substantially increase the computation burden of the Gibbs sampling approach and it encourages the use of the EP distribution for statistical inference via uniform scale mixtures.

6 Concluding Remarks

This paper aims to adopt the class of EP distributions for SV models. The EP family provides both heavier-than and lighter-than normal tails. Moreover, we can also consider the EP-EP SV model, using two different kurtosis parameters for the EP distributions of y_t and h_t. In this case, the full conditional distributions are also of standard forms. The normal distribution can be recovered by setting the kurtosis parameter equal to 1. Furthermore, we can assume the kurtosis parameters to be unknown and suitable prior distributions can be assigned. If vague priors are chosen, then we let the data to determine the two kurtosis parameters.

Regarding to the Gibbs sampling algorithm, we adopt a single-move sampler in drawing h_i's and u_i's although some researchers, for example, Shephard (1994) and Kim *et al.* (1998), suggest using multi-move to speed up the rate of convergence. The reason is that the full conditional distributions of h_i and u_i are truncated normal and truncated exponential, respectively which may make the multi-move sampler difficult to run. However, we believe that it is worthy to consider the multi-move sampler.

In this paper, it is novel to use the EP distribution via uniform scale mixtures in SV models. Whether we should use the EP distribution instead of the normal and Student-t distributions is a model selection problem and in Section 4, we have demonstrated the possible advantage of using the EP family in SV models.

Acknowledgment

This work was partially supported by a grant for the Research Grants Council of HKSAR, China (Project No. HKU 133/98H). The authors would like to thank Luk Chi Ho for carrying out some simulation works.

References

1. Andrews, D.F. and Mallows, C.L. (1974), "Scale mixtures of normal distribution", *Journal of the Royal Statistics Society, Series B*, **36**, 99-102.

2. Bollerslev, T., Chou, R.Y. and Kroner, K.F. (1992), "ARCH Modeling in Finance: A Selective Review of the Theory and Empirical Evidence", *Journal of Econometrics*, **52**, 5-59.

3. Box, G.E.P. and Tiao, G.C. (1973), "*Bayesian Inference in Statistical Analysis*". Massachusettes: Addison Wesley.

4. Choy, S.T.B. and Smith, A.F.M. (1997), "Hierarchical models with scale mixtures of normal distributions", *TEST*, **6**, 205-211.

5. Choy, S.T.B. and Walker, S.G. (1998), "The extended exponential power distribution and Bayesian Robustness", Submitted for publication.

6. Engle, R.F. (1982), "Autoregressive conditional heteroskedasticity with estimates of the variance of the United Kingdom inflation", *Econometrica*, **50**, 987-1007.

7. Fernandez, C. and Steel, M.F.J. (1998), "Bayesian Regression Analysis with Scale mixtures of normals", Technical Report. University of Bristol.

8. Gelfand, A.E. and Smith, A.F.M. (1990), "Sampling-based approaches to calculating marginal densities", *Journal of the American Statistical Association*, **85**, 398-409.

9. Harvey, A.C., Ruiz, E. and Shephard, N. (1994), "Multivariate stochastic variance models", *Rev. Economic Studies*, **61**, 247-264. Reprinted as 256-276.

10. Hull, J. and White, A. (1987), "The pricing of options on assets with stochastic volatilities", *Journal of Finance*, **42**, 281-300.

11. Jacquier, E., Polson, N.G. and Rossi, P.E. (1994), "Bayesian analysis of stochastic volatility models (with discussion)", *Journal of Business and Economic Statistics*, **12**, 371-417.

12. Kim, S., Shephard, N. and Chib, S. (1998), "Stochastic volatility: likelihood inference and comparison with ARCH models", *Review of Economic Studies*, **65**, 361-393.

13. Philippe, A. (1997), "Simulation of right and left truncated gamma distributions by mixtures", *Statistics and Computing*, **7**, 173-181.

14. Pitt, M.K. and Walker, S.G. (1998), "Marginal construction of stationary time series with application to volatility models", Technical Report. Imperial College London.

15. Robert, C.P. (1995), "Simulation of truncated normal variables", *Statistics and Computing*, **5**, 121-125.

16. San Martini, A. and Spezzaferri, F. (1984) "A predictive model selection criterion", *Journal of the Royal Statistical Society, Series B*, **46**, 296-303.

17. Shephard, N. (1994), "Partial non-Gaussian state space", *Biometrika*, **81**, 115-131.

18. Smith, A.F.M. and Roberts, G.O. (1993), "Bayesian computations via the Gibbs sampler and related Markov Chain Monta Carlo Methods", *Journal of the Royal Statistical Society, Series B*, **55**, 3-23.

19. Tierney, L. (1994), "Markov Chain for exploring posterior distributions (with discussion)", *The Annals of Statistics*, **22**, 1701-1762.

20. Walker, S.G. and Gutierrez-Pena, E. (1999), "Robustifying Bayesian Procedures", In *Bayesian Statistics 6* (Bernardo, J.M., Berger, J.O., Dawid, A.P. and Smith, A.F.M. eds.). New York: Oxford University Press, 85-710.

21. Wakefield, J.C., Gelfand, A.E. and Smith, A.F.M. (1991), "Efficient generation of random variate via the ratio-of-uniforms methods", *Statistics and Computing*, **1**, 129-133.

ON A SMOOTH TRANSITION DOUBLE THRESHOLD MODEL

Y.N. LEE and W.K. LI
Department of Statistics and Actuarial Science
The University of Hong Kong
Hong Kong

This paper considers a generalization of the double threshold ARCH model by using smooth transition functions as links between different regimes in the conditional mean and variance of the time series. The model can cope with the situation where both specifications of the mean and variance of a financial time series change with respect to the market condition. Lagrange multiplier tests for linearity are derived and a modelling procedure for the proposed new class of models is proposed. An application to real data is considered.

1 Introduction

Recently, several useful classes of non-linear time series models have emerged. A popular class of non-linear time series model is the threshold autoregressive (TAR) model (Tong, 1978; Tong and Lim, 1980). The basic idea is a local linear approximation over states which results in a piecewise linear model. The threshold autoregressive model can capture features such as limit cycles, jumps and time irreversibility. On the other hand, the autoregressive conditional heteroscedastic (ARCH) model of Engle (1982) is an important time series tool in modelling changing variance and is popular in financial applications (Christie, 1982; Engle and Bollerslev, 1986). As these models have found useful applications, many researchers bring in new models by combining these basic non-linear models. Following Tong (1990), the basic threshold and ARCH models will be called first generation models. The hybrids resulted by combining the first generation models are referred to as second generation models.

Li and Li (1996) proposed a double-threshold autoregressive heteroscedastic time series (DTARCH) model which may be thought of as a second generation model. It is a model where both the conditional mean and variance can switch from one regime to another. It is motivated by the observation that for financial time series the variance specification conditional on previous information probably changes according to the market condition. For instance, Schwert (1989) found that for financial assets volatility is usually higher during recession. Black (1976) noted that volatility tends to grow in reaction to bad news and to fall in response to good news. This suggests that such asymmetric behaviour in volatility could be a characteristic of financial time series. In a related development, Pesaran and Potter (1997) considered a so called "floor and ceiling" model for the business cycle which can be treated as a type of double threshold model.

In reality, changes may develop slowly and some fuzziness in the change of regimes may be desirable (Tong, 1983, p 276). In this connection, Chan and Tong (1986a) suggested that a smooth transition threshold autoregressive model may be more attractive than the traditional threshold model in many applications. Teräsvirta and Anderson (1992) and Teräsvirta (1994) developed this theme further for the TAR models. It seems therefore worthwhile to consider a double smooth transition time series (DST) model. One can think of it as a generalization of the double threshold ARCH (DTARCH) model (Li and Li, 1996) because in the DST model, both the conditional mean and the conditional variance can switch from one regime to another smoothly. In particular, a steep transition function for the mean will give the traditional threshold autoregressive model. Extending Lee and Li (1998), Lundbergh and Teräsvirta (1998) considered a double smooth generalized ARCH (GARCH) model.

The organisation of the paper is as follows. In section 2, the definition and assumptions of the DST model are given. In section 3, we discuss the problem of testing for the DST models. Lagrange multiplier tests are considered because they are easy to apply and they only require estimation under the null model. The empirical size and power of the tests will be discussed. Section 4 considers the problem of specification and estimation. Ordinary least squares method does not work well with the DST model because of the existence of heteroscedasticity. Besides, high correlation among the parameters makes estimation difficult. A Newton-Raphson method is proposed to deal with the problems. In section 5, we consider an application of the DST model to some financial time series. With all the supporting tools developed in the paper, it is not difficult to apply the DST model to the real data.

2 Model Definition and Assumptions

Let $\{x_t\}$, $t = 1, 2, \ldots$ be the given time series. We define a double smooth transition model (DST) of order $(q_1, q_2; p_1, p_2)$ as follows,

$$x_t = \beta_0^{(1)} + \sum_{j=1}^{q_1} \beta_j^{(1)} x_{t-j} +$$

$$\left\{1 + e^{-\gamma(x_{t-d}-c)}\right\}^{-1} \left\{\beta_0^{(2)} + \sum_{j=1}^{q_2} \beta_j^{(2)} x_{t-j}\right\} + \varepsilon_t \quad , \tag{1}$$

$$h_t = \alpha_0^{(1)} + \sum_{s=1}^{p_1} \alpha_s^{(1)} \varepsilon_{t-s}^2 +$$

$$\left\{1 + e^{-\kappa(x_{t-b}-r)}\right\}^{-1} \left\{\alpha_0^{(2)} + \sum_{s=1}^{p_2} \alpha_s^{(2)} \varepsilon_{t-s}^2\right\} \tag{2}$$

where ε_t follows a normal distribution with mean zero and conditional variance h_t given information set $\Im_{t-1}$; $\Im_{t-1}$ is the information set $\{\varepsilon_{t-1}, \varepsilon_{t-2}, \ldots\}$.

There are two sets of parameters both in the conditional mean and the conditional variance which are distinguished by the superscripts (1) and (2). We refer to those parameters with superscript (1) the first regime parameters and those with superscript (2) the second regime parameters. The logistic functions $\left\{1 + e^{-\gamma(x_{t-d}-c)}\right\}^{-1}$ and $\left\{1 + e^{-\kappa(x_{t-b}-r)}\right\}^{-1}$ are used for linking up the two regimes smoothly. Chan and Tong (1986) suggested that any sufficiently smooth function with a rapidly decaying tail will suffice for that purpose. In this paper, we focus mainly on the logistic function. Here d and b are called the delay parameters while c and r are called the transition parameters. Depending on different values of the smoothness parameters (γ and κ), it is clear that ARCH, smooth transition autoregressive (STAR) and double threshold ARCH (DTARCH) models are special cases of the DST model. In the above model, the transition variables are set to be x_{t-d} and x_{t-b} respectively. However, they are not necessarily restricted to the $x_t's$. In real life situations, abrupt changes in the observations $\{x_t\}$ are often accompanied by those in the disturbances $\{\varepsilon_t\}$. Hence, the delay variables determining the changes in the conditional mean or the conditional variance can either be x_t or ε_t or other measurable functions of the two. For instance, by allowing parameters in the conditional variance other than those in the first regime to be negative and replacing x_{t-b} by ε_{t-b}, we can model the phenomenon pointed out by Rabemanajara and Zakoian (1993) that high negative shocks produce a stronger impact on future volatility than positive shocks.

We will assume that (i) The time series $\{x_t\}$ is at least second order stationary and ergodic. (ii) All the parameters in the first regime of the conditional variance are either positive or non-negative, in particular, $\alpha_0^{(1)} > 0$ and $\alpha_j^{(1)} \geq 0$ for $j = 1, 2, \ldots, p_1$. Further let $\alpha_j = \alpha_j^{(1)} + \alpha_j^{(2)}$. Then all the α_j must also be non-negative and (iii) the two regimes in both the conditional mean and the conditional variance are distinct.

For simplicity we write the equation in vector form and assume $q_1 = q_2 = q$ and $p_1 = p_2 = p$. That is,

$$x_t = \{\Omega_{10} + \Omega_1^T Z_t\} + \left\{1 + e^{-\gamma(x_{t-d}-c)}\right\}^{-1} \{\Omega_{20} + \Omega_2^T Z_t\} + \varepsilon_t, \quad (3)$$

and

$$h_t = \{\Pi_{10} + \Pi_1^T W_t\} + \left\{1 + e^{-\kappa(x_{t-b}-r)}\right\}^{-1} \{\Pi_{20} + \Pi_2^T W_t\} \quad (4)$$

where $Z_t = (x_{t-1}, x_{t-2}, \cdots, x_{t-q})^T$, $(\Omega_{10}, \Omega_1^T) = \left(\beta_0^{(1)}, (\beta_1^{(1)}, \beta_2^{(1)}, \cdots, \beta_q^{(1)})\right)$, $(\Omega_{20}, \Omega_2) = \left(\beta_0^{(2)}, (\beta_1^{(2)}, \beta_2^{(2)}, \cdots, \beta_q^{(2)})\right)^T$, $W_t = (\varepsilon_{t-1}^2, \varepsilon_{t-2}^2, \cdots, \varepsilon_{t-p}^2)^T$, $(\Pi_{10}, \Pi_1) = \left(\alpha_0^{(1)}, (\alpha_1^{(1)}, \alpha_2^{(1)}, \cdots, \alpha_p^{(1)})\right)^T$, $(\Pi_{20}, \Pi_2) = \left(\alpha_0^{(2)}, (\alpha_1^{(2)}, \alpha_2^{(2)}, \cdots, \alpha_p^{(2)})\right)^T$. Extensions using other smooth distribution functions and the inclusion of exogenous variables as in Pesaran and Potter (1997) should not be difficult.

3 Linearity Tests

Before going into the section of specification and estimation, we would like to derive several Lagrange multiplier (LM) tests for testing non-linearity. When one wants to specify a model on a time series, the first thing one needs to determine is whether a linear model is adequate or not. However, in financial time series, it is natural to consider an autoregressive conditional heteroscedastic model as the first model. Therefore, linearity in the conditional variance here means that it has a fixed autoregressive conditional heteroscedasticity specification.

Chan and Tong (1990) discuss the possibility of using a likelihood ratio test statistic for testing linearity against SETAR models. The null distribution of the statistic has been determined by Chan (1991). Within this section, LM tests suggested in Luukkonen, Saikkonen, and Teräsvirta (1988) and Teräsvirta (1994) are considered for the DST models. Following the notation in the last section and assuming that, for simplicity, $1 \le d \le q$ and $1 \le b \le p$, the DST model can be written as

$$x_t = \left\{\Omega_{10} + \Omega_1^T\ Z_t\right\} + \left\{(1 + e^{-\gamma(x_{t-d}-c)})^{-1} - \frac{1}{2}\right\} \cdot$$

$$\left\{\Omega_{20} + \Omega_2^T\ Z_t\right\} + \varepsilon_t\ , \tag{5}$$

and

$$h_t = \left\{\Pi_{10} + \Pi_1^T\ W_t\right\} + \left\{(1 + e^{-\kappa(x_{t-b}-r)})^{-1} - \frac{1}{2}\right\} \cdot$$

$$\left\{\Pi_{20} + \Pi_2^T\ W_t\right\}\ . \tag{6}$$

Note that there is a one-half subtracted from the logistic function as it will be useful in deriving linearity tests. The DST models that we estimate after this section do not contain this term. There are several ways of defining linearity in (5) and (6). For instance, if $\gamma = 0$ and $\kappa = 0$ both the logistic functions in (5) and (6) are equal to zero as a result of subtracting the one-half. Then (5) will be an AR process and (6) will be an ARCH process. The hypotheses of interests are therefore $H_0^\gamma : \gamma = 0$ and $H_0^\kappa : \kappa = 0$. Under the null H_0^γ, Ω_{20}, Ω_2 and c can assume any value in (5). And if H_0^κ holds, Π_{20}, Π_2 and r would be nuisance parameters in (6). In a similar way, if $\Omega_{20} = 0$ and $\Omega_2 = 0$, γ and c would become nuisance parameters whereas κ and r are nuisance parameters if $\Pi_{20} = 0$ and $\Pi_2 = 0$. For the moment, the delay parameters d and b are assumed to be known and this restriction will be relaxed later. The tests with this assumption will be useful in the determination of the delay parameters. In addition, if (5) and (6) are linear, we assume that the resulting time series is stationary and ergodic. Under $\gamma = 0$ and $\kappa = 0$, we define $\theta = (m^T, v^T)^T$ with $m = (\beta_0^{(1)}, \beta_1^{(1)}, \cdots, \beta_q^{(1)}, r)^T$ and $v = (\alpha_0^{(1)}, \alpha_1^{(1)}, \cdots, \alpha_p^{(1)}, c)^T$.

Given the nuisance parameters, the general form of the Lagrange multiplier test for testing $H_0^\gamma : \gamma = 0$ and $H_0^\kappa : \kappa = 0$ against $H_1^\gamma : \gamma \neq 0$ or $H_1^\kappa : \kappa \neq 0$ is :

$$LM = n^{-1}\ \left(\hat{S}_\theta\right)^T \left(\hat{H}_{\theta\theta}\right)^{-1} \left(\hat{S}_\theta\right)\quad ,$$

where $\hat{H}_{\theta\theta}$ and $\hat{S}_\theta$ are the estimated information matrix and score function respectively. Both are evaluated under the null. The information matrix can be seen to be block-diagonal by theorem 4 of Engle (1982). That is, $\hat{H}_{\theta\theta} =$ diag $(\hat{H}_{mm}, \hat{H}_{vv})$. Hence the Lagrange multiplier test, LM_θ, can be split into two sub-tests, LM_m and LM_v where

$$LM_m = n^{-1}\left(\hat{S}_m\right)\left(\hat{H}_{mm}\right)^{-1}\left(\hat{S}_m\right)$$

and

$$LM_v = n^{-1}\left(\hat{S}_v\right)\left(\hat{H}_{vv}\right)^{-1}\left(\hat{S}_v\right)$$

For LM_m, the hypotheses are $H_0^{\gamma}: \gamma = 0$ given $\kappa = 0$ vs $H_1^{\gamma}: \gamma \neq 0$ while for LM_v, the hypotheses are $H_0^{\kappa}: \kappa = 0$ given $\gamma = 0$ vs $H_1^{\kappa}: \kappa \neq 0$. By direct differentiation, we have

$$LM_m = \left(\sum_{t=1}^{n} \hat{s}_t\, \hat{\varepsilon}_t\, \frac{\partial \hat{\varepsilon}_t}{\partial m}\right)^T \left(\sum_{t=1}^{n} \frac{\partial \hat{\varepsilon}_t}{\partial m}\, \hat{r}_t^2\, \frac{\partial \hat{\varepsilon}_t}{\partial m^T}\right)^{-1} \left(\sum_{t=1}^{n} \hat{s}_t\, \hat{\varepsilon}_t\, \frac{\partial \hat{\varepsilon}_t}{\partial m}\right) \tag{7}$$

where

$$\hat{r}_t = \left\{\frac{1}{\hat{h}_t} + 2\hat{\varepsilon}_t^2 \sum_j (\hat{\alpha}_j^{(1)})^2 \hat{h}_{t+j}^{-2}\right\}^{\frac{1}{2}}$$

and

$$\hat{s}_t = \left(\sum_j \hat{\alpha}_j^{(1)} \hat{h}_{t+j}^{-2} (\hat{\varepsilon}_{t+j}^2 - \hat{h}_{t+j})\right) - \frac{1}{\hat{h}_t}$$

Note that all the quantities $\hat{\varepsilon}_t$ and $\hat{h}_t$ are evaluated under the null throughout this section. For simplicity, the hat above the variables will be omitted as it is understood that they are estimates. We derive a similar expression for LM_v,

$$LM_v = \left(\sum_{t=1}^{n} \frac{1}{2\hat{h}_t}\left(\frac{\hat{\varepsilon}_t^2}{\hat{h}_t} - 1\right)\frac{\partial h_t}{\partial v}\right)^T \left(\sum_{t=1}^{n} \frac{1}{2\hat{h}_t^2}\frac{\partial h_t}{\partial v}\frac{\partial h_t}{\partial v^T}\right)^{-1}$$

$$\left(\sum_{t=1}^{n} \frac{1}{2\hat{h}_t}(\frac{\hat{\varepsilon}_t^2}{\hat{h}_t} - 1)\frac{\partial h_t}{\partial v}\right).$$

The two statistics LM_m and LM_v are functions of nuisance parameters. Davies (1977) suggested a conservative statistic to deal with this problem. The standard test statistics are then the suprema of LM_m and LM_v over the respective sets of nuisance parameters. Denote these suprema by LM_m^1 and LM_v^1 respectively. The distribution of these statistics are generally not known. However, in the present case we can overcome this by bringing in a simple and direct auxiliary regression technique in the evaluation of LM_m^1 and LM_v^1 (Luukkonen, Saikkonen and Teräsvirta (1988) and Teräsvirta (1994)). The asymptotic distributions of LM_m^1 and LM_v^1 are then given by standard chi-square distributions. See also Granger and Teräsvirta (1993). One can refer to the technical report by Lee and Li or Lee's University of Hong Kong M.Phil thesis for details. These works also consist of a small simulation study on the size and power of the proposed test statistics. Below we only report the algorithm for calculating LM_m^1 and LM_v^1.

The calculation of LM_m^1.

(1) Regress $\varepsilon_t s_t / r_t$ on r_t and $r_t x_{t-j}$ for $j = 1, \ldots, q$. Form the residuals $\hat{a}_t$ $(t = 1, \ldots, n)$ and the residual sum of squares $SSR_0 = \sum \hat{a}_t^2$.

(2) Regress $\varepsilon_t s_t / r_t$ on r_t, $r_t\ x_{t-j}$ and $r_t\ x_{t-j}\ x_{t-d}$; $j = 1, \ldots, q$. Form the residuals $\hat{a'}_t$ and SSR $= \sum \hat{a'}_t^2$.

(3) Compute the test statistic,

$$LM_m^1 = n\ \frac{SSR_0 - SSR}{SSR_0} \overset{asy}{\sim} \chi_q^2$$

under H_0^γ.

Similarly we have the following steps for the calculation of LM_v^1.

(1) Regress $\varepsilon_t^2/h_t - 1$ on $1/h_t$ and ε_{t-j}^2/h_t; $j = 1, ..., p$. Form the residuals $\hat{v}_t$ $(t = 1, ..., n)$ and the residuals sum of squares $SSR_0 = \Sigma \hat{v}_t^2$.

(2) Regress $\varepsilon_t^2/h_t - 1$ on $1/h_t$, ε_{t-j}^2/h_t, x_{t-b}/h_t and $x_{t-b}\varepsilon_{t-j}^2/h_t$; j=1,...,p. Form the residuals $\hat{v}_t'$ and SSR $= \Sigma \hat{v'}_t^2$.

(3) Compute the test statistic,

$$LM_v^1 = n\ \frac{SSR_0 - SSR}{SSR_0} \overset{asy}{\sim} \chi_{p+1}^2$$

under H_0^κ.

4 Model Specification and Parameter Estimation

4.1 Determining the order and the delay parameters

We can use the plot of autocorrelation function (ACF) and partial autocorrelation function (PACF) as a guide to set the upper bounds for q and p. Tsay (1989) suggested a procedure to select the delay parameter d in threshold AR models and his method is applied here. His idea is to vary the values of d and choosing those values that minimize the p-values of the linearity tests. In other words, suppose $p_m(d)$ and $p_v(b)$ are the p-values of the test statistics LM_m^1 and the modified LM_v^1 respectively. We choose the delay parameters $\hat{d}$ and $\hat{b}$ such that $p_m(\hat{d})=min_{\{1\leq d\leq D_m\}}\ p_m(d)$ and $p_v(\hat{b})=min_{\{1\leq b\leq D_v\}}\ p_v(b)$ where D_m and D_v are the collection of all possible values of d and b.

4.2 Estimating other model parameters

In practice, the joint estimation of $\{\gamma,c,\Omega_1,\Omega_2\}$ and $\{\kappa,r,\Pi_1,\Pi_2\}$ presents some difficulties. The reason is that the estimators of γ, κ, c and r tend to be heavily negatively correlated with those of Ω_2 and Π_2. When γ and κ are large, the transition function would be very steep and hence, it will take many observations in the neighborhood of the transition parameters to estimate the values of smoothness parameters accurately. Even with relatively large smoothness parameters, the corresponding logistic transition functions, change only slightly. As a result the convergence rates of the smoothness parameter estimates are relatively slow. Haggan and Ozaki(1980) suggested a method in the exponential AR model where the final model is chosen over a grid of values of $\{\gamma, \kappa\}$. Another problem is that, in case the smoothness parameter values are large and the transition parameters are close to zero, negative definite Hessian may not be obtained for numerical reasons. Teräsvirta (1994) suggested to rescale γ and κ by dividing them with $\hat{\sigma}^2(x)$, the sample variance of x_t and $\hat{\sigma}^2(\varepsilon)$ respectively after we have a preliminary estimation of the model. After standardization the value of one can be a reasonable initial value for these parameters in the Newton-Raphson algorithm.

Here, maximum likelihood estimation based on the Newton-Raphson method is summarized as follows:

(1) Estimate $\{\Omega, \Pi\}$ by fixing some value to the smoothness parameters and transition parameters temporarily.

(2) Use the value of $\{\hat{\Omega}, \hat{\Pi}\}$ obtained in step (1) to estimate the transition parameters, c, r with some fixed value on the smoothness parameters.

(3) Compute the smoothness parameters based on the values of $\{\hat{\Omega}, \hat{\Pi}\}$ and $\hat{c}, \hat{r}$ obtained from steps (1) and (2).

(4) Repeat the process (1) - (3) until convergence is reached.

The scoring algorithm is applied to calculate estimates of Ω, Π, c, r, γ and κ. Since according to theorem 4 in Engle(1982), the Hessian matrix is block diagonal; we can estimate the mean parameters and conditional variance parameters individually. The technical report of Lee and Li contains some simulation results on the estimation procedure.

5 An Application of the Double Smooth Transition Model

Linear time series model have been found not adequate in the modeling of daily time series especially daily financial series. Recently, many researchers have tried to apply different kinds of non-linear time series models to the financial data and economic data. Li and Li (1996) suggested a DTARCH model and applied it to the Hong Kong Hang Seng Index daily returns. The results demonstrated that a threshold structure could be present in both the conditional mean and conditional variance of the index. As an illustration, we apply the DST$(q_1, q_2; p_1, p_2)$ model to the Daily Hong Kong Hang Seng Index (HSI) return from year 1970 to year 1991. The return, R_t is defined as the difference of the logarithm of the index and multiplied by 100, i.e. $R_t = 100 \times (ln\ P_t - ln\ P_{t-1})$ where P_t is the daily closing index observed. Note that the logarithm differences are very small, in the order of 10^{-2} and the order of magnitude of the smooth function can also be very small, even lower than 10^{-4}. Accordingly, there could be a significant error in calculating the inverse of the information matrices. Therefore, we enlarge their values by multiplying them by 100. During the last two decades, the Hong Kong financial market has undergone many structural changes. It seems reasonable to divide the 22 years observations into 11 non-overlapping sub-series with two years data each. This will also help to sort out the effect of the different economical changes existed in different periods. Moreover, the number of regimes can be reasonably specified to be two in a short series.

The procedure of specification and estimation of the DST models follow the steps stated in section 4. Smoothness parameters were estimated by grid searching. The trick is that we first keep γ at 100 and search for a value of κ that maximizes the log likelihood function. Then we vary γ and obtain $\hat{\gamma}$ that maximizes the likelihood function. The tried values were 1, 5, 10, 20, 50 and 100. Too large a value could introduce error in the calculation of the transition parameters. Four different DST models were considered in fitting the HSI. The first model is a DST($q_1, q_2; p_1, p_2$) with both γ and κ not equal to zero. The second one is called a STM($q_1, q_2; p$) [Smooth Transition exists in the conditional Mean only] model which is a special case of the DST model with κ equals zero. The third one is called a STV($q; p_1, p_2$) [Smooth Transition exists in the conditional Variance only] model which is a special case of the DST model with γ equals zero. The last model is the popular autoregressive conditional heteroscedastic model ARCH(q;p) where both γ and κ are zero. The STM is a model with smooth transition only in the mean while the STV has smooth transition only in the conditional variance.

The first step is to determine the order in the mean and the conditional variance using autocorrelation and autocorrelation of the squared observations. The results are reported in Table 5. The linearity tests, LM_m^1 and LM_v^1, are used to detect the existence of non-linearity in the series. We assume that $q_1 = q_2 = q$, $p_1 = p_2 = p$, $1 \leq d \leq q$ and $1 \leq b \leq p$ so that the number of non-linear models is limited. The results are summarized in Table 6. The tests with the corresponding delay parameters are significant if the probability (underlined) value is less than 0.1. Corresponding DST models are suggested after nonlinearities are detected. If no nonlinearity is detected for both the conditional mean and the conditional variance, an ARCH model is entertained. With the proposed order and delay parameters, we try to check the adequacy of the fitted model using goodness of fit statistics $Q_m(M)$ and $Q_v(M)$ as proposed in Li and Mak (1994) and Li and Li (1996). The degrees of freedom for the tests are set at 5. The p-values of the $Q_m(M)$ and $Q_v(M)$ statistics are reported in Table 8. In the estimation, we have restricted $\beta_0^{(1)}$ and $\beta_0^{(2)}$ to zero. After fine tuning the final models are summarized in the Appendix. The values shown in the brackets are the standard errors of the estimates.

For periods 80-81, 82-83 and 84-85, a large intercept α_0 in the conditional variance was obtained. The reason may be due to the existence of stock crisis in these two periods. For example, from the climax in 1973 to the depression in 1974, HSI dropped nearly 92% within 21 months. And in 1982, China and England started to bring up the sovereignty issue of Hong Kong leading to confidence crisis. HSI again dropped 62.7%, from 1,810 points to 676 points in Dec, 1982. A high value of α_0 implies that the value of the error is dominated by the overall variance of the data. Consequently, abnormal fluctuations of HSI resulted in poor model fitting.

Several features are observed from the DST models fitted to the HSI. First in the mean equations, the sign of the parameters of the same lag are largely opposite in the two regimes. It suggests that perhaps the reason why it is difficult to reject the efficiency market hypothesis since the positive (negative) $\beta_i^{(1)}$ and the negative (positive) $\beta_i^{(2)}$ may cancel each other out in a linear model. In Li and Lam (1995), a similar phenomenon is reported. Second, some parameter estimates of the conditional variance in the second regime are negative. Note that when $x_t > r$, the smoothing function approaches 1. While as $x_t < r$, the smoothing function tends to zero. Therefore, the negative property of the parameters support the fact that volatility tends to be higher on the arrival of bad news and lower on the arrival of good news (Black, 1976). Liu, Li and Li (1997) also have similar findings. Third, the transition parameter, c, in the mean component are always different from zero. It seems to suggest that the relationship between the expected return and its past values is related to the magnitude of the previous return and is not just related to their signs. Fourth, the transition parameter, r, in the conditional variance is always greater than zero. Hence, a small positive return may also have a larger volatility than a big positive return. A possible explanation is that people may fear that a small increase in return is due to short term fluctuation and the price will not keep going again. Therefore, volatility may increase.

6 Conclusion

The DST model with smooth transition both in the mean and the variance is an extension of Li and Li (1996) DTARCH model. In this paper, model estimation, linearity tests and applications are discussed. The Newton-Raphson method is used in the estimation of the series. The number of possible models can be very large. These models should find themselves useful in financial time series. The derived linearity tests seem effective in detecting nonlinear property in a series. These tests are also useful in determining the orders q and p of a DST model. We applied the DST model to the HSI Index and found that some smooth transition in the conditional variance or the conditional mean do exist.

Further extension to the smooth transition GARCH model have been considered by Lundbergh and Teräsvirta (1998). Engle and Gonzalez-Rivera (1991) introduced the semi-parametric ARCH model to deal with the violation of the normality assumption in the error terms. Li and Li (1994) generalize the DTARCH model to the Extended DTARCH model by assuming a conditional error distribution of the Gram-Charlier type. This type of distribution allows for unknown skewness and leptokurtosis. Generalization in this direction should be useful in applications. It is hoped that the DST model can be a useful tool in financial time series modeling.

Acknowledgement

The research was partially supported by the Hong Kong Research Grants Council. The constructive comments of a referee is gratefully acknowledged.

Table 5: Tentative determination of order in the mean and the variance

Year	conditional mean q	conditional variance p	number of observations
70 - 71	2	5	479
72 - 73	1	5	481
74 - 75	3	5	483
76 - 77	1	2	488
78 - 79	3	2	484
80 - 81	1	2	482
82 - 83	1	2	485
84 - 85	1	3	484
86 - 87	1	4	488
88 - 89	1	4	487
90 - 91	1	3	486

Table 6: Linearity test for HSI

Year	order(q;p)	delay parameters d & b	p-value LM_m^1	LM_v^1	proposed model
70 - 71	(2;5)	1	0.6670	0.5954	ARCH
		2	0.4934	0.6597	
		3		0.6064	
		4		0.9760	
		5		0.8072	
72 - 73	(1;5)	1	0.5937	0.5287	ARCH
		2		0.8289	
		3		0.2572	
		4		0.7142	
	5		0.5395		
74 - 75	(3;5)	1	0.0323	0.5932	STM
		2	0.3203	0.4343	
		3	0.0987	0.6535	
		4		0.7297	
		5		0.4318	
76 - 77	(1;2)	1	0.2131	0.7008	ARCH
		2		0.2358	
78 - 79	(3;2)	1	0.0373	0.1739	STM
		2	0.1941	0.7383	
		3	0.1077		
80 - 81	(1;2)	1	0.1087	0.0696	STV
		2		0.0155	
82 - 83	(1;2)	1	0.3774	0.7075	ARCH
		2		0.3274	
84 - 85	(1;3)	1	0.1435	0.0735	STV
		2		0.2082	
		3		0.4567	
86 - 87	(1;4)	1	0.0043	0.0559	DST
		2		0.0837	
		3		0.3732	
		4		0.2449	
88 - 89	(1;4)	1	0.2829	0.8159	ARCH
		2		0.1780	
		3		0.8558	
		4		0.5427	
90 - 91	(1;3)	1	0.6215	0.7523	STV
		2		0.0731	
		3		0.7386	

Table 7: Proposed model and p-values of the corresponding goodness of fit tests for HSI

Year	model with order(q;p)	p-values $Q_m(5)$	$Q_v(5)$
70 - 71	ARCH (2;5)	0.2678	0.4341
72 - 73	ARCH (1;5)	0.1628	0.2929
74 - 75	STM (3,3;5)	0.3483	0.2705
76 - 77	ARCH (1;2)	0.9205	0.4674
78 - 79	STM (3,3;2)	0.8353	0.2169
80 - 81	STV (1;2,2)	0.8039	0.1064
82 - 83	ARCH (1;2)	0.1251	0.2119
84 - 85	STV (1;3,3)	0.1120	0.2183
86 - 87	DST (1,1;4,4)	0.7805	0.7721
88 - 89	ARCH (1;4)	0.1523	0.9829
90 - 91	STV (1;3,3)	0.6990	0.8560

Appendix

Estimation result for the HSI

70–71 Model	ARCH (2;5)	p-values:	$Q_m(5) = 0.2678$	$Q_v(5) = 0.4341$
$\beta^{(1)}$	0.3084 (0.5398×10^{-1})	-0.1011 (0.5517×10^{-1})		
$\alpha^{(1)}$	0.3785 (0.7195×10^{-1})	0.1964 (0.6950×10^{-1})	0.2140 (0.7569×10^{-1})	0.1357 (0.6789×10^{-1})
$\alpha^{(1)}$ (continue)	0.1499 (0.6714×10^{-1})	0.1221 (0.6058×10^{-1})		

72–73 Model	ARCH (1;5)	p-values:	$Q_m(5) = 0.1628$	$Q_v(5) = 0.2929$
$\beta^{(1)}$	0.2114 (0.4979×10^{-1})			
$\alpha^{(1)}$	1.0660 (0.2606)	0.1216 (0.5995×10^{-1})	0.2387 (0.7354×10^{-1})	0.2180 (0.7269×10^{-1})
$\alpha^{(1)}$ (continue)	0.2224 (0.7325×10^{-1})	0.1753 (0.6729×10^{-1})		

74–75 Model STM (3,3;5) p-values: $Q_m(5) = 0.3483$ $Q_v(5) = 0.2705$

$\gamma = 10$ $d = 1$ $c = -0.8711$ (0.2642)

$\beta^{(1)}$	0.2256 (0.5592×10^{-1})	−0.1278 (0.7757×10^{-1})	−0.0249 (0.7220×10^{-1})	
$\beta^{(2)}$	−0.2060 (0.8415×10^{-1})	0.1879 (0.9874×10^{-1})	0.2374 (0.9236×10^{-1})	
$\alpha^{(1)}$	0.7491 (0.2147)	0.0601 (0.4896×10^{-1})	0.3776 (0.8282×10^{-1})	0.1856 (0.6621×10^{-1})
$\alpha^{(1)}$ (continue)	0.1439 (0.6182×10^{-1})	0.2354 (0.7083×10^{-1})		

76–77 Model ARCH (1;2) p-values: $Q_m(5) = 0.9205$ $Q_v(5) = 0.4674$

$\beta^{(1)}$	0.1357 (0.5387×10^{-1})		
$\alpha^{(1)}$	0.4556 (0.5489×10^{-1})	0.2605 (0.7101×10^{-1})	0.3686 (0.8201×10^{-1})

78–79 Model STM (3,3;3) p-values: $Q_m(5) = 0.8353$ $Q_v(5) = 0.2169$

$\gamma = 100$ $d = 1$ $c = 0.2612$ (0.7953×10^{-1})

$\beta^{(1)}$	0.1019 (0.7555×10^{-1})	−0.1192 (0.6361×10^{-1})	0.2884 (0.6310×10^{-1})	
$\beta^{(2)}$	0.1138 (0.9525×10^{-1})	0.0997 (0.9368×10^{-1})	−0.2443 (0.9389×10^{-1})	
$\alpha^{(1)}$	0.8221 (0.1187)	0.1426 (0.6349×10^{-1})	0.2620 (0.7537×10^{-1})	0.2012 (0.7063×10^{-1})

80–81 Model STV (1;2,2) p-values: $Q_m(5) = 0.8039$ $Q_v(5) = 0.1064$

$\beta^{(1)}$	0.1057 (0.4073×10^{-1})		

$\kappa = 50$ $b = 2$ $r = 1.2612\ (0.4890 \times 10^{-1})$

$\alpha^{(1)}$	1.6773 (0.2137)	0.0285 (0.3666×10^{-1})	0.2012 (0.6457×10^{-1})
$\alpha^{(2)}$	2.9956 (0.9966)	0.5062 (0.3039)	−0.1337 (0.1568)

82–83 Model ARCH (1;2) p-values: $Q_m(5) = 0.1251$ $Q_v(5) = 0.2119$

$\beta^{(1)}$	0.1767 (0.5005×10^{-1})		
$\alpha^{(1)}$	2.1928 (0.2428)	0.1332 (0.5800×10^{-1})	0.2573 (0.7254×10^{-1})

84–85 Model STV (0;3,0) p-values: $Q_m(5) = 0.1120$ $Q_v(5) = 0.2183$

$\kappa = 100$ $b = 1$ $r = 1.0970\ (0.2439 \times 10^{-1})$

$\alpha^{(1)}$	0.9728 (0.1294×10^{-1})	0.0133 (0.3332×10^{-1})	0.0559 (0.3926×10^{-1})	0.1002 (0.4608×10^{-1})
$\alpha^{(2)}$	2.8112 (0.5426)			

86–87 Model DST (1,1;4,4) p-values: $Q_m(5) = 0.7805$ $Q_v(5) = 0.7721$

$\gamma = 100$ $d = 1$ $c = -0.2432\ (0.1897)$

$\beta^{(1)}$	-0.0828 (0.7985×10^{-1})
$\beta^{(2)}$	0.4182 (0.9597×10^{-1})

$\kappa = 50$ $b = 1$ $r = 0.1458\ (0.8310 \times 10^{-1})$

$\alpha^{(1)}$	0.6832 (0.1745)	0.2170 (0.1013)	0.1597 (0.1006)	0.0801 (0.8361×10^{-1})	0.4320 (0.1352)
$\alpha^{(2)}$	0.0468 (0.2355)	-0.1608 (0.1191)	-0.0673 (0.1225)	0.0786 (0.1113)	-0.3381 (0.1589)

88–89 Model ARCH (1;4) p-values: $Q_m(5) = 0.1523$ $Q_v(5) = 0.9829$

$\beta^{(1)}$	0.1654 (0.5218×10^{-1})			
$\alpha^{(1)}$	0.6195 (0.8593×10^{-1})	0.2026 (0.6684×10^{-1})	0.0122 (0.4270×10^{-1})	0.1881 (0.6505×10^{-1})
$\alpha^{(1)}$ (continue)	0.1027 (0.5548×10^{-1})			

90–91 Model STV (1;3,1) p-values: $Q_m(5) = 0.6990$ $Q_v(5) = 0.8560$

$\beta^{(1)}$	0.1523 (0.4481×10^{-1})

$\kappa = 100$ $b = 2$ $r = 0.7032\ (0.6263 \times 10^{-1})$

$\alpha^{(1)}$	0.3524 (0.5220×10^{-1})	0.1555 (0.5629×10^{-1})	0.0563 (0.4034×10^{-1})	0.1335 (0.4967×10^{-1})
$\alpha^{(2)}$	0.4805 (0.1633)	-0.15057 (0.1339)		

References

Black, F. (1976), Studies of stock price volatility changes, *Proceedings of the Business & Economic Statistics Section, American Statistical Association*, 177-181.

Chan, K. S. (1991), Percentage points of likelihood ratio tests for threshold autoregression, *J. R. Statist. Soc. B*, **53**, 691-696.

Chan, K. S. and H. Tong (1986), On estimating thresholds in autoregressive models, *Journal of Time Series Analysis*, **7**, 179-194.

Chan, K. S. and H. Tong (1990), On the likelihood ratio tests for threshold autoregression, *J. R. Statist. Soc. B*, **52**, 469-476.

Christie, A. A. (1982), The stochastic behavior of common stock variances, *Journal of Financial Economics*, **10**, 407-432.

Davies, R. B. (1977), Hypothesis testing when a nuisance parameter is present only under the alternative, *Biometrika*, **64**, 247-257.

Engle, R. F. (1982), Autoregressive conditional heteroskedasticity with estimates of the variance of UK inflation, *Econometrica*, **50**, 987-1008.

Engle, R. F. and T. Bollerslev (1986), Modeling the persistence of conditional variance, *Econometric Reviews*, **5**, 1-87.

Engle, R. F. and Gonzalez-Rivera, G. (1991), Semiparametric ARCH models, *Journal of Business and Economic Statistics*, **9**, 345-359.

Granger, C. W. J. and T. Teräsvirta (1993), *Modelling nonlinear economic relationships*. Oxford University Press.

Haggan, V. and T. Ozaki (1980), Amplitude-dependent exponential AR model fitting for nonlinear random vibrations, In *Time Series*, O.D. Anderson, ed., North-Holland Publishing Company.

Haggan, V. and T. Ozaki (1981), Modelling nonlinear random vibrations using an amplitude-dependent autoregressive time series model, *Biometrika*, **68**, 1, 189-196.

Lee, Y. N. and W. K. Li (1998), On smooth transition double threshold models. Research Report #198, Department of Statistics, The University of Hong Kong, July 1998.

Li, C. W. and W. K. Li (1994), Semiparameteric modelling of a double threshold autoregressive heteroscedastic time series model. Research Report #64, Department of Statistics, The University of Hong Kong.

Li, C. W. and W. K. Li (1996), On a double threshold autoregressive heteroscedastic time series model, *Journal of Applied Econometrics*, **11**, 253-274.

Li, W. K. (1992), On the asymptotic standard errors of residual autocorrelations in nonlinear time series modeling, *Biometrika*, **79**, 2, 435-437.

Li, W. K. and K. Lam (1995), Modelling asymmetry in stock returns by a threshold autoregressive conditional heteroscedastic model, *The Statistician*, *44*, 333-341.

Li, W. K. and T. K. Mak (1994), On the squared residual autocorrelations in conditional heteroskedastic time series modeling, *Journal of Time Series Analysis*, **15**, 627-636.

Liu, J, W. K. Li and C. W. Li (1997), On a threshold autoregression with conditional heteroscedastic variances, *Journal of Statistical Planning and Inference*, **62**, 279-300.

Lundbergh, S. and T. Teräsvirta (1998), Modelling economic high frequency time series with STAR-STGARCH models. Working Paper #291, Dec., 1998, Stockholm School of Economics, The Economic Research Institute.

Luukkonen, R., P. Saikkonen and T. Teräsvirta (1988), Testing linearity against smooth transition autoregressive models, *Biometrika*, **75**, 491-499.

Pesaran, H. and S. M. Potter (1997), A floor and ceiling modeling of US output, *Journal of Economic Dynamics and Control*, **21**, 661-695.

Rabemananjara, R. and J. M. Zakoian (1993), Threshold ARCH models and asymmetries in volatility. *J. Appl. Econ.*, **8**, 31-49.

Schwert, G. W. (1989), Why do stock market volatility change over time?, *Journal of Finance*, **44**, 5, 1115-1153.

Teräsvirta, T. (1994), Specification, estimation, and evaluation of smooth transition autoregressive models, *Journal of the American Statistical Association*, **89**, 202-218.

Teräsvirta, T. and H. M. Anderson (1992), Characterizing nonlinearities in business cycles using smooth transition autoregressive models, *Journal of Applied Econometrics*, **7**, S119-S139.

Tong, H. (1978), On a threshold model, In *Pattern Recognition and Signal Processing*. (C. H. Chen ed. 575-586), Sijhoff and Noordhoff, Amsterdam.

Tong, H. (1983), *Threshold models in non-linear time series analysis*. Springer Lecture Notes in Statistics, **21**, Springer: New York.

Tong, H. (1990), *Non-linear time series: A dynamical system approach*, Oxford University Press.

Tong, H. and K. S. Lim (1980), Threshold autoregressive, limit cycles and cyclical data, *Journal of the Royal Statistical Association*, **B 42**, 245-292.

Tsay, R. S. (1986), Non-linearity tests for time series, *Biometrika*, **73**, 461-466.

Weiss, A. A. (1986), Asymptotic theory for ARCH models: estimation and testing, *Econometric Theory*, **2**, 107-131.

TESTING GARCH VERSUS E-GARCH

SHIQING LING, MICHAEL MCALEER
Department of Economics,
The University of Western Australia, Nedlands, Perth,
Western Australia 6009, Australia
E-mail: sling@ecel.uwa.edu.au, Michael.McAleer@uwa.edu.au

This paper develops non-nested tests of the GARCH and E-GARCH models against each other, based on a weighted function of the competing conditional variances. The asymptotic distributions and power functions of the non-nested tests are derived. Two novel joint tests of the ARCH and E-ARCH models against their GARCH and E-GARCH counterparts are analysed. Non-nested tests based on the weighting scheme in an L_λ−family are also examined. It is shown that the non-nested test based on a linear weighting of the competing conditional variances is optimal in the L_λ−family.

1 Introduction

Various volatilities, such as asset returns, stock returns and exchange rates, are believed to change over time. Modelling time-varying volatility has been one of the most important research topics in various economic and financial applications over the last fifteen years. The first development to capture such volatility was the autoregressive conditional heteroskedasticity (ARCH) model of Engle (1982). Following Engle's seminal contribution, many different ARCH-type models have been proposed; see, for example, ARMA-ARCH (Weiss, 1984), GARCH (Bollerslev, 1986), CHARMA (Tsay, 1987), E-GARCH (Nelson, 1989), Threshold ARCH (Zakoian, 1994), and double threshold ARCH (Li and Li, 1996), among others (for a survey of recent theoretical results, see Li et al. (1999)). Without doubt, two of the more widely used models in the ARCH family are Bollerslev's GARCH and Nelson's E-GARCH.

The GARCH model has two quite attractive features. First, it can capture the persistence of volatility. A substantial body of empirical evidence has helped to explain various economic and financial phenomena (see, for example, Engle and Bollerslev (1986a, b), Bollerslev et al. (1992, 1994), and Bollerslev and Mikkelsen (1996)). Second, GARCH is mathematically and computationally straightforward, as compared with some other ARCH-type models. Many theoretical results, including the statistical properties of the model and the large sample properties of some estimation methods, are now available, and these provide a solid foundation for applications of the model. However, as argued by Nelson (1989), the GARCH model has several drawbacks, including an inability to capture asymmetric volatility and to impose nonnegativity restrictions.

In order to avoid these shortcomings, Nelson (1989) proposed the E-GARCH model. The GARCH and E-GARCH models are non-nested (or separate) and the volatilities modelled by these two models should be substantially different from each other. However, as the true feature of the volatility is not known in practice, it is also not known whether the true model is GARCH or E-GARCH when a series of economic or financial data are observed. This suggests the motivation for exploring an approach to test the GARCH and E-GARCH models against each other.

A primary aim of this paper is to develop and examine the asymptotic properties of non-nested tests of the GARCH and E-GARCH models. The non-nested testing methodology was developed almost four decades ago, and has been demonstrated to be a powerful tool for testing such models (see McAleer (1995) for a recent review). However, virtually all non-nested tests have been developed for the functional forms of the regression, or for the conditional means. This paper adapts the non-nested procedure for testing the conditional variances of different models, in particular, to develop non-nested tests of the GARCH and E-GARCH models. The asymptotic distributions and power functions of the non-nested tests are also derived. Two novel joint tests of the ARCH and E-ARCH models against their GARCH and E-GARCH counterparts are developed, and their asymptotic distributions are derived. Alternative weighting schemes are also examined.

The paper is organised as follows. Section 2 presents the GARCH and E-GARCH models, and the non-nested testing procedures. Section 3 develops the non-nested tests of the GARCH and E-GARCH models against each other, and derives their asymptotic distributions and power functions. Section 4 develops two joint Lagrange multiplier tests of the ARCH and E-ARCH models against the GARCH and E-GARCH counterparts, and derives their asymptotic distributions. Section 5 discusses non-nested tests based on the weighting scheme in an L_λ−family, and shows that the non-nested test based on a linear weighting of the competing conditional variances is optimal in the L_λ−family. Concluding remarks are given in Section 6.

2 Non-nested Testing Procedures

Suppose that $\{\varepsilon_t\}$ is the time series process of interest. One possible specification for ε_t is the GARCH (p, q) model, namely:

$$H_0 : \ \varepsilon_t = z_{0t} h_t^{1/2}, \ \ h_t = \alpha_0 + \sum_{i=1}^{p} \alpha_i \varepsilon_{t-i}^2 + \sum_{i=1}^{q} \beta_i h_{t-i}, \tag{1}$$

where $\alpha_0 > 0$, $\alpha_i \geq 0$ and $\beta_i \geq 0$; $\{z_{0t}\}$ is a series of independently and identically distributed (i.i.d.) random variables with mean zero and variance one; and $t = 1, \cdots, n$. It is assumed that $\sum_{i=1}^{p} \alpha_i + \sum_{i=1}^{q} \beta_i < 1$, which ensures that the GARCH model is strictly stationary and ergodic, and $E\varepsilon_t^2 < \infty$ (see Bollerslev (1986) and Ling and Li (1997)).

A popular alternative specification to GARCH is the E-GARCH (r, s) model, which is defined by:

$$H_1 : \varepsilon_t = z_{1t} g_t^{1/2}, \ \ln g_t = \omega + (1 - \sum_{i=1}^{r} \phi_i B^i)^{-1} (1 + \sum_{i=1}^{s} \psi_i B^i) u(\varepsilon_{t-1}), \quad (2)$$

where $u(\varepsilon_t) = \theta \varepsilon_t / g_t^{1/2} + \gamma[|\varepsilon_t|/g_t^{1/2} - E(|\varepsilon_t|/g_t^{1/2})] = \theta z_{1t} + \gamma[|z_{1t}| - E(|z_{1t}|)]$, and $\{z_{1t}\}$ is a series of i.i.d. random variables with mean zero and variance one. It is assumed that γ and θ are not both equal to zero, $1 - \sum_{i=1}^{r} \phi_i B^i = 0$ and $1 + \sum_{i=1}^{s} \psi_i B^s = 0$ have no common root, and all the roots of $1 - \sum_{i=1}^{r} \phi_i B^i = 0$ lie outside the unit circle. This is the condition for the strict stationarity, ergodicity, and covariance stationarity of $\ln g_t^2$ (see Nelson (1989)). From (1)-(2), it is clear that H_0 and H_1 are non-nested, in that neither can be obtained from the other by the imposition of suitable parametric restrictions.

In a similar spirit to that of Davidson and MacKinnon (1981), MacKinnon et al. (1983), and Bera and McAleer(1989), we can construct the auxiliary ARCH-type model given by the following linear weighting of the competing conditional variances:

$$H_L : \varepsilon_t = z_t f_t^{1/2}, \ f_t = (1 - \delta) h_t + \delta g_t, \quad (3)$$

where $\{z_t\}$ is a series of i.i.d. random variables with mean zero and variance one. If H_0 is true, then $\delta = 0$, and $\delta = 1$ if H_1 is true. Denote $\beta = (\omega, \phi_1, \cdots, \phi_r, \psi_1, \cdots, \psi_s, \theta, \gamma)'$, where A' denote the transpose of the vector or matrix A. Now let $\hat{\beta}_n$ be the maximum likelihood estimator (MLE) of β in (2) and denote $g_t(\hat{\beta}_n)$ by $\hat{g}_t$. H_L can be approximated by the following model:

$$H_{0L} : \varepsilon_t = z_t f_{0t}^{1/2}, \ f_{0t} = (1 - \delta) h_t + \delta \hat{g}_t. \quad (4)$$

It can be seen that $\hat{g}_t$ is a function of $\{\varepsilon_{t-1}, \cdots, \varepsilon_0, \varepsilon_{-1}, \cdots\}$, and hence is independent of z_t because the influence of any particular error term on the estimates tends to zero as the sample size approaches infinity (this argument is similar to that in Davidson and MacKinnon (1981)). In practice, it is usually assumed that the pre-sample values, i.e. ε_t with $t \leq 0$, are zero. This assumption does not affect the asymptotic properties of the estimators or tests (see Bollerslev (1986) and Weiss (1986)). Under H_0, ε_t is strictly stationary and ergodic, and has a finite unconditional variance.

Using maximum likelihood estimation, we can obtain the joint estimators, $\hat{\delta}_n$ and $\hat{\alpha}_n$, of δ and α, where $\alpha = (\alpha_0, \alpha_1, \cdots, \alpha_p, \beta_1, \cdots, \beta_q)'$. Under H_0, we can derive the asymptotic distribution of $\hat{\delta}_n$ in order to test H_0.

Under H_1, the estimator of δ in (4) will converge to 1 (see the following section for details). However, as the $t-$statistic for testing $\delta = 0$ in (4) is conditional on the truth of H_0, it is valid only for testing H_0. In order to derive a test for H_1, consider the following auxiliary ARCH-type model:

$$H_{1L}: \ \varepsilon_t = z_t f_{1t}^{1/2}, \ f_{1t} = (1-\delta)g_t + \delta \hat{h}_t, \tag{5}$$

where $\hat{h}_t$ denotes $h_t(\hat{\alpha}_n)$ and $\hat{\alpha}_n$ is the MLE of α. The MLE of δ in (5) can be used to test H_1 as the null, namely $\delta = 0$.

Unlike the typical linear and nonlinear regression models considered in the literature, the estimation of (4) or (5) is more complicated. Consider the quasi-maximum likelihood estimation of (4). Under the regularity conditions given in Bollerslev and Wooldridge (1992, Theorem 2.1) or White (1994, Theorem 6.2), there exist a series of consistent MLEs which are asymptotically normal or satisfy condition (7) of the following section. Unfortunately, the verification of these regularity conditions can be difficult, even for the GARCH and E-GARCH models. A weaker regularity condition is available only for the GARCH (1,1) model (see Lee and Hansen (1994) and Lumsdaine (1996)). For the general ARCH (p) model, Ling and McAleer (1999) showed asymptotic normality under the second moment condition, while the finite fourth-moment condition is required for the general GARCH(p, q) model (see Ling and Li (1997)). The latter is a strong condition and may not alway be satisfied in practical applications. For the E-GARCH model, no regularity condition has yet been established. In what follows, it is assumed that all of the appropriate regularity conditions are satisfied.

3 Asymptotic Properties of the Non-nested Tests

In this section, we derive the asymptotic distribution of the non-nested test under the null, H_0:GARCH, and the corresponding power function under the alternative, H_1:E-GARCH. The problem is symmetric, and the case of testing the E-GARCH null against the GARCH alternative is also examined.

Consider the (conditional) quasi-log-likelihood function of model (4), i.e.

H_{0L}:

$$L(\delta, \alpha) = -\frac{1}{2n}\sum_{t=1}^{n} \log f_{0t} - \frac{1}{2n}\sum_{t=1}^{n}\frac{\varepsilon_t^2}{f_{0t}}. \tag{6}$$

Suppose that $(\hat{\delta}_n, \hat{\alpha}'_n)$ are a series of MLEs of (δ, α'), such that:

$$\sqrt{n}\begin{pmatrix} \hat{\delta}_n - \delta \\ \hat{\alpha}_n - \alpha \end{pmatrix} = -\sqrt{n}B^{-1}\begin{pmatrix} \frac{\partial L(\delta,\alpha)}{\partial \delta} \\ \frac{\partial L(\delta,\alpha)}{\partial \alpha} \end{pmatrix} + o_p(1), \tag{7}$$

where $\partial L(\delta, \alpha)/\partial \delta$ and $\partial L(\delta, \alpha)/\partial \alpha$ are the first-order derivatives of $L(\delta, \alpha)$ with respect to δ and α, respectively, and $B = \partial L^2(\delta, \alpha)/\partial (\delta, \alpha)\partial (\delta, \alpha)'$ is the corresponding second-order derivative of $L(\delta, \alpha)$.

Under H_0, we obtain the score functions:

$$\frac{\partial L(0, \alpha)}{\partial \delta} = \frac{1}{2n}\sum_{t=1}^{n}\frac{(\hat{g}_t - h_t)}{h_t}\left(\frac{\varepsilon_t^2}{h_t} - 1\right), \frac{\partial L(0, \alpha)}{\partial \alpha} = \frac{1}{2n}\sum_{t=1}^{n}\frac{1}{h_t}\frac{\partial h_t}{\partial \alpha}\left(\frac{\varepsilon_t^2}{h_t} - 1\right), \tag{8}$$

and the information blocks:

$$\frac{\partial^2 L(0, \alpha)}{\partial \delta^2} = -\frac{1}{n}\sum_{t=1}^{n}\frac{(\hat{g}_t - h_t)^2}{2h_t^2} + o_p(1), \tag{9}$$

$$\frac{\partial^2 L(0, \alpha)}{\partial \delta \alpha} = -\frac{1}{n}\sum_{t=1}^{n}\frac{(\hat{g}_t - h_t)}{2h_t^2}\frac{\partial h_t}{\partial \alpha} + o_p(1), \tag{10}$$

$$\frac{\partial^2 L(0, \alpha)}{\partial \alpha\, \partial \alpha'} = -\frac{1}{n}\sum_{t=1}^{n}\left[\frac{1}{2h_t^2}\frac{\partial h_t}{\partial \alpha}\frac{\partial h_t}{\partial \alpha'}\right] + o_p(1), \tag{11}$$

where equations (9)-(11) hold by the ergodic theorem and the strict stationarity and ergodicity of the process ε_t, and $\partial h_t/\partial \alpha = \tilde{\varepsilon}_t + \sum_{i=1}^{q}\beta_i(\partial h_{t-i}/\partial \alpha)$, with $\tilde{\varepsilon}_t = (1, \varepsilon_{t-1}^2, \cdots, \varepsilon_{t-p-1}^2, h_{t-1}, \cdots, h_{t-q-1})'$. Let $\sigma_{\delta\delta}$, $\sigma_{\delta\alpha}$ and $\sigma_{\alpha\alpha'}$ be the first terms of the right-hand sides of (9)-(11). From (7), we have:

$$\sqrt{n}\begin{pmatrix} \hat{\delta}_n - 0 \\ \hat{\alpha}_n - \alpha \end{pmatrix} = -\begin{pmatrix} \sigma_{\delta\delta} & \sigma'_{\delta\alpha} \\ \sigma_{\delta\alpha} & \sigma_{\alpha\alpha'} \end{pmatrix}^{-1}\begin{pmatrix} \sqrt{n}\frac{\partial L(0,\alpha)}{\partial \delta} \\ \sqrt{n}\frac{\partial L(0,\alpha)}{\partial \alpha} \end{pmatrix} + o_p(1). \tag{12}$$

Denote $F = (a_1, \cdots, a_n)'$, with $a_t = (\partial h_t/\partial \alpha)/\sqrt{2}h_t$; $e = (e_1, \cdots, e_n)'$, with $e_t = (\varepsilon_t^2/h_t - 1)/\sqrt{2}$; and $R = (r_1, \cdots, r_n)'$, with $r_t = (\hat{g}_t - h_t)/\sqrt{2}h_t$. It follows that:

$$\sqrt{n}\hat{\delta}_n = \sqrt{n}R'M_0 e/||M_0 R||^2 + o_p(1), \tag{13}$$

where

$$M_0 = I - F(F'F)^{-1}F'.$$

Let β^* be the limit of $\hat{\beta}_n$ in probability under H_0 as n goes to infinity. Note that:

$$\frac{1}{\sqrt{n}}\sum_{t=1}^{n}\frac{(\hat{g}_t - h_t)}{h_t}(\frac{\varepsilon_t^2}{h_t} - 1) = \frac{1}{\sqrt{n}}\sum_{t=1}^{n}\frac{(g_t(\beta^*) - h_t)}{h_t}(\frac{\varepsilon_t^2}{h_t} - 1)$$
$$+(\hat{\beta}_n - \beta^*)'\left[\frac{1}{\sqrt{n}}\sum_{t=1}^{n}\frac{1}{h_t}\frac{\partial g_t(\bar{\beta})}{\partial\beta}(\frac{\varepsilon_t^2}{h_t} - 1)\right] \quad (14)$$

and

$$\frac{1}{n}\sum_{t=1}^{n}\frac{(\hat{g}_t - h_t)}{h_t^2}\frac{\partial h_t}{\partial\alpha} = \frac{1}{n}\sum_{t=1}^{n}\frac{(g_t(\beta^*) - h_t)}{h_t^2}\frac{\partial h_t}{\partial\alpha}$$
$$+(\hat{\beta}_n - \beta^*)'\left[\frac{1}{n}\sum\frac{1}{h_t^2}\frac{\partial g_t(\bar{\beta})}{\partial\beta}\frac{\partial h_t}{\partial\alpha}\right], \quad (15)$$

where $\bar{\beta}$ is an intermediate point between $\hat{\beta}_n$ and β^*. Under some suitable conditions, $\sum_{t=1}^{n}||\partial g_t(\bar{\beta})/\partial\beta||^2/nh_t^2$ is bounded in probability in a neighbourhood of β^*, and $\sum_{t=1}^{n}||\partial h_t/\partial\alpha||/nh_t^2$ is finite in probability.[1] Thus, the second terms in (14)-(15) will vanish as n goes to infinity, that is, $\hat{g}_t$ in $R'M_0e/\sqrt{n}$ can be replaced asymptotically by $g_t(\beta^*)$. Similarly, we can replace $\hat{g}_t$ in $||M_0R||^2/n$ by $g_t(\beta^*)$. By the martingale central limit theorem (see Theorem 4.4 in Hall and Heyde (1980)), it is straightforward to show that $\sqrt{n}\hat{\delta}_n$ is asymptotically normal with mean zero and variance c/σ^2:

$$\sigma^2 = \lim_{n\to\infty}||\frac{1}{\sqrt{n}}M_0R||^2 \text{ (in probability)} = \left\{E\left(\frac{h_t - g_t^*}{h_t}\right)^2\right.$$
$$\left.-E\left[\frac{(h_t - g_t^*)}{h_t^2}\frac{\partial h_t}{\partial\alpha'}\right]E^{-1}\left(\frac{1}{h_t^2}\frac{\partial h_t}{\partial\alpha}\frac{\partial h_t}{\partial\alpha'}\right)E\left[\frac{(h_t - g_t^*)}{h_t^2}\frac{\partial h_t}{\partial\alpha}\right]\right\}^{-1}, \quad (16)$$

where $g_t^* = g_t(\beta^*)$ and $c = (Ez_t^4 - 1)/2$. In particular, when z_t is normal, $c = 1$. The asymptotic variance may be estimated by:

$$nc/||\hat{M}_0\hat{R}||^2,$$

[1]The condition for $\sum_{t=1}^{n}||\partial h_t/\partial\alpha||/nh_t^2$ being finite in probability is $E\varepsilon_t^4 < \infty$, and can be weakened to $E\ln(\alpha_1 z_{1t}^2 + \beta_1) < \infty$ if $p = q = 1$ and $\beta_1 \neq 0$. However, we cannot give a simple explicit condition for $\sum_{t=1}^{n}||\partial g_t(\bar{\beta})/\partial\beta||^2/nh_t^2$ being bounded in probability in a neighbourhood of β^* since E-GARCH is a misspecified model under H_0, so that g_t is very complicated.

where $\hat{M}_0$ and $\hat{R}$ are, respectively, M_0 and R with all the parameters replaced by their corresponding estimates. Thus, we have the following theorem.

Theorem 3.1. The $t-$statistic for $\hat{\delta}_n$ generated by (4) is asymptotically distributed as $N(0,1)$ if H_0 is true.

Denote α^* as the limit of $\hat{\alpha}_n$ in probability under H_1 as n goes to infinity. Under H_1:

$$e_t = \frac{1}{\sqrt{2}}(\frac{\varepsilon_t^2}{h_t^*} - 1) = \frac{z_{1t}^2(g_t - h_t^*)}{\sqrt{2}h_t^*} + \frac{1}{\sqrt{2}}(z_{1t}^2 - 1), \tag{17}$$

where $g_t = g_t(\beta)$ and $h_t^* = h_t(\alpha^*)$. Denote $\tilde{e} = (\tilde{e}_1, \cdots, \tilde{e}_n)$, with $\tilde{e}_t = (z_{1t}^2 - 1)/\sqrt{2}$. It follows that:

$$\sqrt{n}\hat{\delta}_n = \sqrt{n} + \sqrt{n}R_1' M_{10}\tilde{e}/||M_{10}R_1||^2 + o_p(1), \tag{18}$$

and hence $\hat{\delta}_n \to 1$ as n goes to infinity, where R_1 and M_{10} are defined as R and M_0 with α and β^* replaced by α^* and β, respectively. Let $N_\delta = ||M_{10}R_1/(\sqrt{c}||\hat{\delta}_n)$, i.e. the $t-$statistic for $\hat{\delta}_n$ from (4). It follows that:

$$N_\delta = ||M_{10}R_1/\sqrt{c}|| + R_1' M_{10}\tilde{e}/(\sqrt{c}||M_{10}R_1||) + o_p(1). \tag{19}$$

By the martingale central limit theorem, $R_1' M_{10}\tilde{e}/(\sqrt{c}||M_{10}R_1||)$ is asymptotically distributed as $N(0,1)$. Thus, we have the theorem.

Theorem 3.2. The $t-$statistic for $\hat{\delta}_n$ generated by (4) is asymptotically distributed as $N(||M_{10}R_1/\sqrt{c}||, 1)$ under H_1. The power function is asymptotically given by:

$$\Phi(\zeta_\mu + ||M_{10}R_1/\sqrt{c}||) + o(1),$$

where $\Phi(\cdot)$ is the cumulative distribution function of the standardised normal and ζ_μ is the 100μ percentile of the standardised normal distribution.

For purposes of testing E-GARCH (that is, $\delta = 0$) in (5) as the null, the estimation procedure is similar to the above and hence is omitted. In this case, the asymptotic distribution of the $t-$statistic of $\hat{\delta}_n$ is the same as that given for testing GARCH (that is, $\delta = 0$) in (4), with h_t, $\hat{g}_t$ and $\partial h_t/\partial\alpha$ replaced by g_t, $\hat{h}_t$ and $\partial g_t/\partial\beta$, respectively, where

$$\partial g_t/\partial\beta = g_t^{-1}[\tilde{\varepsilon}_t + \sum_{i=1}^{r} \phi_i g_{t-i}^{-1}(\partial g_{t-i}/\partial\beta)], \tag{20}$$

$$\tilde{\varepsilon}_t = (1, u(\varepsilon_{t-1}), \cdots, u(\varepsilon_{t-r}), \ln g_{t-1}, \cdots, \ln g_{t-s}, \sum_{i=0}^{s} \psi_i\varepsilon_{t-i-1}/g_{t-i-1}^{1/2}, \xi_t)',$$

and

$$\xi_t = \sum_{i=0}^{s} \psi_i [|\varepsilon_{t-i-1}|/g_{t-i-1}^{1/2} - E(|\varepsilon_{t-i-1}|/g_{t-i-1}^{1/2})],$$

with $\psi_0 = 1$. Similarly, we can evaluate the power function of the $t-$statistic based on (5).

Remark. From Theorem 3.2, the $t-$statistic from (4) rejects H_0 against H_1 with probability 1 when H_1 is true. A similar conclusion holds for the $t-$statistic from (5). In practice, as it is possible that both H_0 and H_1 are rejected using the two non-nested tests from (4) and (5), the appropriate inference would be to reconsider both models. Thus, the non-nested tests from (4) and (5) are intended as simple and useful diagnostic tools. The results from both tests should provide some guidance for further empirical analysis.

4 Joint Lagrange Multiplier Tests of ARCH and E-ARCH Against GARCH and E-GARCH

It should be noted that, in (4) and h_t in (1), if $\delta = \beta_1 = \cdots = \beta_q = 0$, then the true model should be the ARCH rather than the GARCH model. Under ARCH, the estimator $\hat{\delta}_{0n}^*$ of $\delta_0^* = (\delta, \beta_1, \cdots, \beta_q)'$ can be used to construct a test of $\delta^* = \mathbf{0}$. Denote $\alpha^* = (\alpha_0, \alpha_1, \cdots, \alpha_p)'$, $F_0^* = [(\partial h_1/\partial \alpha^*)/(\sqrt{2}h_1), \cdots, (\partial h_n/\partial \alpha^*)/(\sqrt{2}h_n)]'$, and $R_0^* = [(\partial h_1/\partial \delta^*)/(\sqrt{2}h_1), \cdots, (\partial h_n/\partial \delta^*)/(\sqrt{2}h_n)]'$. From (12), it follows that:

$$\sqrt{n}(\delta_{0n}^* - \mathbf{0}) = (\frac{1}{n} R_0^{*'} M_{00}^* R_0^*)^{-1} (\frac{1}{\sqrt{n}} R_0^{*'} M_{00}^* e), \tag{21}$$

where e is defined as in (13), and

$$M_{00}^* = I - F_0^* (F_0^{*'} F_0^*)^{-1} F_0^{*'}.$$

Similarly, $\hat{g}_t$ in (21) can be replaced by $g_t(\beta^*)$, where β^* is defined as in (14). By the Gramme-advice and the martingale central limit theorem, $\sqrt{n}\delta_{0n}^*$ is an asymptotically normal vector with mean zero and covariance $n(R_0^{*'} M_{00}^* R_0^*)^{-1}$. Thus, we have the following theorem.

Theorem 4.1. Under $\delta = \beta_1 = \cdots = \beta_q = 0$, the Lagrange multiplier test, $L_{0n} = \delta_{0n}^{*'}(\hat{R}_0^{*'} \hat{M}_{00}^* \hat{R}_0^*)\delta_{0n}^*$, has an asymptotic χ^2 distribution with $q+1$ degrees of freedom, where $\hat{R}_0^*$ and $\hat{M}_{00}^*$ are, respectively, R_0^* and M_{00}^* with all parameters replaced by their corresponding estimates.

Remark. Bollerslev (1986) developed a Lagrange multiplier test for testing ARCH against the GARCH model, which has an asymptotic χ^2 distribution with q degrees of freedom under the ARCH null. Here, we provide a novel joint test of the ARCH model against the nested GARCH model and the non-nested E-GARCH model.

In (5) and $\ln g_t$ in (2), if $\delta = \phi_1 = \cdots = \phi_r = 0$, then the true model should be E-ARCH rather than the E-GARCH model. Under E-ARCH, the estimator, δ_{1n}^*, of $\delta_1^* = (\delta, \phi_1, \cdots, \phi_r)'$ can be used to test the E-ARCH model jointly against the non-nested GARCH model and the nested E-GARCH model. In a similar manner to (21), we can obtain:

$$\sqrt{n}(\delta_{1n}^* - \mathbf{0}) = (\frac{1}{n} R_1^{*'} M_{10}^* R_1^*)^{-1} (\frac{1}{\sqrt{n}} R_1^{*'} M_{10}^* e), \tag{22}$$

where R_1^* and M_{10}^* are defined as R^* and M_0^*, respectively, with h_t, $\hat{g}_t$, $\partial h_t / \partial \alpha^*$ and $\partial h_t / \partial \delta^*$ replaced by g_t, $\hat{h}_t$, $\partial g_t / \partial \psi^*$ and $\partial g_t / \partial \delta_1^*$, where $\psi^* = (\omega, \psi_1, \cdots, \psi_s, \theta, \gamma)'$. Along the lines of Theorem 4.1, by (22), we have the following theorem.

Theorem 4.2. Under $\delta = \phi_1 = \cdots = \phi_r = 0$, the Lagrange multiplier test, $L_{1n} = \delta_{1n}^{*'} (\hat{R}_1^{*'} \hat{M}_{10}^* \hat{R}_1^*) \delta_{1n}^*$, has an asymptotic χ^2 distribution with $r + 1$ degrees of freedom, where $\hat{R}_1^*$ and $\hat{M}_{10}^*$ are, respectively, R_1^* and M_{10}^* with all parameters replaced by their corresponding estimates.

Remark. Nelson (1989) did not propose a test of the E-ARCH model against the E-GARCH model. Theorem 4.2 not only provides such a test, but also gives a novel joint test of the E-ARCH model against the nested E-GARCH model and the non-nested GARCH model.

5 The Optimal Non-nested Test in an L_λ−Family

It is clear that auxiliary ARCH-type models can be constructed using different weighting functions. Consider the following two alternative forms:

$$H_{0L}' : \varepsilon_t = z_t f_{0t}^{1/2}, \quad f_{0t} = h_t^{1-\delta} \hat{g}_t^{\delta} \tag{23}$$

and

$$H_{0L}'' : \varepsilon_t = z_t f_{0t}^{1/2}, \quad f_{0t}^{1/2} = (1 - \delta) h_t^{1/2} + \delta \hat{g}_t^{1/2}. \tag{24}$$

In (23), the auxiliary ARCH-type model is given as a linear combination of the logarithms of the competing conditional variances, i.e. $\ln f_{0t} = (1 - \delta) \ln h_t + \delta \ln \hat{g}_t$, whereas in (24), it is given as a linear combination of the competing conditional standard deviations. If H_0 is true, then $\delta = 0$ in each case, and if H_1 is true, then correspondingly $\delta = 1$. The asymptotic distribution of the MLE of δ can be obtained in each case for testing H_0.

First, consider the non-nested test of H_0:GARCH (that is, $\delta = 0$) based on (23). The (conditional) quasi-log-likelihood function of (23) is the same as that of (6), with f_{0t} defined in (23). Suppose that $(\hat{\delta}_n, \hat{\alpha}'_n)$ are a series of MLEs of (δ, α'), such that:

$$\sqrt{n}\begin{pmatrix} \hat{\delta}_n - \delta \\ \hat{\alpha}_n - \alpha \end{pmatrix} = -\sqrt{n}B^{-1}\begin{pmatrix} \frac{\partial L(\delta,\alpha)}{\partial \delta} \\ \frac{\partial L(\delta,\alpha)}{\partial \alpha} \end{pmatrix} + o_p(1), \tag{25}$$

where $\partial L(\delta, \alpha)/\partial \delta$, $\partial L(\delta, \alpha)/\partial \alpha$ and B are defined analogously to (7). Under H_0, we can obtain:

$$\begin{aligned} \frac{\partial L(0,\alpha)}{\partial \delta} &= \frac{1}{2n}\sum_{t=1}^{n} \ln\left(\frac{\hat{g}_t}{h_t}\right)\left(\frac{\varepsilon_t^2}{h_t} - 1\right), \\ \frac{\partial L(0,\alpha)}{\partial \alpha} &= \frac{1}{2n}\sum_{t=1}^{n} \frac{1}{h_t}\frac{\partial h_t}{\partial \alpha}\left(\frac{\varepsilon_t^2}{h_t} - 1\right) \end{aligned} \tag{26}$$

and

$$\sum_{t=1}^{n} \frac{\partial^2 L(0,\alpha)}{\partial(\delta,\alpha)} = -\hat{\Omega} + o_p(1), \tag{27}$$

where

$$\hat{\Omega} = \frac{1}{2n}\sum_{t=1}^{n}\begin{pmatrix} \ln^2\left(\frac{\hat{g}_t}{h_t}\right) & \ln\left(\frac{\hat{g}_t}{h_t}\right)\frac{1}{h_t}\frac{\partial h_t}{\partial \alpha'} \\ \ln\left(\frac{\hat{g}_t}{h_t}\right)\frac{1}{h_t}\frac{\partial h_t}{\partial \alpha} & \frac{1}{h_t^2}\frac{\partial h_t}{\partial \alpha}\frac{\partial h_t}{\partial \alpha'} \end{pmatrix}.$$

Denote $F = (a_1, \cdots, a_n)'$, with $a_t = (\partial h_t/\partial\alpha)/(\sqrt{2}h_t)$; $e = (e_1, \cdots, e_n)'$, with $e_t = (\varepsilon_t^2/h_t - 1)/\sqrt{2}$; and $\tilde{R} = (\tilde{r}_1, \cdots, \tilde{r}_n)'$, with $\tilde{r}_t = \ln(\hat{g}_t/h_t)/\sqrt{2}$. By (26)-(27), it follows that, under H_0:

$$\sqrt{n}\hat{\delta}_n = \sqrt{n}\tilde{R}'M_0 e/||M_0\tilde{R}||^2 + o_p(1), \tag{28}$$

where

$$M_0 = I - F(F'F)^{-1}F'.$$

Let β^* be the limit of $\hat{\beta}_n$ in probability under H_0 as n goes to infinity. Note that:

$$\ln\frac{\hat{g}_t}{h_t} - \ln\frac{g_t^*}{h_t} = \ln\frac{\hat{g}_t}{g_t^*} = \ln\{1 + O_p[\frac{\partial g_t^*(\bar{\beta})}{g_t\partial\beta'}(\hat{\beta}_n - \beta^*)]\}, \tag{29}$$

where $g_t^* = g_t(\beta^*)$ and $\bar{\beta}$ is an intermediate point between $\hat{\beta}_n$ and β^*. In a similar manner to (14)-(15), by (29), $\hat{g}_t$ in (28) can be replaced asymptotically by g_t^*. Similarly, by the martingale central limit theorem, $\sqrt{n}\hat{\delta}_n$ is asymptotically normal with mean zero and variance $c/\tilde{\sigma}^2$:

$$\tilde{\sigma}^2 = \lim_{n\to\infty} ||\frac{1}{\sqrt{n}} M_0 \tilde{R}||^2 \text{ (in probability) } = \tag{30}$$

$$\left\{E\left[\ln^2\left(\frac{g_t^*}{h_t}\right)\right] - A'E^{-1}\left(\frac{1}{h_t^2}\frac{\partial h_t}{\partial \alpha}\frac{\partial h_t}{\partial \alpha'}\right)A\right\}^{-1},$$

where c is defined as in (16) and $A = E\left[(1/h_t)\ln\left(g_t^*/h_t\right)(\partial h_t/\partial\alpha)\right]$. The asymptotic variance can be estimated by

$$nc/||\hat{M}_0\hat{\tilde{R}}||^2,$$

where $\hat{M}_0$ and $\hat{\tilde{R}}$ are, respectively, M_0 and $\tilde{R}$ with all the parameters replaced by their corresponding estimates. Thus, we have the following theorem.

Theorem 5.1. The $t-$statistic for $\hat{\delta}_n$ generated by (23) is asymptotically distributed as $N(0,1)$ if H_0 is true.

Denote α^* as the limit of $\hat{\alpha}_n$ in probability under H_1 as n goes to infinity. Under H_1:

$$e_t = \frac{1}{\sqrt{2}}\left(\frac{\varepsilon_t^2}{h_t^*} - 1\right) = \frac{z_{1t}^2(g_t - h_t^*)}{2h_t^*} + \frac{1}{\sqrt{2}}(z_{1t}^2 - 1), \tag{31}$$

where $g_t = g_t(\beta)$ and $h_t^* = h_t(\alpha^*)$, with α^* being defined as in (21). Denote $\tilde{e} = (\tilde{e}_1, \cdots, \tilde{e}_n)$, with $\tilde{e}_t = (z_{1t}^2 - 1)/\sqrt{2}$. It follows that:

$$\sqrt{n}\hat{\delta}_n = \sqrt{n}\tilde{R}_1' M_{10} R_1/||M_{10}\tilde{R}_1||^2 + \sqrt{n}\tilde{R}_1' M_{10}\tilde{e}/||M_{10}\tilde{R}_1||^2 + o_p(1), \tag{32}$$

where $\tilde{R}_1$, M_{10} and R_1 are defined as $\tilde{R}$, M_0 and R with α and β^* replaced by α^* and β, respectively. Note that $\hat{\delta}_n$ does not converge to 1 as n goes to infinity since $\tilde{R}_1' M_{10} R_1/||\tilde{R}_1 M_{10}||$ is not equal to 1, so that the estimator of δ is not consistent under H_1. Let $N_\delta = ||M_{10}\tilde{R}_1/\sqrt{c}||\hat{\delta}_n$, i.e. the $t-$statistic given in Theorem 5.1. It follows that:

$$N_\delta = \tilde{R}_1' M_{10} R_1/(\sqrt{c}||M_{10}\tilde{R}_1||) + \tilde{R}_1' M_{10}\tilde{e}/(\sqrt{c}||M_{10}\tilde{R}_1||) + o_p(1). \tag{33}$$

By the martingale central limit theorem, $\tilde{R}_1' M_{10}\tilde{e}/(\sqrt{c}||M_{10}\tilde{R}_1||)$ is asymptotically distributed as $N(0,1)$. Thus, we have the following theorem.

Theorem 5.2. The $t-$statistic for $\hat{\delta}_n$ generated by (23) is asymptotically distributed as $N(\tilde{R}_1' M_{10} R_1/(\sqrt{c}||M_{10}R_1||), 1)$ under H_1. The power function is asymptotically given by:

$$\Phi(\zeta_\mu + |\tilde{R}_1' M_{10} R_1|/(\sqrt{c}||M_{10}\tilde{R}_1||)) + o(1),$$

where $\Phi(\cdot)$ and ζ_μ are defined as in Theorem 3.2.

Remark. From Theorem 5.2, the $t-$statistic based on (23) still has asymptotic power of unity under H_1 if $|\tilde{R}_1' M_{10} R_1| \neq 0$. However, $|\tilde{R}_1' M_{10} R_1|$ may be zero, and in this case, the power function of the $t-$statistic based on (23) will be asymptotically $N(0,1)$. It is expected that the $t-$statistic from (23) will not be robust. Since M_{10} is an orthogonal projection matrix, $|\tilde{R}_1' M_{10} R_1| \leq ||\tilde{R}_1' M_{10}|| \cdot ||M_{10} R_1||$. Thus, from Theorems 3.2 and 5.2, it follows that the $t-$statistic from (4) is more powerful than that from (23).[2]

Now consider the alternative weighting scheme given as H_{0L}'' in (24). In a similar manner to (28)-(30), we can show that $\sqrt{n}\hat{\delta}_n$ is asymptotically normal with mean zero and variance $c/\tilde{\tilde{\sigma}}^2$, under H_0:

$$\tilde{\tilde{\sigma}}^2 = \lim_{n\to\infty} ||\frac{1}{\sqrt{n}} M_0 \tilde{\tilde{R}}||^2 \text{ (in probability) } =$$

$$\left\{ 4E\left(\frac{h_t^{1/2} - g_t^{1/2}}{h_t^{1/2}}\right)^2 - A'E^{-1}\left(\frac{1}{h_t^2}\frac{\partial h_t}{\partial \alpha}\frac{\partial h_t}{\partial \alpha'}\right) A \right\}^{-1},$$

where M_0 and c are defined as in (16), $A = E\{[2(h_t^{1/2} - g_t^{1/2})/h_t^{1/2}](h_t^{-1}\partial h_t/\partial\alpha)\}$, and $\tilde{\tilde{R}} = (\tilde{\tilde{r}}_1, \cdots, \tilde{\tilde{r}}_n)'$, with $\tilde{\tilde{r}}_t = [\sqrt{2}(h_t^{1/2} - g_t^{1/2})/\sqrt{h_t}]$. The asymptotic variance can be estimated by

$$nc/||\hat{M}_0 \hat{\tilde{\tilde{R}}}||^2,$$

where $\hat{M}_0$ and $\hat{\tilde{\tilde{R}}}$ are, respectively, M_0 and $\tilde{\tilde{R}}$ with all the parameters replaced by their corresponding estimates. Thus, we have the following theorem.

Theorem 5.3. The $t-$statistic for $\hat{\delta}_n$ generated by (24) is asymptotically distributed as $N(0,1)$ if H_0 is true.

In a similar manner to (33), it can be shown that, under H_1:

$$\sqrt{n}\hat{\delta}_n = \sqrt{n}\tilde{\tilde{R}}_1' M_{10} R_1/||M_{10}\tilde{\tilde{R}}_1||^2 + \sqrt{n}\tilde{\tilde{R}}_1' M_{10}\tilde{e}/||M_{10}\tilde{\tilde{R}}_1||^2 + o_p(1), \quad (34)$$

[2] Here we exclude the case where $\hat{\alpha}_n$ from (4) and (5) has different limits under H_1.

where $\tilde{\tilde{R}}_1$, M_{10} and R_1 are defined as $\tilde{\tilde{R}}$, M_0 and R with α and β^* replaced by α^* and β, respectively, and α^* is defined as in (17). The estimator of δ is not consistent under H_1. Letting $N'_\delta = ||M_{10}\tilde{\tilde{R}}_1/\sqrt{c}||\hat{\delta}_n$, it follows that:

$$N'_\delta = \tilde{\tilde{R}}_1' M_{10} R_1/(\sqrt{c}||M_{10}\tilde{\tilde{R}}_1||) + \tilde{\tilde{R}}_1' M_{10}\tilde{e}/(\sqrt{c}||M_{10}\tilde{\tilde{R}}_1||) + o_p(1), \quad (35)$$

and $\tilde{\tilde{R}}_1' M_{10}\tilde{e}/(\sqrt{c}||M_{10}\tilde{\tilde{R}}_1||)$ is asymptotically distributed as $N(0,1)$. Thus, we have the following theorem.

Theorem 5.4. The t–statistic for $\hat{\delta}_n$ generated by (24) is asymptotically distributed as $N(\tilde{\tilde{R}}_1' M_{10}R_1/(\sqrt{c}||M_{10}R_1||), 1)$ under H_1. The power function is asymptotically given by:

$$\Phi(\zeta_\mu + |\tilde{\tilde{R}}_1' M_{10}R_1|/(\sqrt{c}||M_{10}\tilde{\tilde{R}}_1||)) + o(1),$$

where $\Phi(\cdot)$ and ζ_μ are defined as in Theorem 3.2.

Remark. In a similar manner to the weighting scheme in (23), the power function of the t–statistic from (24) is asymptotically $N(0,1)$ if $\tilde{\tilde{R}}_1' M_{10}R_1 = 0$, and 1 in probability if $\tilde{\tilde{R}}_1' M_{10}R_1 \neq 0$. Thus, it is also not robust as compared with the linear weighting scheme given in (4). Similarly, since $\tilde{R}_1' M_{10}R_1 \leq ||\tilde{R}_1' M_{10}|| \cdot ||M_{10}R_1||$, the t–statistic from (4) is more powerful than that from (24).

Consider the following more general weighting scheme:

$$H_{0L}^\lambda : \varepsilon_t = z_t f_{0t}^{1/2}, \quad f_{0t}^{1/\lambda} = (1-\delta)h_t^{1/\lambda} + \delta\hat{g}_t^{1/\lambda}, \quad (36)$$

where $\lambda \neq 0$. It is clear that, when $\lambda = 1$, (36) reduces to (4); as $\lambda \to \infty$, (36) reduces to (23); and when $\lambda = 2$, (36) reduces to (24). Thus, (36) will be referred to as the L_λ–family. If H_0 is true, then $\delta = 0$ in the L_λ–family, and if H_1 is true, then correspondingly $\delta = 1$.[3] In a similar manner to (28)-(30), we can show that, under H_0, $\sqrt{n}\hat{\delta}_n$ is asymptotically normal with mean zero and variance c/σ_λ^2:

$$\sigma_\lambda^2 = \lim_{n\to\infty} ||\frac{1}{\sqrt{n}} M_0 R_\lambda||^2 \text{ (in probability) } =$$

$$\left\{\lambda^2 E\left(\frac{h_t^{1/\lambda} - g_t^{1/\lambda}}{h_t^{1/\lambda}}\right)^2 - A_\lambda' E^{-1}\left(\frac{1}{h_t^2}\frac{\partial h_t}{\partial \alpha}\frac{\partial h_t}{\partial \alpha'}\right) A_\lambda\right\}^{-1},$$

[3] Although the conditional heteroskedasticity in (36) is free with respect to λ under H_0: $\delta = 0$, the power of the non-nested test will depend on the choice of λ when $\delta \neq 0$.

where M_0 and c are defined as in (16), $A_\lambda = E\{[\lambda(h_t^{1/\lambda} - g_t^{1/\lambda})/h_t^{1/\lambda}](h_t^{-1}\partial h_t/\partial\alpha)\}$, and $R_\lambda = (r_{\lambda 1}, \cdots, r_{\lambda n})'$, with $r_{\lambda t} = [\lambda(h_t^{1/\lambda} - g_t^{1/\lambda})/(\sqrt{2}h_t^{1/\lambda})]$. The asymptotic variance can be estimated by

$$nc/||\hat{M}_0\hat{R}_\lambda||^2,$$

where $\hat{M}_0$ and $\hat{R}_\lambda$ are, respectively, M_0 and R_λ with all the parameters replaced by their corresponding estimates.

In a similar manner to (32), it can be shown that, under H_1:

$$\sqrt{n}\hat{\delta}_n = \sqrt{n}R'_{\lambda 1}M_{10}R_1/||M_{10}R_{\lambda 1}||^2 + \sqrt{n}R'_{\lambda 1}M_{10}\tilde{e}/||M_{10}R_{\lambda 1}||^2 + o_p(1), \quad (37)$$

where $R_{\lambda 1}$, M_{10} and R_1 are defined as R_λ, M_0 and R with α and β^* replaced by α^* and β, respectively, and α^* is defined as in (17). Under H_1, the estimator of δ is not consistent unless $\lambda = 1$. Letting $N'_\delta = ||M_{10}R_{\lambda 1}/\sqrt{c}||\hat{\delta}_n$, it follows that:

$$N'_\delta = R'_{\lambda 1}M_{10}R_1/(\sqrt{c}||M_{10}R_{\lambda 1}||) + R'_{\lambda 1}M_{10}\tilde{e}/(\sqrt{c}||M_{10}R_{\lambda 1}||) + o_p(1), \quad (38)$$

where $R'_{\lambda 1}M_{10}\tilde{e}/(\sqrt{c}||M_{10}R_{\lambda 1}||)$ is asymptotically distributed as $N(0,1)$. Note that $|R'_{\lambda 1}M_{10}R_1|/(\sqrt{c}||M_{10}R_{\lambda 1}||) \leq ||M_{10}R_1/\sqrt{c}||$, which determines the non-nested test with maximum power in finite samples when $\lambda = 1$. The above results are given in the following theorem.

Theorem 5.5. (a) The $t-$statistic for $\hat{\delta}_n$ generated by (36) is asymptotically distributed as $N(0,1)$ if H_0 is true.

(b) The $t-$statistic for $\hat{\delta}_n$ generated by (36) is asymptotically distributed as $N(R'_{\lambda 1}M_{10}R_1/(\sqrt{c}||M_{10}R_1||), 1)$ under H_1. The power function is asymptotically given by:

$$\Phi(\zeta_\mu + |R'_{\lambda 1}M_{10}R_1|/(\sqrt{c}||M_{10}R_{\lambda 1}||)) + o(1),$$

where $\Phi(\cdot)$ and ζ_μ are defined as in Theorem 3.2.

(c) The test from (4) (that is, (36) with $\lambda = 1$) is the optimal non-nested test of H_0: $\delta = 0$ in the $L_\lambda-$family with respect to maximum power under H_1 in finite samples.

Remark: Note that $|R'_{\lambda 1}M_{10}R_1|$ may be zero unless $\lambda = 1$. In a similar manner to that given in the Remark for Theorem 5.4, the $t-$statistic for $\hat{\delta}_n$ from (36) may not be robust unless $\lambda = 1$. Moreover, it is clear that the $t-$statistic for $\hat{\delta}_n$ from (36) has asymptotic power of unity for all $\lambda \neq 0$ if $|R'_{\lambda 1}M_{10}R_1| \neq 0$. It should be noted that the optimal property of the non-nested test in the $L_\lambda-$family given in Theorem 5.5 relates to differences in finite samples. The optimal property of the non-nested test of H_1 from (5) can be obtained from a similarly defined $L_\lambda-$family.

6 Concluding Remarks

This paper has developed non-nested tests of the GARCH and E-GARCH models against each other. It was shown that the $t-$statistic based on a linear weighting of the competing conditional variances is asymptotically normal and has asymptotic power of unity. The corresponding power function was also derived. Two novel joint LM tests were developed for the ARCH and E-ARCH models against their nested and non-nested GARCH and E-GARCH counterparts, and the corresponding asymptotic distributions were established. In addition, the non-nested tests based on the weighting schemes in an $L_\lambda-$family were evaluated. The asymptotic distributions and power functions of the corresponding $t-$ statistics were presented. It was demonstrated that the $t-$statistic based on a linear weighting of the competing conditional variances is robust and yields the maximum power in the $L_\lambda-$family in finite samples. Thus, the $t-$statistic based on the linear weighting scheme is recommended for practical purposes.

Acknowledgements

The authors wish to acknowledge the helpful comments of seminar participants at the Catholic University of Leuven, Chinese University of Hong Kong, Curtin University of Technology, Edith Cowan University, Erasmus University Rotterdam, National University of Singapore, Niigata University, Osaka University, Tilburg University, Tohoku University, University of Amsterdam, University of Melbourne, University of Western Australia and Yokohama National University, and the financial support of the Australian Research Council. An earlier version of the paper was presented at the Kansai Econometrics Conference, Osaka, November 1998, and at the International Workshop on Statistics in Finance, Hong Kong, July 1999.

References

1. A.K. Bera and M. McAleer, Nested and non-nested procedures for testing linear and log-linear regression models. *Sankhya B* **51**, 212–224 (1989).
2. T. Bollerslev, Generalized autoregressive conditional heteroskedasticity. *Journal of Econometrics* **31**, 307–327 (1986).
3. T. Bollerslev, R.Y. Chou and K.F. Kroner, ARCH modelling in finance. *Journal of Econometrics* **52**, 5–59 (1992).
4. T. Bollerslev, R.F. Engle and D.B. Nelson, ARCH models, in R.F. Engle and D. McFadden (eds.). *Handbook of Econometrics*, Vol. 4, pp. 2959-

3038. (North-Holland, Amsterdam, 1994).

5. T. Bollerslev, R.F. Engle and J.M. Woodridge, A capital asset pricing model with time varying covariance. *Journal of Political Economy* **96**, 116–131 (1988).
6. T. Bollerslev and H.O. Mikkelsen, Modelling and pricing long memory in stock market volatility. *Journal of Econometrics* **73**, 151–184 (1996).
7. T. Bollerslev and J.M. Woodridge, Quasi-maximum likelihood estimation and inference in dynamic models with time-varying covariances. *Econometric Reviews* **11**, 143–173 (1992).
8. R. Davidson and J.G. MacKinnon, Several tests for model specification in the presence of alternative hypotheses. *Econometrica* **49**, 781–793 (1981).
9. R.F. Engle, Autoregressive conditional heteroscedasticity with estimates of the variance of United Kingdom inflation. *Econometrica* **50**, 987–1007 (1982).
10. R.F. Engle and T. Bollerslev, Modelling the persistence of conditional variance. *Econometric Reviews* **5**, 1–50 (1986a).
11. R.F. Engle and T. Bollerslev, Modelling the persistence of conditional variance: reply. *Econometric Reviews* **5**, 81–88 (1986b).
12. P. Hall and C.C. Heyde, *Martingale Limit Theory and Its Applications.* (Academic Press, New York, 1980).
13. S.-W. Lee and B.E. Hansen, Asymptotic theory for the GARCH (1,1) quasi-maximum likelihood estimator. *Econometric Theory* **10**, 29–52 (1994).
14. C.W. Li and W.K. Li, On a double threshold autoregressive with heteroskedasticity time series model. *Journal of Applied Econometrics* **11**, 253–274 (1996).
15. W.K. Li, S. Ling and M. McAleer, A survey of recent theoretical results for time series models with GARCH errors (submitted) (1999).
16. S. Ling and M. McAleer, Asymptotic theory for a new vector ARMA-GARCH model (submitted) (1999).
17. S. Ling and W.K. Li, On fractionally integrated autoregressive moving-average time series models with conditional heteroskedasticity. *Journal of the American Statistical Association* **92**, 1184-1194 (1997).
18. R.L. Lumsdaine, Consistency and asymptotic normality of quasi-maximum likelihood estimator in IGARCH (1,1) and covariance stationary GARCH(1,1) models. *Econometrica* **64**, 575–596 (1996).
19. J.G. MacKinnon, H. White and R. Davidson, Tests for model specification in the presence of alternative hypotheses: some further results. *Journal of Econometrics* **21**, 53–70 (1983).

20. M. McAleer, The significance of testing empirical non-nested models. *Journal of Econometrics* **67**, 149–171 (1995).
21. D.B. Nelson, Conditional heteroskedasticity in asset returns: a new approach. *Econometrica* **59**, 347–370 (1989).
22. R.S. Tsay, Conditional heteroskedastic time series model. *Journal of the American Statistical Association* **81(7)**, 590-604 (1987).
23. A.A. Weiss, ARMA models with ARCH errors. *Journal of Time Series Analysis* **5**, 129–143 (1984).
24. A.A. Weiss, Asymptotic theory for ARCH models: estimation and testing. *Econometric Theory* **2**, 107–131 (1986).
25. H. White, *Estimation, Inference, and Specification Analysis.* (Cambridge University Press, New York, 1994).
26. J.-M. Zakoian, Threshold heteroskedastic model. *Journal of Economic Dynamics and Control*, **18**, 931-955 (1994).

Part IV: Forecasting

INTERVAL PREDICTION OF FINANCIAL TIME SERIES

B. CHENG
Institute of Applied Mathematics,
The Chinese Academy of Science,
Beijing, 100008, China
E-mail: chb@mail.musoft.com

H. TONG
Department of Statistics & Actuarial Science
University of Hong Kong, Hong Kong
E-mail: htong@hku.hk

In this paper, we introduce a percentile-based method to predict return distribution of financial time series and provide a way to calculate value at risk under non-normal portfolio changes.

1 Statistical characteristics of changes of market factors and portfolio value

1.1 Non-normality and quasi-stable volatility

A long held assumption is that stock market data are normally distributed.

Table 1 summarizes the moments of the distributions of daily returns for some equity data from the London Stock Exchange. The standard deviations taken across the returns for different stocks are reasonably similar, with the exception of the Mirror Group, which has an estimated standard deviation nearly two and a half times that for BT, and is considerably larger than any of the other standard deviations.

The return distributions tend to be very leptokurtic and positively skewed, but some are considerably more so than the others. See, for example, the return for the Mirror Group (Figure 1.1). In this case, the shares were suspended during December 1991 following the death of Robert Maxwell. They were then requoted on 17th July 1992 and suffered a 57.8% loss in value on that day!

Normality is therefore a questionable assumption. The following statement is cited from Candace F. Daly (*Asia Risk* journal, December 1997, page 41). "A study in April 1997 showed the standard deviation of US dollar/Thai baht daily spot moves was 0.12%, compared with the typical larger US dollar/Deutschmark daily spot moves of 0.55%. However, the daily percentage change in US dollar/Thai baht spot was greater than two standard deviations (0.24%) on 11 days out of 22 trading days in May 1997. With all the usual

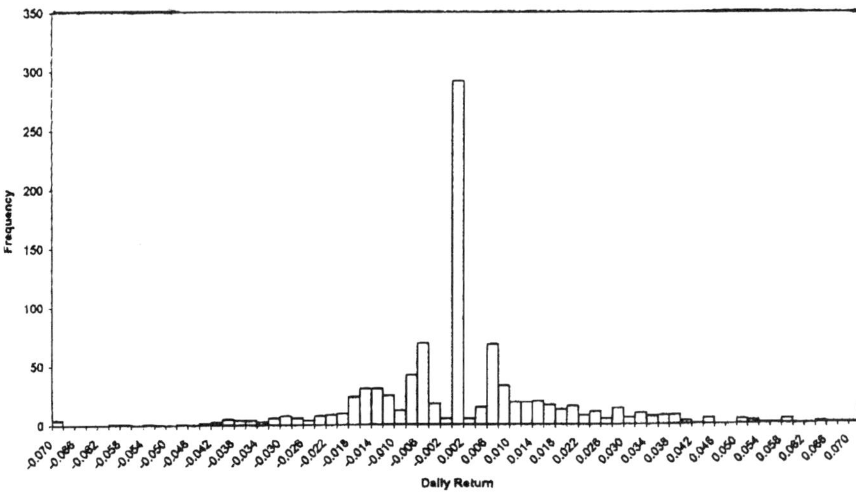

Figure 1.1 Return distribution of daily Mirror Croup

assumptions about normality, one would expect this to happen on only one day out of 20. This large number of extreme moves during May gave a strong indication that Thailand's exchange rate was under attack due to the Market sentiment that the currency was overvalued." See Table 2 for details.

As would be expected, the frequency distribution of market factor changes over a long period reflects some very large market movements. This feature is illustrated in Figure 1.2, which shows the distribution of monthly changes in 3-month Libor for the period from 1973 to 1991. In the 1980s, the Federal Reserve Bank attempted to reduce inflation by abandoning its long-standing policy of maintaining stable interest rates. Instead they increased short-term interest rates significantly, leading to large movements in interest rates in the 1980s. The plot reflected the significant increase in market volatility.

Also evident in the long-term distribution in Figure 1.2 are two distinctly different regions: the bell-shaped portion and the outliers. The bell-shaped portion includes the majority of events and describes the so-called business-as-usual type of market volatility. This type is associated with a stable relationship between the magnitude of market changes and their frequency. It is the **quasi-stability** of this type of market volatility in the short run which enables us to forecast future distributions and to perform risk measurement and capital allocation.

The outliers, on the other hand, usually represent structural disruptions of

Table 1: Moment of return distribution

Stock	SD of return	Skewness	Kurtosis
Barclays	1.6%	0.48	9.83
BT	1.2%	0.048	0.33
Glaxo	1.6%	0.09	1.20
ICI	1.3%	0.52	2.61
Next	1.5%	0.47	1.91
Argos	1.3%	1.15	9.76
Courtaulds	1.6%	0.35	5.47
Delta	0.012	-0.26	7.26
Hardy O&G	1.6%	2.03	23.02
Mirror Gp	2.8%	-9.09	193.98

Table 2: Volatility study on Thai Baht Markets

	Thai baht spot	1 week implied rate	1 month implied rate
May 13, 1996 - April 3, 1997			
Volatility (1 standard deviation)	0.12%	1.1%	0.5%
May 1, 1997 - May 30, 1997			
Highest price/yield	26.12%	168%	62.3%
Lowest price/yield	25.20%	7.8%	8.5%
Maximum positive change	1.80%	149.8%	47.7%
Maximum negative change	-0.84%	-122.4%	-38.2%
Largest move (number of SD)	15	139	103
number of days $\geq$ 2 SD	11	22	21

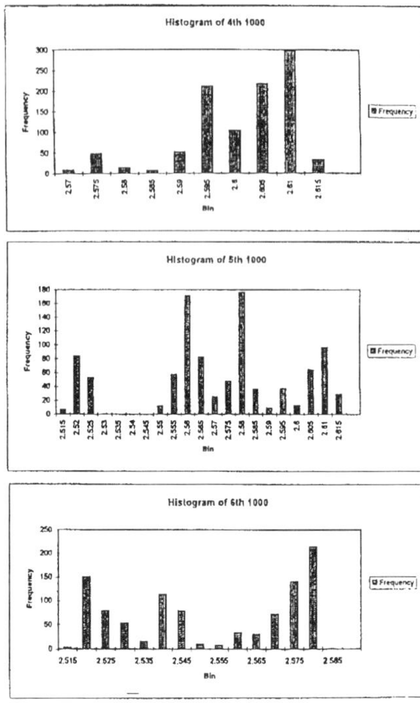

Figure 1.2 three-month Libor distribution on monthly changes.

the markets. It seems that no clear relationship could be established between the magnitudes of outlying events and their probabilities.

We need to find ways to handle both *business-as-usual risk* and *catastrophic risks.*

1.2 Non-stationarity

Each business day produces a new distribution of market rates, giving rise to a dynamic of distributions over the time horizon. As an illustration, we use the time series of tick bid rates of the Deutschmark against the Sterling over a period of about 3 months. In total there are 40,000 observations. However, 9 exceedingly large observations have been removed by using a data cleaning method. The tick series are divided into non-overlapping groups of 1000 observations. Due to the removal of the 9 irregular data, several groups contain only 999 or 998 observations. The 40 groups are labelled as *1st 1000*, *2nd 1000* and so on. Histograms for the groups are given in Figures 1.3 - 1.4. We can see that they are different from one another.

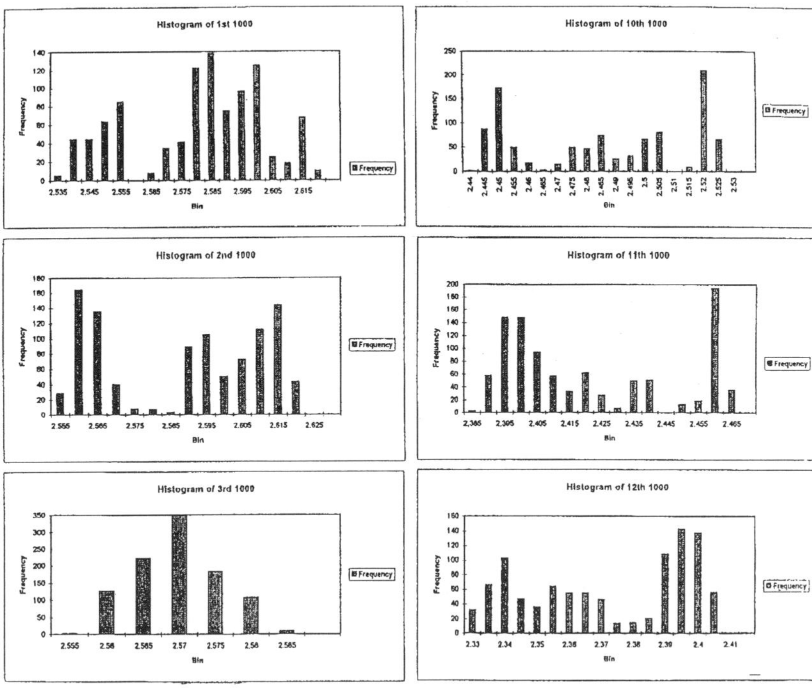

Figures 1.3 and 1.4 about here: Histograms of tick bid rates of the Deutschmark against the Sterling.

2 Term structure of volatility and correlation and the mean-reverting property

2.1 Term structure of market volatility

Volatility measures the intensity of random or unpredictable changes in a market value. We commonly visualize it by plotting return against time and watching the fluctuation of the amplitude of the return over time. The episodes of high and low volatility are often called "volatility clusters". These clusters show the possibility of forecasting volatility because high-volatility periods tend to persist for some time but eventually decay to periods of low volatility.

Thus we build a volatility model to describe the typical historical pattern of volatility and to forecast future episodes. Volatility over time can be thought of as a stochastic process, which we seek to uncover.

Historical data tend to reveal that the duration of volatility clusters can vary between several hours and a decade. These differences are commonly seen to be driven by different economic processes. The primary source of changes in market prices is *news* about the fundamental value of the related asset. With the news arriving in bunches, the volatility of returns tends to cluster. High frequency volatility is often associated with *noise*, the most likely sources of which are the pressures and turbulence induced through trading. Lower frequency volatility is most likely due to macro-economic and institutional changes.

A common definition of volatility at time t over time horizon d is given by the standard deviation, σ_t^d, of return series over a delay of d units; The unit could be tick, minute, hour, day, week, month or year. (A precise mathematical definition of σ_t^d is given later.) The plot of volatility σ_t^d against d is called the **term structure of volatility**.

Most popular risk models make the assumption that volatility forecasts follow the so-called *square-root-of-time rule.* Under this rule, if returns are calculated over, say 10 days, and daily prices are recorded, then a 10-day return has a mean value 10 times the daily mean and its variance is 10 times the daily variance. Similarly, if the current estimate of the monthly variance of returns of the US dollar-Deutschemark exchange rate is 9%, then the variance over next year (i.e. the next 12 months), or the annualized variance, is 12 times 9%, or 108%. In general, the volatility forecast over the next T periods is simply $\sqrt{T}$ times the unit period (*e.g.* day) volatility. Figure 2.1 is the term structure of volatility following the square root of time rule.

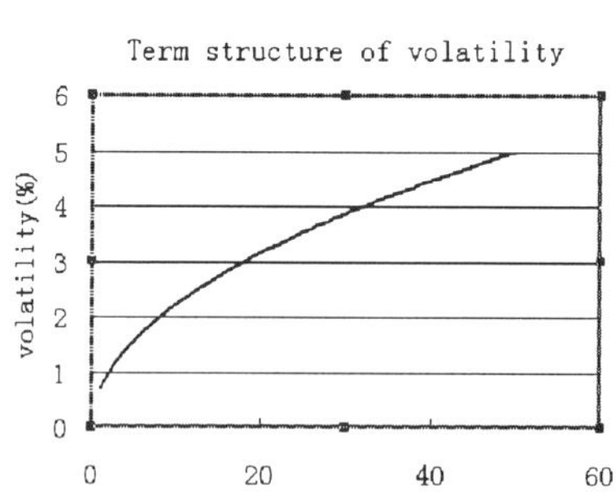

Figure 2.1 Term structure of square-root volatility.

Since the volatility forecasts 'grow' with the square root of time, there is no limit on their size. In other words, the volatility forecasts do not converge to a constant long-run value. In this sense, there is no **mean-reverting**. However, in practice, the term structures are typically found to be mean-reverting. This phenomenon implies that as the horizon broadens, the limit of the forecast is a constant and does not depend on the current information. Specifically

$$\sigma_{t,\ d} \rightarrow_{d \rightarrow \infty} \sigma. \tag{2.1}$$

Figure 2.2 is a plot of the term structure of daily volatility of the UK's FTSE100 index data.

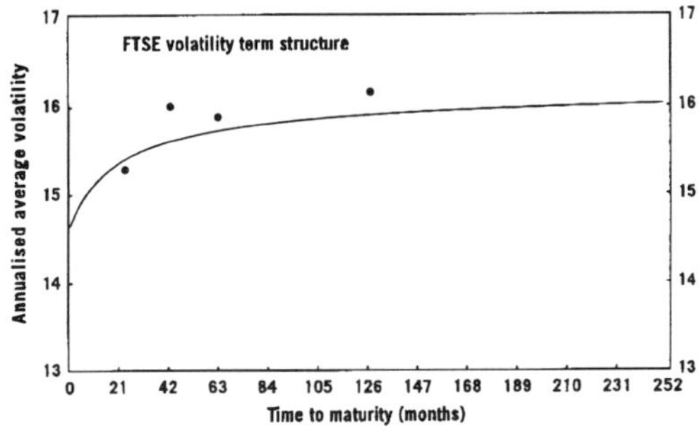

Figure 2.2

A few comments are in order.

1. Risk modelling over the d horizon has to be dealt with individually.
2. Shapes and patterns of the movements in the term structure are useful for hedging volatility exposure.
3. Similar to volatility, correlation also has a term structure.

2.2 A temporal covariance matrix method for the derivation of the term structure of volatility and correlation

Let P_t be a market time series and d be a positive integer. The return of d-th delay is defined by

$$R_t^d = \ln\left\{\frac{P_t}{P_{t-d}}\right\}. \tag{2.2}$$

Then it is easy to see $R_t^d = \sum_{i=1}^{d} \ln\left\{\frac{P_{t-i+1}}{P_{t-i}}\right\} = \sum_{i=1}^{d} R_{t-i}^1$.

Therefore the variance of R_t^d is given by

$$\mathrm{Var}(R_t^d) \;=\; \sum_{i=1}^{d}\sum_{j=1}^{d} \mathrm{Cov}(R_{t-i}^1, R_{t-j}^1). \tag{2.3}$$

In particular, if the return series $\{R_t^1\}$ is independent, which is true under the assumption of the efficient market hypothesis, then $\mathrm{Cov}(R_{t-i}^1, R_{t-j}^1) = 0$, so that

$$\mathrm{Var}(R_t^d) \;= d \times \mathrm{Var}(R_t^1), \tag{2.4}$$

which is the so-called squared-root-of-time rule.

Define a variance-covariance matrix Σ_d by $\Sigma_d \;=\; (\sigma_{i,j})_{d\times d}$ with $\sigma_{i,j} = cov(R_{t-i}^1, R_{t-j}^1)$, and a unit vector $I_d = (1, \ldots, 1)$. Hence

$$\sigma_t^d \;=\; \sqrt{\mathrm{Var}(R_t^d)} \;=\; \sqrt{I_d' \Sigma_d I_d}. \tag{2.5}$$

In order to produce a term structure of volatility by $\{(d,\ \sigma_t^d)\}_{d=1}^{D}$, all we need to do is to calculate the matrix Σ_D once. Furthermore by examining values of the elements of the matrix, we can see how the market rate links itself backwards and forwards over the D horizon.

Similarly, to get a term structure of correlation between two market rates A and B, we simply replace $\mathrm{Cov}(R_{t-i}^1, R_{t-j}^1)$ by $\mathrm{Cov}(R_{A,t-i}^1, R_{B,t-j}^1)$.

3 A universal interval prediction model for portfolio market value

We agree with the view that the risk of a position can be measured if a relationship can be established between all possible future losses (and gains) of the position and their likelihood over the holding period, i.e. the distribution of changes in its market values, or equivalently the risk profile of the position. In addition to risk measurement, the distribution can be used to assess various aspects of the position's future performance, including the likelihood and magnitude of large gains and amount of expected change in market value over the holding period. An accurate assessment of the distributions is the primary task of interval prediction models.

3.1 A percentile-based functional distribution model

A percentile is the real number which divides the data under the probability density function of a random variable, say X, into two parts of specified amounts. For $0 < p < 1$, the pth (or $100p\%$) percentile $\xi(p)$ is defined as

$$\text{Prob}(X < \xi(p)) \leq p \textit{ and } \text{Prob}(X > \xi(p)) \leq 1 - p. \tag{3.1}$$

The estimation of percentiles is related to order statistics. Let $X_1, \cdots, X_n$ be an independent sample from the distribution F of X and order them ascendingly to give the order statistics $X_{(1)} \leq \cdots \leq X_{(n)}$. Then the estimator of $\xi(p)$ is given by

$$\hat{\xi}(p) = X_{([np])} + (n+1)\left\{p - \frac{[np]}{n+1}\right\}(X_{([np]+1)} - X_{([np])}), \tag{3.2}$$

where $[a]$ is the largest integer which is not bigger than a. We have a limiting distribution for $\hat{\xi}(p)$ given below.

Theorem 1 Let f be the density function of F and continuous at $\xi(p)$, then

$$\sqrt{n}\{\hat{\xi}(p) - \xi(p)\} \rightarrow^{d}_{n\rightarrow\infty} N\left(0, \frac{p(1-p)}{f^2(\xi(p))}\right). \tag{3.3}$$

Proof: See [1].

Suppose that $\xi_A(p)$ and $\xi_B(p)$ are the p-th percentiles of the distributions F_A and F_B, respectively. Varying the probability p over a regular grid from 0 to 1, say $0 < p_1 < \ldots < p_M < 1$, produces two seqences of percentiles $\xi_A(p_i)$ and $\xi_B(p_i)$, $i = 1, \ldots, M$. For example, if we are interested in the low bound prediction with 5 percent, simply take $M = 100/5 = 20$.

Consider a linear regresion by

$$\xi_B(p_i) = \alpha - \beta\xi_A(p_i) + \epsilon_i, \tag{3.4}$$

where ϵ_i is a 'residual' variable with a zero mean and an unknown variance σ^2.

Obviously, when two distributions are identical, the plot, called the *percentile plot*, of ξ_B against ξ_A will resemble a straight line with slope 1, passing through the origin. By looking at various transformations of this basic straight line, we gain some basic insights into how the model works.

Case 1: Suppose that this line is shifted upwards by an amount c. This effectively means that for each ξ_A, the corresponding value for ξ_B is larger by the amount c. It then follows that the distribution F_B has been shifted a distance c to the right.

Case 2: Suppose that this line is shifted downwards by an amount c. This effectively means that for each ξ_A, the corresponding value for ξ_B is larger by the amount c. It then follows that the distribution F_B has been shifted a distance c to the left.

Case 3: Suppose that there is no vertical displacement but the slope of the line is increased C times, i.e. the line is now steeper. Graphically the distribution F_B is stretched where the degree of stretching depends on the value of C. However, the stretching need not be symmetrical. In fact, symmetrical stretching only occurs when the percentiles are evenly spread on both sides of zero. If there are more positive precentiles than negative ones, then the distribution is stretched more to the right than the left and vice-versa.
Case 4: Suppose that there is no vertical displacement but the slope of the line is reduced by a factor of c, *i.e.* the line is less steep. The percentiles of F_B are now c times smaller. The converse of stretching has occurred. That is the distribution F_B is now squeezed c times. This means that the range of the distribution F_B is c times smaller than that of F_A and has a high peak. Once again the squeezing need not be symmetric except in the case when there are equal numbers of positive and negative percentiles.

Some theoretical results can be established.

Theorem 2: A linear transformation on the percentiles of a Normal distribution yields the percentiles of another distribution within the class of Normal distributions, with a variance that is a multiple of the original variance and a mean that is a linear function of the original mean.

Given percentile sequences $\{\xi_A(p_i)\}_{i=1}^M$ and $\{\xi_B(p_i)\}_{i=1}^M$, we can estimate unknown parameters α and β by the least-squares:

$$\hat{\beta} = \frac{\sum_{i=1}^M(\xi_A(p_i) - \bar{\xi}_A)\xi_B(p_i)}{\sum_{i=1}^M(\xi_A(p_i) - \bar{\xi}_A)^2} \tag{3.5}$$

and

$$\hat{\alpha} = \bar{\xi}_B - \hat{\beta}\bar{\xi}_A, \tag{3.6}$$

where $\bar{\xi}_A = \frac{1}{M}\sum_{i=1}^M \xi_A(p_i)$ and $\bar{\xi}_B = \frac{1}{M}\sum_{i=1}^M \xi_B(p_i)$.

Figure 3.1 gives percentile plot of model (3.4) for the four cases and Figures 3.2a and 3.2b give the corresponding changes of the distribution. Figure 3.3 gives the precentile plots for the patches 1 - 7 of 1000 tick US dollar to Deutschemark exchange rates.

3.2 Forecasting the distribution of changes of portfolio market values.

Let $\{P_t\}_{i=1}^T$ be the market values of portfolio P with instruments $I_1, \cdots I_L$ and d be a positive integer. See Figure 3.5 for example. Define the return of market

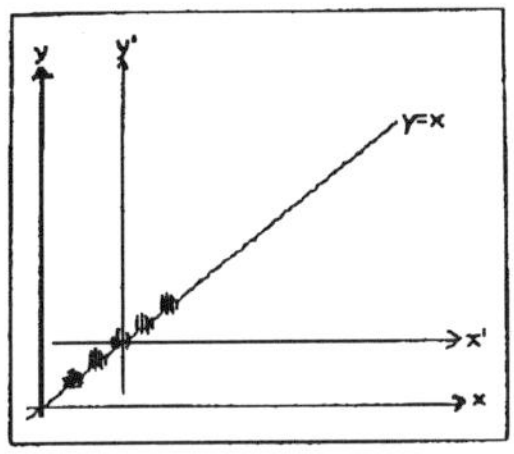

Diagram 1

The original axis, y and x, have been shifted to new positions, y' and x' respectively. Hence, there are now an equal number of positive and negative percentiles, represented by the striped circles.

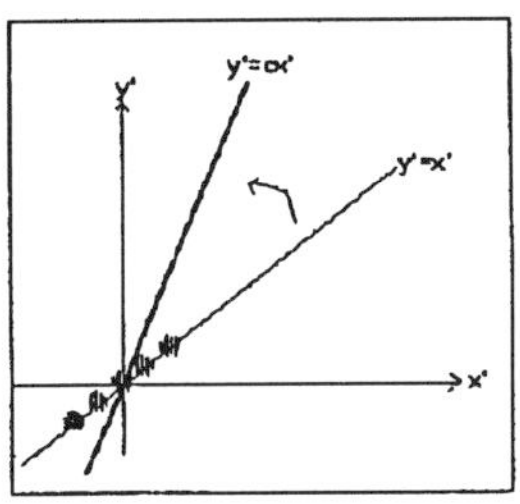

Diagram 2

The line, y' = x', which has a slope of one and a zero intercept, is now rotated so that it's slope is increased c times. Note that the percentiles will be shifted vertically so that they lie on the line y' = cx'.

Figure 3.1: Four cases of the percentile plot.

value over time horizon d by

$$R_t^d = \ln\left(\frac{P_t}{P_{t-d}}\right).$$

Let $\xi_t^d(p)$ be p-th (or $100p\%$) percentile for R_t^d. The percentile $\xi_t^d(p)$ is unknown. There are two ways to estimate $\xi_t^d(p)$: (1) historic simulation and (2) Monte-Carlo. In this paper, we use a hybrid method. Based on historic observations $R_{t-\Delta}^d, \cdots, R_t^d$, we generate scenarios $\{\tilde{R}_s^d\}$ of portfolio changes by

$$\tilde{R}_s^d \;=\; Mean(R_s^d) \;+\; \sqrt{Var(R_s^d)} \;\times\; \epsilon_s, \tag{3.7}$$

where

$$\begin{aligned} Mean(R_s^d) \;&=\; (1.0-\lambda_1)\times R_s^d \;+\; \lambda_1 * Mean(R_{s-1}^d), \\ Var(R_s^d) \;&=\; (1.0-\lambda_2)\times (R_s^d - Mean(R_s^d))^2 \;+\; \lambda_2 * Var(R_{s-1}^d), \end{aligned}$$

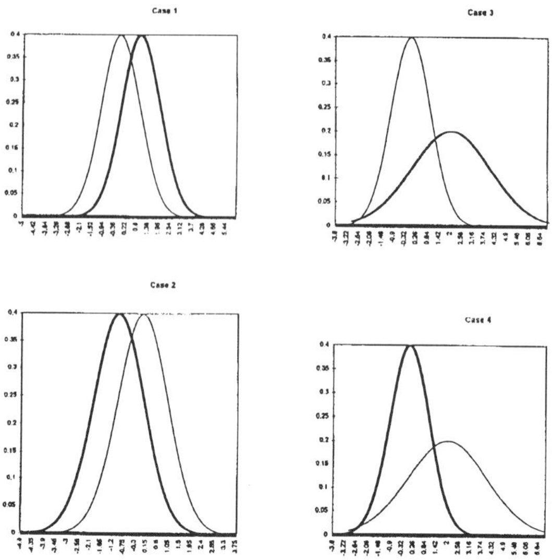

Figures 3.2a, 3.2b: The corresponding changes of the distribution for the four cases.

$0 < \lambda_1, \lambda_2 < 1$ and the random variable $\epsilon_s \sim GEV(\rho)$. Here, $GEV(\xi)$ is a truncated version (to unit variance) of the generalized extreme value distribution defined by Frechet, Gumbel and Weibull:

$$P(\epsilon_s \leq x) \;=\; exp[-(1+\rho x)^{-1/\rho}].$$

In this paper, we simply take $\lambda_1 = \lambda_2 = 0.85$ and $\rho = 1$.

Therefore, based on the scenarios $\{\tilde{R}_s^d\}$, we can estimate $\xi_t^d(p)$ by using formulae (3.2). The estimate is denoted by $\hat{\xi}_t^d(p)$.

For time horizon d, and given $s = \Delta, \Delta + d, \Delta + 2d, \cdots T - d, T$, consider the percentile regression model

$$\hat{\xi}_{s+d}^d(p_i) \;=\; \alpha_s \;+\; \beta_s \hat{\xi}_s^d(p_i) \;+\; \epsilon_i. \tag{3.8}$$

Since the data are available up to and including time T, we can estimate $(\alpha_s,\ \beta_s)$ up to and including $s = T - d$ by the least squares estimators (3.5) and (3.6). Then we can predict $\xi_{T+d}^d(p)$ by

$$\hat{\xi}_{T+d}^d(p) \;=\; \hat{\alpha}_{T-d} \;+\; \hat{\beta}_{T-d}\hat{\xi}_T^d(p). \tag{3.9}$$

Procedure: Given the confidence level $(1-q)100\%$, *say* $q = 0.05$ or $q = 0.25$,

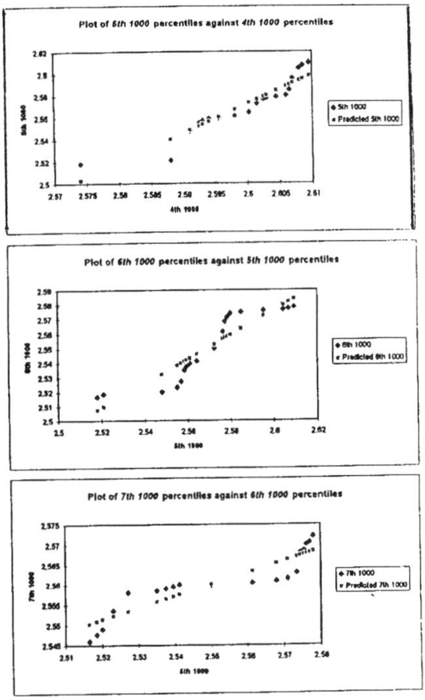

Figure 3.3: The percentile plot of the FX tick data (US dollar to Deutschmark).

1. for the portfolio P over time horizon d at time t we denote

$$-\min\{\hat{\xi}^d_{t+d}(q), 0\} \tag{3.10}$$

by $\text{VAR}_{q,d}$ and call it the value at risk for brevity;

2. for the portfolio P over time horizon d at time t we denote

$$\max\{\hat{\xi}^d_{t+d}(1-q), 0\} \tag{3.11}$$

by $\text{VAB}_{q,d}$ and call it value at best for brevity.

In practice, we may sometimes treat $(\text{VAR}_{q,d}, \text{VAB}_{q,d})$ if it were an interval predictor for the porfolio P.

3.3 The selection of look-back block size

One of the most important factors in the accurate measurement of interval prediction is the selection of the length, Δ, of the historical time series of market factors to be used in risk modelling. It is also called the look-back block size.

In selecting this parameter, we must consider two conflicting requirements. The first requirement is related to the fact that we are using statistics as a tool to measure risk. Thus, a fourfold increase in the sample size will double the accuracy of statistical calculation, *provided the changes in the market factors are stationary.*

The second requirement is related to the fact that the stochastic nature of a financial time series (*i.e.* market volatility) does change with time.

Empirical analyses of market volatility suggest that volatility clusters, yielding extended periods of relatively "stable" volatility. The whole process of risk measurement is based on this relative stability of volatility of the current cluster; it can be used to forecast the volatility of the market, in other words the nature of the risk.

For this purpose, the optimal look-back block size is given by the length of the current volatility cluster. As a result, this approach effectively relates the look-back block size to the nature of current market volatility. If a certain level of volatility has been stable for a long period (a large volatility cluster), the look-back block size should be increased as this will increase the accuracy of statistical calculations. On the other hand, any significant change in the nature of market volatility would indicate that it is prudent to reduce the look-back block size.

In practice, however, the current volatility cluster is not always easy to identify. Another difficulty is that clusters for different markets can be of different duration. In the following we describe a way to determine Δ automatically.

The idea is to count the number of times that the interval prediction (with the confidence level fixed) underpredicts future losses. We use the term 'violation' to refer to the case where an interval prediction underpredicts a future price move. Specifically, let p be a fixed real number between 0 and 1. Let Num_p be the number of $P\&L$ series that violate its interval prediction bound, *i.e.*,

$$\mathrm{Num}_p = \#\{R^d_{s+d} < \hat{\xi}^d_{s+d}(p) \mid s = T-\Delta, T-\Delta+d, \cdots, T-2d, T-d\}. \quad (3.12)$$

Define the violation ratio by

$$\text{Ratio}(\Delta) = \frac{\text{Num}_p}{\#\{T-\Delta, T-\Delta+d, \cdots, T-2d, T-d\}}. \tag{3.13}$$

Choose that look-back block size Δ which minimises

$$\min_{\Delta} \{|\text{Ratio}(\Delta) - p|\}. \tag{3.14}$$

In the following example, we first use the daily UK's FTSE100 close index to calculate the mean and variance series $\{Mean(R_s^d)\}$ and $\{Var(R_s^d)\}$ which we set initial mean = 0 and variance = 0. Then we generate daily return series by 100 replications, so that we can calculate the value at risk at the 99% confidence level for the return series from the UK's FTSE100 index data over one-day horizon (in Figure 3.4) and over five-day horizon (in Figure 3.5). From Figure 3.4, we can see that the violation ratio is 1.8%, which compares reasonably with the nominal level of 1%, but increases to 3.2% in Figure 3.5. This indicates that prediction accuracy decreases as the time horizon increases.

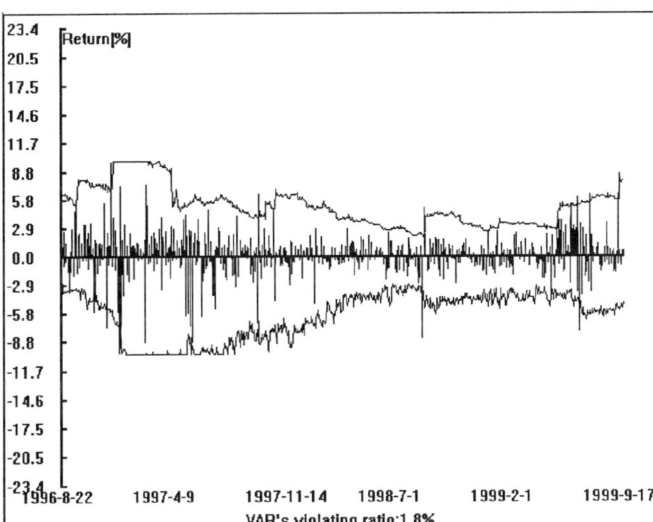

Figure 3.4

References

1. X.R. Chen (1992) *Nonparametric Statistics* (in Chinese). The Eastern University Press, Shanghai, China.

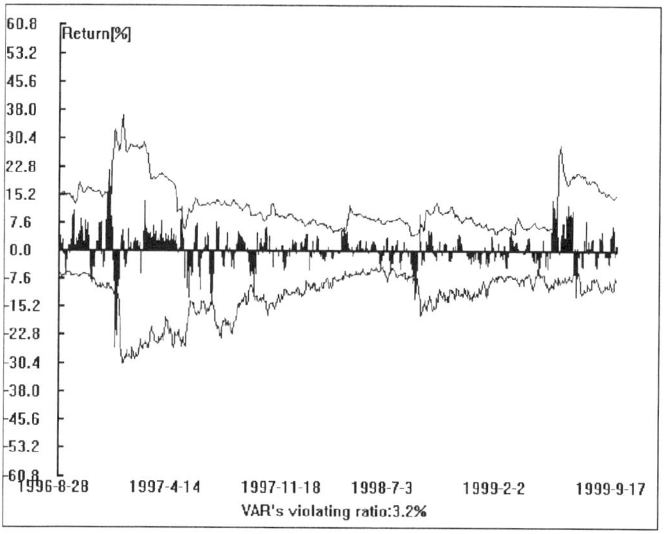

Figure 3.5

2. Engle, R. F. (1982). Autoregressive conditional heteroscedasticity with estimates of the variance of UK inflation *Econometrica* **50** pp. 987-1007.
3. Muller, U.A. *et al* (1996). Heavy tails in high-frequency finanical data. Tech. Report, Olsen & Associates.

A DECISION THEORETIC APPROACH TO FORECAST EVALUATION

C.W.J. GRANGER
University of California at San Diego

M. HASHEM PESARAN
Trinity College, Cambridge

This paper addresses the problem of forecast evaluation in the context of a simple but realistic decision problem, and proposes a procedure, for the evaluation of forecasts based on their average realized value to the decision maker. It is shown that by concentrating on probability forecasts stronger theoretical results can be achieved than if just event forecasts were used. A possible generalisation is considered concerning the use of the correct, conditional predictive density function when forming forecasts.

1 Introduction

In econometric applications forecasts are generally presented as point or interval forecasts. But focusing on point forecasts is justified only when the underlying decision problems are linear in constraints and quadratic in the loss function. However, in most decision making problems where the loss function is asymmetric and/or the constraints are non-linear point forecasts will not be sufficient and probability forecasts will be needed.[1] This paper argues in favor of probability forecasting and a closer integration of forecast evaluation and the decision making process.[2]

To illustrate the type of decision/forecasts problem that concerns us in this paper, suppose that one is considering making a forecast of the variable X_t at time $t-1$, having available an information set, Ω_{t-1} for use in a particular decision problem. It is usual to concentrate on the forecast of the mean of X_t conditional on Ω_{t-1}, or of some other function of X_t such as another measure of location. With each forecast there will be linked a value or cost function, as making a forecast error will cause a cost to some decision maker, but the link

[1] The use of probability forecasts in macroeconometric applications have been emphasized, for example, by Fair (1993). More recently, the Bank of England, also routinely publishes a range of inflation forecasts, see Britton, Fisher and Whitley (1998). Interval forecasts have also been suggested by some investigators. See, or example, Chatfield (1993). But the forecast uncertainty characterized by means of interval forecasts is only indirectly informative in context of decision making. Evaluation of interval forecasts also present new difficulties. See Christoffersen (1998).

[2] See also the companion paper, Granger and Pesaran (1999).

between the forecasts and the decisions is usually left vague. In this paper, a specific forecast/decision problem, of a simple but realistic nature is considered, with a general cost or value function. Some earlier literature discussing evaluation of forecasts within a decision framework, is Murphy and Winkler (1987), Ehrendorfer and Murphy (1988), and Katz and Murphy (1990). Optimum forecasts are considered both at any given moment of time and also by considering averaged realized values over time. A simple procedure for comparison of forecasts, based on the average realized value of the forecasts to the decision maker, is proposed.

2 The Simple Model

Consider a situation in which there are two "states" of the world, which for ease will be called "bad" and "good". Examples would be "freezing" or not, "high winds" or not, and "high inflation" or not. Suppose also that a sequence of forecasts are made on day $t-1$ of the events to occur on day t. Let $\widehat{\pi}_t$ be the forecast that the bad event will occur on day t (the correct notation should be ${}_t\widehat{\pi}_{t-1}$ or $\widehat{\pi}_{t-1,1}$ but $\widehat{\pi}_t$ is used for convenience). Thus the forecast of the good event is $1-\widehat{\pi}_t$. Note that these are not point forecasts or even an interval forecast but that the forecast of the whole distribution is given for all possible outcomes, in this very simple situation comprising just two possibilities. (Later more than two possibilities (states) are considered). Given the probability forecast, $\widehat{\pi}_t$, a decision maker will then decide whether to take action, by comparing the expected benefit of taking the action with its cost. Let the values of the activities and the cost of taking action be as shown in the following matrix

		State	
		Bad	Good
Action	Yes	$Y_{11}-C$	$Y_{12}-C$
	No	Y_{21}	Y_{22}

where Y_{11} is the value in the bad state if preventative action is taken, Y_{21} is the value of the activity in the bad state when no preventative action is taken (clearly $Y_{11} > Y_{21}$) and Y_{12}, Y_{22} are the values of the activities in the good state, irrespective whether action is taken or not, and C is the cost of taking the preventative action. In the previous example, if the bad event is an icy road, the action will be laying down grit which will have a cost C. If the forecast is wrong and the roads are not icy, so the good event actually occurs, there still will be a cost.

From the perspective of someone accepting the forecasts then one can form the following expected values:

$$\begin{aligned} &\text{Expected value of } \textbf{taking} \text{ action} &&= (Y_{11} - C)\,\widehat{\pi}_t + (Y_{12} - C)\,(1 - \widehat{\pi}_t) \\ &\text{Expected value of } \textbf{not taking} \text{ action} &&= Y_{21}\widehat{\pi}_t + Y_{22}\,(1 - \widehat{\pi}_t) \end{aligned}$$

and so action is taken if the first of these exceeds the second, which gives the condition:

$$\widehat{\pi}_t\,(Y_{11} - Y_{12} + Y_{22} - Y_{21}) > C + Y_{22} - Y_{12}.$$

Under the simplifying assumption, which will be used here, that $Y_{12} = Y_{22}$ this gives

$$\widehat{\pi}_t > C/\,(Y_{11} - Y_{21}) = q. \tag{1}$$

In other words, preventative action will be taken if the forecast, or the perceived probability of a bad state, occurring ($\widehat{\pi}_t$) exceeds q, as defined by (1), which is the ratio of the cost of prevention, C, relative to the economic benefit that results from taking preventative action, $Y_{11} - Y_{21}$. q may be called the cost-benefit ratio. It is clear that $q > 0$. In order not to rule out the possibility of preventative action being taken, we also assume that $q < 1$.

The definition of the states can be somewhat arbitrary, but it will be convenient to assume that there is a stochastic process x_t and a critical level, b, which together define the "state determination" procedure:

the bad state occurs if $x_t > b$.

Thus, x_t could be the average temperature over the last hour and b some specific temperature, such as -5°C. The bad state may then correspond to frozen, icy roads or to damaged agricultural crops.

On any particular occasion, or date t, the realized value of the economic benefit of the decision rule based on the probability forecast $\widehat{\pi}_t$, which we shall denote by V_t, will also depend on the actual outcome, that is whether a good or bad event occurs. This can be displayed in terms of indicator functions as follows: (where $I(w) = 1$ if $w \geq 0$, $I(w) = 0$ if $w < 0$)

$$\begin{aligned} V_t = &\,(Y_{11} - C)\,I\,(x_t - b)\,I\,(\widehat{\pi}_t - q) \\ &+ (Y_{12} - C)\,\{1 - I\,(x_t - b)\}\,I\,(\widehat{\pi}_t - q) \\ &+ Y_{21}I\,(x_t - b)\,\{1 - I\,(\widehat{\pi}_t - q)\} \\ &+ Y_{22}\,\{1 - I\,(x_t - b)\}\,\{1 - I\,(\widehat{\pi}_t - q)\} \end{aligned} \tag{2}$$

which can be simplified into (recall that $Y_{12} = Y_{22}$).

$$V_t = A_t + (Y_{11} - Y_{21})\,\{I\,(x_t - b) - q\}\,I\,(\widehat{\pi}_t - q)\,, \tag{3}$$

where

$$A_t = Y_{21} I(x_t - b) + Y_{22}\{1 - I(x_t - b)\}. \qquad (4)$$

Suppose that an information set Ω_{t-1} is available at time $t-1$, then the following holds:

Proposition 1 *The optimum solution set of forecasts is given by all $\widehat{\pi}_t$, with values $\widehat{\pi}_t \geq q$ if $\pi_t \geq q$, or $\widehat{\pi}_t < q$ if $\pi_t < q$ with the "supreme" solution given by*

$$\widehat{\pi}_t(\text{sup}) = \pi_t$$

where $\pi_t = \text{Prob}(x_t > b \,|\Omega_{t-1})$

Proof The proof is straightforward. Since A_t does not depend on π_t in (3), only the second term is of concern. Taking conditional expectations throughout (3) gives

$$E[V_t \,|\Omega_{t-1}] = E[A_t \,|\Omega_{t-1}] + (Y_{11} - Y_{21})(\pi_t - q) I(\widehat{\pi}_t - q). \qquad (5)$$

It is clear that the last term can be negative for some values of π_t, $\widehat{\pi}_t$ and q, unless $\widehat{\pi}_t = \pi_t$ giving $\widehat{\pi}_t(\text{sup})$. It should also be noted that $E[V_t \,|\Omega_{t-1}]$ is not altered if $(\widehat{\pi}_t - q)(\pi_t - q) > 0$.

Thus, all solutions in the optimum solution set give the same (conditional) expected value of V_t, and in that sense are equal to each other. However, the supreme solution has an advantage over other solutions as it does not depend on the cost function, whereas the other optimum solutions depend on the cost/benefit ratio q, and on a knowledge of the region in which π_t lies.

It should be noted that the supreme solution provides the complete forecast distribution, conditional on a particular information set. If there are several users, with different values of q, they can all use the supreme forecasts, but this is not true for other solutions in the optimal set. To obtain the supreme solution one will need to know the "true" conditional probability distribution function of the event; in practice the proposition suggests that the forecaster should attempt to obtain a good estimate of this probability. One can view the forecasters as producers of goods and the users of forecasts as the consumers of these goods. Without a precise knowledge of the cost function of the users of the forecasts, the most appropriate course open to the forecasters is to do their best to obtain the supreme solution.

It may be noted that if a forecaster is undecided between offering an *event forecast* ("tomorrow will be bad") or *probability event forecast* ("probability that tomorrow is bad is 0.6"), where the former may be based on a rule such as "bad if $\widehat{\pi} > d$" for some d, then the *Proposition 1* suggests that the event forecast will be sub-optimal unless both $\widehat{\pi}_t = \pi_t$ and $d = q$. Probabilistic event forecasts are more useful to customers than just event forecasts.

3 Comparisons of Forecasts

Apart from the additive and multiplicative terms A_t and $(Y_{11} - Y_{21})$, both of which are positive and are the same for a forecast using the same cost function, the essential component of the value V_t given by (3) can be written as

$$v_t = (z_t - q)\, I\, (\widehat{\pi}_t - q) \tag{6}$$

where $z_t = 1$ if the "bad" event occurs (namely $x_t > b$) and is 0 otherwise. We can imagine having a span of dates $t = 1, ..., T$ for which z_t is observed and also probability forecasts from two competing models giving $\widehat{\pi}_t^{(1)}$, $\widehat{\pi}_t^{(2)}$. Thus, average values

$$\overline{v}_T^{(i)} = \frac{1}{T} \sum_{t=1}^{T} (z_t - q)\, I \left(\widehat{\pi}_t^{(i)} - q\right), \qquad i = 1, 2 \tag{7}$$

can be formed and the forecasts providing the largest value will be preferred. It is seen that only the term q derived from the cost matrix is relevant.

The two forecasts could be combined in any relevant fashion, such as linearly

$$\widehat{\pi}_t^{(c)} = \theta \widehat{\pi}_t^{(1)} + (1 - \theta)\, \widehat{\pi}_t^{(2)} \qquad 0 \leq \theta \leq 1$$

giving an average value $\overline{v}^{(c)}(\theta)$ and one could search over θ to obtain the maximum average value available from such combinations. Clearly, from its method of construction, the optimum average value will be no less than Max $\left(\overline{v}_T^{(1)}, \overline{v}_T^{(2)}\right)$, as one could select $\theta = 0$ or 1.

It is easy to compare the values achieved by a particular forecasting model with those from two very simple models, one very naive and the other in which perfect forecasts are achieved. The naive model simply sets $\widehat{\pi}_t =$ constant, say p, for every t and so ignores the contents of the information set. It is easily seen that if $p \geq q$ then the expected value of this forecast is Prob $(z_t = 1) - q$, which could be negative, and if $p < q$, the value is zero. A rather more interesting case assumes that the information set Ω_{t-1} can eventually be expanded sufficiently so that x_t, and thus z_t, can be forecast virtually perfectly. Whether or not this is actually possible is debatable. The value becomes

$$\overline{v}_T^{opt} = \frac{1}{T} \sum_{t=1}^{T} (z_t - q)\, I\, (z_t - q) \tag{8}$$

which is necessarily positive. As q lies in the range (0, 1), $I\,(z_t - q)$ will only be non-zero when $z_t = 1$ and so

$$\overline{v}_T^{opt} = (1-q)\left\{\frac{1}{T}\sum_{t=1}^{T} I\left(z_t - q\right)\right\}, \tag{9}$$

where $T^{-1}\sum_{t=1}^{T} I\left(z_t - q\right)$ is the fraction of the times the "bad" event occurs in the sample. In the limit as $T \to \infty$, assuming $\{x_t\}$ is a strictly stationary process then $\overline{v}_T^{opt}$ tends to $(1-q)\,\text{Prob}\,(z_t = 1)$, which sets an absolute upper bound to the expected value of forecasts, and presents the supreme optimum.

In practice, however, the supreme optimum (9), which is based on the perfect event forecast, $\widehat{\pi}_t = z_t$, is unlikely to be attainable.

4 A Particular Example

In the previous section it was pointed out that for each value of t there could be many optimal forecasts and that all forecasts would be compared by their average values over a period of time. In this section, using a very particular example, it is shown that there may be a unique optimum for the average value, which is the supreme forecast.

For illustrative purposes suppose the values of x_t, which determine whether a bad event occurs, are generated according to the following stationary AR(1) process:

$$x_t = \rho x_{t-1} + \epsilon_t, \qquad t = 1, 2, ..., T$$

where e_t are independently and identically distributed with the distribution function $F_\epsilon(\cdot)$, $\rho \in (0,1)$, and the 'bad' event occurs if $x_t > b$. For this simple example,

$$\begin{aligned}\pi_t &= \text{Prob}\left(x_t > b \,|\Omega_{t-1}\right) \\ &= \text{Prob}\left(\rho x_{t-1} + \epsilon_t > b\right) \\ &= 1 - F_\epsilon\left(b - \rho x_{t-1}\right).\end{aligned} \tag{10}$$

Suppose now that instead of π_t a decision maker base his/her action on $\widehat{\pi}_t$ given by

$$\widehat{\pi}_t = 1 - F_\epsilon\left(b - r x_{t-1}\right), \tag{11}$$

where r is an 'estimate' of ρ. Using (3) and (4) the average realized loss over the period $t = 1, 2, ..., T$ arising from using the estimate r, instead of ρ is given by

$$L_T\left(\rho, r\right) = CqT^{-1}\sum_{t=1}^{T}\left(z_t - q\right)\left(I\left(\pi_t - q\right) - I\left(\widehat{\pi}_t - q\right)\right),$$

where $z_t = I(x_t - b)$. In what follows for convenience we normalize $L_T(\rho, r)$ by setting $Cq = Y_{11} - Y_{21} = 1$. Since x_t is strictly stationary and ergodic, the limit of $L_T(\rho, r)$ as $T \to \infty$ exists and is given by

$$L(\rho, r) = \lim_{T\to\infty} T^{-1} \sum_{t=1}^{T} E[(z_t - q) I(\pi_t - q)] - \lim_{T\to\infty} T^{-1} \sum_{t=1}^{T} E(v_t), \quad (12)$$

where as before $v_t = (z_t - q) I(\widehat{\pi}_t - q)$. The first term of $L(\rho, r)$ does not depend on r (or $\widehat{\pi}_t$), and to minimize $L(\rho, r)$ it is sufficient to consider values of r that maximize

$$V(\rho, r) = \lim_{T\to\infty} T^{-1} \sum_{t=1}^{T} E(v_t). \quad (13)$$

It is useful to note that

$$\begin{aligned} E(v_t) &= E\{E(v_t \,|\Omega_{t-1})\} \\ &= E((\pi_t - q) I(\widehat{\pi}_t - q)), \end{aligned}$$

where expectations are taken with respect to the unconditional distribution of x_{t-1}. Also from (11) we note that $\widehat{\pi}_t > q$ if $rx_{t-1} > b - F_\epsilon^{-1}(1-q) = \theta$, which defines θ. For $r > 0$, $\widehat{\pi}_t > q$ if $x_{t-1} > \theta/r$. Hence

$$E(v_t) = \int_{\theta/r}^{\infty} [1 - q - F_\epsilon(b - \rho x_{t-1})] h_x(x_{t-1}) dx_{t-1}, \quad (14)$$

where $h_x(x_t)$ stands for the unconditional density function of x_t. Since x_t is strictly stationary, $E(v_t)$ is time-invariant and using (13) and (14) we have

$$V(\rho, r) = E(v_t) = \int_{\theta/r}^{\infty} (1 - q - F_\epsilon(b - \rho x)) h_x(x) dx. \quad (15)$$

It is now easily seen that

$$\frac{\partial V(\rho, r)}{\partial r} = -\frac{\partial L(\rho, r)}{\partial r} = \left(\frac{\theta}{r^2}\right)\left(1 - q - F_\epsilon\left(b - \frac{\rho\theta}{r}\right)\right) h_x\left(\frac{\theta}{r}\right), \quad (16)$$

and the value of r, denoted by r^*, that solves the necessary condition $\partial L(\rho, r)/\partial r = 0$ for the minimization of $L(\rho, r)$ is given by

$$F_\epsilon\left(b - \frac{\rho\theta}{r^*}\right) = 1 - q$$

or

$$\frac{\rho\theta}{r^*} = b - F_\varepsilon^{-1}(1-q) = \theta,$$

which gives the unique solution $r^* = \rho$.[3] Therefore, the optimum value of $\widehat{\pi}_t$ in the multi-shot decision problem is given by the supreme optimum solution of the one-shot decision problem discussed in the previous section.

The extent of the average loss when r is not equal to ρ, can be computed by simulation. One possibility would be to use the form

$$L_{TR}(\rho, r) = \frac{1}{RT}\sum_{j=1}^{R}\sum_{t=1}^{T}\left(z_t^j - q\right)\left(I\left(\pi_t^j - q\right) - I\left(\widehat{\pi}_t^j - q\right)\right), \tag{17}$$

where $z_t^j = I\left(x_t^j - b\right)$, $\pi_t^j = 1 - F_\epsilon\left(b - \rho x_{t-1}^j\right)$, $\widehat{\pi}_t^j = 1 - F_\epsilon\left(b - r x_{t-1}^j\right)$, $x_t^j = \rho x_{t-1}^j + \epsilon_t^j$, and ϵ_t^j is a draw from the distribution of ϵ. Table 1a gives the values of $L_{RT}(\rho, r)$ for $T = 1000$, $R = 100$, $q = 0.2$, $b = 1$, $\rho = 0.0, 0.1, ..., 0.9$, $r = 0.0, 0.1, ..., 0.9, 1.0$, and $\epsilon_t^j \sim N(0,1)$. To minimize the effect of the initialization of the x_t process on the results, the first 100 draws of x_t starting from $x_{-100} = 0$ were discarded.

An alternative, and a more effective, procedure would be to simulate $L(\rho, r)$ given by (12) directly. Since x_t is strictly stationary we have

$$L(\rho, r) = E\left\{(\pi_t - q)\left(I(\pi_t - q) - I(\widehat{\pi}_t - q)\right)\right\},$$

where expectations are now taken with respect to the unconditional distribution of x_t. Under this procedure we have:

$$L_R(\rho, r) = \frac{1}{R}\sum_{j=1}^{R}\left(\pi^j - q\right)\left(I\left(\pi^j - q\right) - I\left(\widehat{\pi}^j - q\right)\right), \tag{18}$$

where $\pi^j = \Phi\left(\rho x^j - b\right)$, $\widehat{\pi}^j = \Phi\left(r x^j - b\right)$, $\Phi(\cdot)$ is the cumulative distribution function of the standard normal, and x^j are drawn from the unconditional density function of x_t, namely $x^j \sim N\left(0, \frac{1}{1-\rho^2}\right)$. Table 1b gives the simulated values of $L_R(\rho, r)$ for $R = 50,000$, and in comparison to the values in Table 1a may be viewed as effectively using an infinite T value and a much larger R.

[3]The second order derivative of $V(\rho, r)$ evaluated at $r = r^* = \rho$ is given by

$$\left.\frac{\partial^2 V(\rho, r)}{\partial r^2}\right|_{r=\rho} = \left(\frac{-\theta^2}{\rho^3}\right) h_x\left(\frac{\theta}{\rho}\right) f_\epsilon(b-\theta) < 0,$$

for $\rho > 0$. $f_\epsilon(\cdot)$ is the density function of ϵ_t.

Table 1a: Empirical Means of 100 $\times L_{TR}(\rho, r)$ Using Relation (17) with $T = 1000$ and $R = 100$

r/ρ	0.0	0.1	0.2	0.3	0.4	0.5	0.6	0.7	0.8	0.9
0.0	0.000	0.056	0.732	1.882	3.359	5.230	7.677	10.756	15.021	21.371
0.1	0.255	0.000	0.311	0.979	1.737	2.606	3.499	4.322	4.862	4.697
0.2	0.911	0.206	0.000	0.100	0.280	0.454	0.641	0.772	0.930	0.892
0.3	1.266	0.415	0.032	0.000	0.024	0.056	0.167	0.192	0.298	0.239
0.4	1.469	0.550	0.140	0.010	0.000	-0.021	0.060	0.053	0.102	0.115
0.5	1.565	0.648	0.174	0.057	0.020	0.000	0.015	0.003	0.017	0.060
0.6	1.614	0.695	0.239	0.082	0.035	0.000	0.000	-0.022	-0.015	0.020
0.7	1.660	0.742	0.280	0.126	0.067	0.023	0.025	0.000	-0.006	0.011
0.8	1.711	0.777	0.319	0.160	0.084	0.046	0.039	0.011	0.000	0.002
0.9	1.728	0.806	0.358	0.169	0.113	0.071	0.046	0.009	0.007	0.000
1.0	1.755	0.838	0.391	0.212	0.130	0.100	0.055	0.011	0.004	-0.001

Table 1b: Empirical Means of 100 $\times L_R(\rho, r)$ Using Relation (18) with $R = 50,000$

r/ρ	0.0	0.1	0.2	0.3	0.4	0.5	0.6	0.7	0.8	0.9
0.0	0.000	0.070	0.767	1.933	3.453	5.360	7.768	10.889	15.129	21.494
0.1	0.234	0.000	0.310	0.961	1.761	2.649	3.548	4.387	4.939	4.788
0.2	0.877	0.194	0.000	0.088	0.273	0.497	0.717	0.906	1.001	0.908
0.3	1.222	0.401	0.062	0.000	0.033	0.104	0.184	0.254	0.303	0.291
0.4	1.422	0.545	0.148	0.026	0.000	0.014	0.046	0.081	0.108	0.114
0.5	1.543	0.639	0.214	0.065	0.012	0.000	0.007	0.022	0.040	0.045
0.6	1.625	0.707	0.268	0.103	0.034	0.006	0.000	0.004	0.012	0.016
0.7	1.687	0.761	0.311	0.136	0.056	0.019	0.004	0.000	0.002	0.006
0.8	1.731	0.799	0.342	0.160	0.075	0.033	0.012	0.002	0.000	0.001
0.9	1.767	0.831	0.370	0.184	0.095	0.047	0.021	0.007	0.001	0.000
1.0	1.799	0.859	0.395	0.204	0.110	0.059	0.030	0.013	0.004	0.001

The two approaches give quantitatively similar results and should converge to the same limits as both R and $T \to \infty$. It is clear that the average cost function is not symmetric in r about its true value ρ over-estimating ρ is less costly than under-estimating it, so long $\rho \geq 0.2$. For example, in Table 1b, when $\rho = 0.5$ the cost of using $r = 0.3$ is almost twice that at $r = 1.0$. Table 1b, may be thought to be preferable as all values are positive and $r = \rho$ gives the minimum cost, as suggested by the theory.

The extent of the asymmetry of the value function in $\rho - r$ can also be seen from Figure 1 which plots the expected values of the benefit function given by (15) as a function of $\rho - r$. These expected values are computed by stochastic simulations using 100,000 replications with $b = 0.2$, $q = 0.3$, and assuming that $x \sim N(0, 1/(1-\rho^2))$. As can be seen from this figure the degree of asymmetry in $V(\rho, r)$ increases sharply as ρ is increased from 0.1 to 0.2.

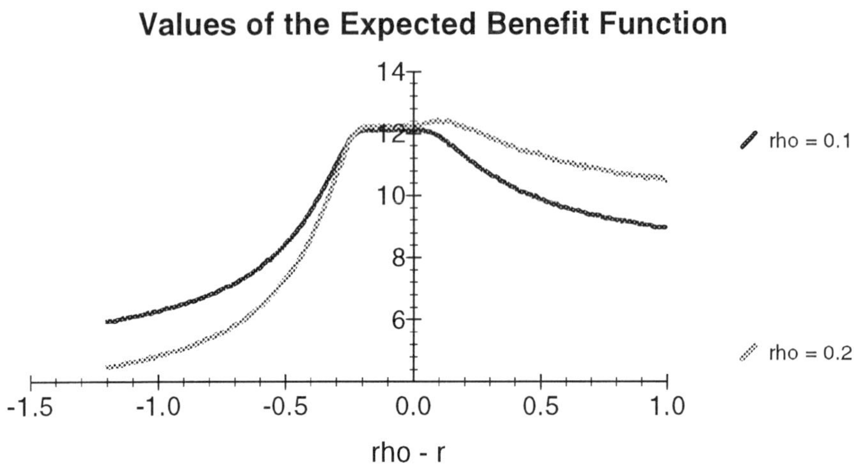

Figure 1: Plot of $V(\rho, r)$ against $\rho - r$, defined by (15) and computed at $b = 0.2$ and $q = 0.3$.

5 The Two-Action, Multi-State Problem

Now consider a more general case, with m states and let $\widehat{\pi}_{it}$ be the forecast probability that the ith state occurs at time t, so that $\sum_{i=1}^{m} \widehat{\pi}_{it} = 1$. The benefit/cost matrix will now take the form

		States				
		1	2	3	$\cdots$	m
Action	Yes	$Y_{11} - C$	$Y_{12} - C$	$Y_{13} - C$	$\cdots$	$Y_{1m} - C$
	No	Y_{21}	Y_{22}	Y_{23}	$\cdots$	Y_{2m}

It is assumed that the cost of taking action is always C, regardless of the state. One should expect that $Y_{1j} \geq Y_{2j}$ for all j, so that for all, or most states, taking action is beneficial. Let

$$\beta_i = Y_{1i} - Y_{2i}$$

and suppose that the states are ranked according to the size of β. Thus, the largest β will correspond to the "worst" state, second largest β to the "next worst" state, and so forth, and with the smallest β, which will be taken to be zero, corresponds to the "good" state. Any pair of states with identical β's will be considered as being collectively identical for our purposes and will be amalgamated, so that it will be assumed that all β's are different. Based on the forecasts, action will be taken if the value of the action ("yes") is greater than no action, which is given by

$$\sum_{i=1}^{m} \widehat{\pi}_{it} \left(Y_{1i} - C\right) > \sum_{i=1}^{m} \widehat{\pi}_{it} Y_{2i}$$

i.e.

$$\sum_{i=1}^{m} \widehat{\pi}_{it} \left(Y_{1i} - Y_{2i}\right) > C$$

or

$$\sum_{i=1}^{m} \widehat{\pi}_{it} \beta_i > C.$$

This can be written more simply in the vector notation as

$$\boldsymbol{\beta}' \widehat{\boldsymbol{\pi}}_{\mathbf{t}} > \mathbf{C} \tag{19}$$

where $\boldsymbol{\beta}' = (\boldsymbol{\beta}_1, \boldsymbol{\beta}_2, \ldots \boldsymbol{\beta}_m)$, $\widehat{\boldsymbol{\pi}}_t' = (\widehat{\pi}_{1t}, \widehat{\pi}_{2t}, \ldots \widehat{\pi}_{mt})$. It should be noted that the constraint (19) has associated with it the other constraints that $\beta_i \geq 0$, $C \geq 0$, $\widehat{\pi}_{it} \geq 0$ and $\sum_{i=1}^{m} \widehat{\pi}_{it} = 1$.

For a given set of costs and probability forecasts, $\widehat{\pi}_{it}$, equation (19), gives the "action rule" of whether action should be taken or not. Given this action rule, it is now possible to determine the properties of the best forecasts. Introducing some further notation, let

$$\begin{aligned} z_{it} &= 1 \text{ if state } i \text{ occurs at time } t \\ &= 0 \text{ if not.} \end{aligned}$$

Therefore, if $z_{it} = 1$, $z_{jt} = 0 \quad j \neq i$ and, for some information set Ω_{t-1} available at time $t-1$ we have

$$E(z_{it} \,|\Omega_{t-1}) = \pi_{it}$$

so that π_{it} is the probability of state i occurring at time t conditional on the information set, Ω_{t-1}. The realized economic value of the decision to act is then

$$\begin{aligned} V_t = &\left\{\sum_{i=1}^{m} \left(Y_{1i} - C\right) z_{it}\right\} I\left(\boldsymbol{\beta}' \widehat{\boldsymbol{\pi}}_{\mathbf{t}} - \mathbf{C}\right) \\ &+ \left\{\sum_{i=1}^{m} Y_{2i} z_{it}\right\} \left(1 - I(\boldsymbol{\beta}' \widehat{\boldsymbol{\pi}}_{\mathbf{t}} - \mathbf{C})\right) \end{aligned} \tag{20}$$

where as before, $I(\cdot)$ is the indicator function. Taking conditional expectations with respect to Ω_{t-1} gives

$$
\begin{aligned}
E\left[V_t \left|\Omega_{t-1}\right.\right] &= \left(\sum_{i=1}^{m} (Y_{1i} - C)\,\pi_{it}\right) I(\boldsymbol{\beta}'\widehat{\boldsymbol{\pi}}_{\mathbf{t}} - \mathbf{C}) \\
&\quad + \sum_{i=1}^{m} (Y_{2i}\pi_{it})\left(1 - I\left(\boldsymbol{\beta}'\widehat{\boldsymbol{\pi}}_{\mathbf{t}} - \mathbf{C}\right)\right) \\
&= \sum_{i=1}^{m} Y_{2i}\pi_{it} + \left(\boldsymbol{\beta}'\boldsymbol{\pi}_{\mathbf{t}} - \mathbf{C}\right) I\left(\boldsymbol{\beta}'\widehat{\boldsymbol{\pi}}_{\mathbf{t}} - \mathbf{C}\right) \qquad (21)
\end{aligned}
$$

The first term is not a function of $\widehat{\boldsymbol{\pi}}_t$. The second term is certainly positive if $\widehat{\boldsymbol{\pi}}_t = \boldsymbol{\pi}_{\mathbf{t}}$ and this solution does not depend on the cost function, and so is the supreme optimum. However, any $\widehat{\boldsymbol{\pi}}_t$ such that the pair of inequalities

$$\boldsymbol{\beta}'\boldsymbol{\pi}_{\mathbf{t}} \geq \mathbf{C}, \quad \boldsymbol{\beta}'\widehat{\boldsymbol{\pi}}_{\mathbf{t}} \geq \mathbf{C}$$

are obeyed will provide the same optimum expected value $E(V_t \left|\Omega_{t-1}\right.)$, but these optima require some knowledge of $\boldsymbol{\pi}_{\mathbf{t}}$ and also involve knowledge of the parameters of the cost function c and β.

As an example, consider the three state case:

	Bad	Medium	Good	
β Values	β_1	β_2	0	; with $\beta = \beta_2$

The constraints are now

$$\beta_1\pi_1 + \beta_2\pi_2 > C$$

and

$$\beta_1\widehat{\pi}_1 + \beta_2\widehat{\pi}_2 > C$$

The supreme forecast is always $\widehat{\pi}_i = \pi_i$, $i = 1, 2, 3$. The optimal set is $(\widehat{\pi}_1, \widehat{\pi}_2)$ anywhere in region A_1, if (π_1, π_2) is in A_1, and $(\widehat{\pi}_1, \widehat{\pi}_2)$ anywhere in region A_2 if (π_1, π_2) is in A_2. (See Figure 2). At any given point of time, it is seen that there are many optimal forecasting possibilities that require a knowledge of the cost-benefit ratios, C/β_1, and C/β_2, and of the location, if not the actual value of (π_1, π_2), but there is only one supreme forecast that is transferable across agents as it is not dependent on the cost function.

Comparisons between forecasts can be made using the obvious generalizations of the procedure discussed in Section 3.

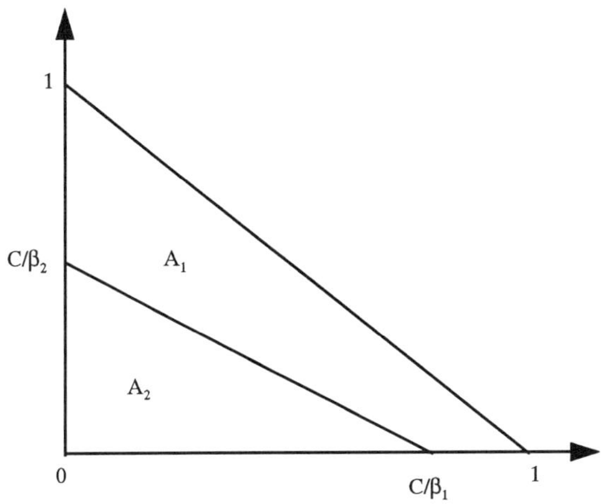

Figure 2

6 Some Further Extensions

Forecasts are often linked with decisions and forecast errors with costs. Here, in a simple example but for a situation that can arise in practice, the full implications of these relationships have been explored. The importance of achieving the complete forecast distribution, that is the distribution of the relevant variable conditional on the available information set, is emphasized. Many of these results readily extend to a more general formulation of the problem.

Let X_t be a stationary process with probability density function conditional on information set Ω_{t-1} denoted by $f(x\,|\Omega_{t-1})$, so that

$$f(x\,|\Omega_{t-1})\,dx = \text{Prob}\,(x < X_t \leq x + dx\,|\Omega_{t-1}).$$

This will be called the "true density function", and let $\widehat{f}(x\,|\Omega_{t-1})$ be some estimate or model of it, also based on Ω_{t-1} which is not completely correct, which will be called the "estimated density function". A cost function $\phi(y_t, y_t^*)$ is considered, where y_t is the actual realization at time t and y_t^* is a forecast of y_t made at time $t-1$. To be "well behaved" it is easier to consider the equivalent form $\phi(e_t, y_t^*) \equiv \overline{\phi}(y_t, y_t^*)$ where $e_t = y_t - y_t^*$ and to assume the properties

(i) $\phi(0, y^*) = 0$ and $\phi(e, y^*) > 0$ for all $e \neq 0$, and all y^*

(ii) $\phi(e, y^*)$ is continuously differentiable in e [4]

(iii) $\phi(e, y^*)$ is convex in e for all y^*, so that it is monotonically non-decreasing as e increases for $e > 0$ and is monotonically non-increasing as e increases for $e < 0$.

Considering the simple case where the cost function depends just on e, and so is given by $\phi(e)$, the above conditions put severe constraints on the analytical form of the function. In particular, the Taylor series expansion of $\phi(e)$ will have zero constant and linear terms.

An important example of an asymmetric cost function is a simple version of the LINEX function introduced in Varian (1975) and analyzed in a Bayesian context by Zellner (1986):[5]

$$\phi(e) = 2\left(\exp(\alpha e) - \alpha e - 1\right)/\alpha^2. \tag{22}$$

This one parameter LINEX function has the interesting property that it reduces to the familiar quadratic loss function for $\alpha = 0$. A pictorial representation of this function for $\alpha = 0.5$ is given in Figure 3. For this particular cost function under-predicting is more costly than over-predicting when $\alpha > 0$. The reverse is true when $\alpha < 0$.

For this cost function the optimal forecast, y_t^* is the solution of

$$E\left(\partial\phi(e_t)/\partial y_t^* \,|\Omega_{t-1}\right) = 0, \tag{23}$$

which is easily seen to be

$$y_t^* = \alpha^{-1}\log\left\{E\left(\exp(\alpha y_t)\,|\Omega_{t-1}\right)\right\},$$

where the expectations are taken with respect to the conditional true density function of y. In the case where this density in normal we have

$$y_t^* = E\left(y_t\,|\Omega_{t-1}\right) + \frac{\alpha}{2}Var\left(y_t\,|\Omega_{t-1}\right),$$

where $E\left(y_t\,|\Omega_{t-1}\right)$, and $Var\left(y_t\,|\Omega_{t-1}\right)$are the conditional mean and variance of y_t. Notice that the higher the degree of asymmetry in the cost function (as measured by the magnitude of α), the larger will be the discrepancy between

[4] Some cost functions discussed in the forecasting literature do not have this property as they are not forecastable on a set of measure zero. However, they can always be arbitrarily well approximated by a function having the property.

[5] This and other cost functions have been considered by Christoffersen and Diebold (1996).

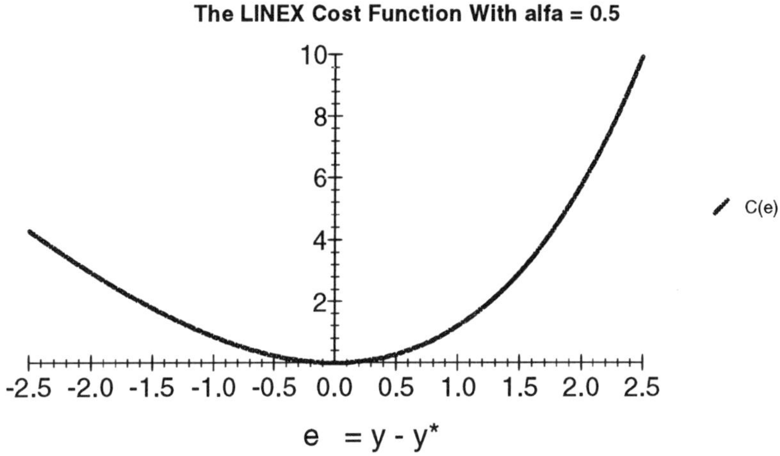

Figure 3: The LINEX Cost Function Defined by (22) for $\alpha = 0.5$

the optimal forecast and $E\left(y_t \left| \Omega_{t-1}\right.\right)$. The average realized value of the LINEX cost function, evaluated at the optimal forecast, is given by

$$E\left(\phi(e_t)\right) = E\left(Var(y_t \left| \Omega_{t-1}\right.)\right),$$

which, interestingly enough, is independent of the degree of asymmetry of the underlying cost function.

In practice, y_t may be $g(x_t)$ for any well behaved function $g(\cdot)$. The optimum forecast of $g(x_t)$ will then be y_t^* which minimizes

$$E\left(\phi\left(e_t, y_t^*\right) \left| \Omega_{t-1}\right.\right) = \int_{-\infty}^{\infty} \phi\left(e_t, y_t^*\right) f\left(x \left| \Omega_{t-1}\right.\right) dx \tag{24}$$

It is important to note that the forecast chosen must not influence the range of the integral. If the estimated density function is used, an alternative forecast is achieved, $\widehat{y}_t$, which minimizes

$$\widehat{E}\left[\phi\left(e_t, \widehat{y}_t\right)\right] = \int_{-\infty}^{\infty} \phi\left(e_t, \widehat{y}_t\right) \widehat{f}\left(x \left| \Omega_{t-1}\right.\right) dx \tag{25}$$

Clearly, as y_t^* globally minimizes (24) and $\widehat{y}_t$ minimizes something else, it follows that[6]

$$E\left[\phi\left(e_t, \widehat{y}_t\right) - \phi\left(e_t, y_t^*\right)\right]$$

[6]The global optimality of the forecasts y_t^* follows from the assumption that the cost function is well-behaved, in the sense set out above.

$$= \int_{-\infty}^{\infty} [\phi(e_t, \widehat{y}_t) - \phi(e_t, y_t^*)] f(x \,|\Omega_{t-1}) \, dx \geq 0$$

with the equality holding only if $\widehat{y}_t = y_t^*$, which occurs only if $\widehat{f}(x\,|\Omega_{t-1}) = f(x\,|\Omega_{t-1})$. We therefore have

Proposition 2 *The forecast of any function of x_t, evaluated by any well behaved cost function, is optimum if the forecast is formed on the basis of the "true conditional density function" $f(x\,|\Omega_{t-1})$ for a given information set Ω_{t-1}. Clearly better forecasts may be achievable by using larger information sets.*

One implication of this result is that it may pay forecasters to concentrate on models for the whole predictive distribution function, from which forecasts for any function, $y_t = g(x_t)$ and for any cost function, $\phi(e_t, y_t^*)$, can be derived using numerical optimization techniques applied to (24). One way to do this would be to estimate the predictive density function using models for quantiles (as in Sin and Granger (1995)) or possibly non-parametrically.

7 Conclusions

It is quite routine to evaluate forecasts by their mean squared errors. However in many realistic circumstances forecasts are used as part of a decision problem where the underlying cost function is asymmetric. The simple model discussed in Section 2 and its extension in Section 5 clearly illustrate the importance of a closer link between the decision and the forecast evaluation problems. The 2×2 action-state formulation, used in the analysis of weather forecasting, has also important applications in economics. For example, over the past few years a number of Central Banks, including the Bank of England, have been setting the nominal interest rate in the light of their forecasts of the inflation rate, thus increasing the interest rate if their prediction of the inflation rate exceeds a politically determined threshold rate. The simple model and its extension are clearly applicable to this problem, and require deriving the predictive distribution function of the inflation rate, rather a point forecast of it. These decision theoretic models also highlight the importance of a complete formulation of the benefits/costs associated with correct/incorrect forecasts. In the case of the inflation problem, immediate cost (benefit) of falsely (correctly) predicting inflation to exceed its threshold value is excessively high interest rates (inflation). In these contexts a satisfactory evaluation of inflation forecasts can be achieved only after a careful formulation of the costs/benefits of the decision problem under consideration.

It is also clear from the analysis that if costs/benefits are quite different for different states, then it is important to take this into account in the forecasting and decision making processes. For example, when considering inflation, the threshold rate could be 2 percent, a "fairly high" inflation would be over 4 percent and "very high" over 8 percent, with substantially different costs and effects of taking actions for each. These would translate into the β values involved in the discussions in Section 5. If these states were incorrectly amalgamated into the two-state system of Section 2, sub-optimal decisions could occur.

Finally, the theory of Section 6 while quite general, has the disadvantage that it is based on a cost function which is difficult to formulate in specific contexts. This should be compared with the discrete state/action formulation where the cost function is an integral component of the decision model.

Acknowledgments

Written whilst the first author was Visiting Fellow Commoner at Trinity College. He would like to thank the College for the excellent hospitality. We are grateful to Yongcheol Shin for carrying out the computations.

References

1. Britton, E., P. Fisher and J. Whitley (1998), "The Inflation Report Projections: Understanding the Fan Chart," *Bank of England Quarterly Bulletin*, 38, No.1, 30-37.
2. Chatfield, C. (1993), "Calculating Interval Forecasts," Journal of Business and Economic Statistics, 11, No.2, 121-139.
3. Christoffersen, P.F. (1998), "Evaluating Interval Forecasts," International Economic Review, Vol 39, No.4, 841-862.
4. Christoffersen, P.F. and F.X. Diebold (1996), "Further Results on Forecasting and Model Selection under Asymmetric Loss," *Journal of Applied Econometrics*, Vol 11, No. 5, 561-571.
5. Ehrendorfer, M. and A.H. Murphy (1988) Comparative evaluation of weather forecasting systems sufficiency, quality and accuracy. *Monthly Weather Review*, 116, 1757-1770.
6. Fair, R.C. (1993), "Estimating Event Probabilities form Macroeconometric Models Using Stochastic Simulations," Chp 3 in *Business Cycles, Indicators, and Forecasting* ed by J.H. Stock and M.W. Watson, National Bureau of Economic Research, Studies in Business Cycles Volume 28, University of Chicago Press.
7. Granger, C.W.J. and M.H. Pesaran (1999), "A Decision Theoretic Approach to Forecast Evaluation," Unpublished manuscript, University of Cambridge, http:\\www.econ.cam.ac.uk\faculty\pesaran\

8. Katz, R.W., and A.H. Murphy (1990) Quality/value relationships for imperfect weather forecasts in a prototype multistage decision-making model. *Journal of Forecasting*, 9, 75-86.
9. Murphy, A.H. and R.L. Winkler (1987) A general framework for forecast verification, *Monthly Weather Review*, 115, 1330-1338.
10. Varian, H.R. (1975) A Bayesian approach to real estate assessment, in *Studies in Bayesian econometrics and statistics in Honor of Leonard J. Savage*, eds. Stephen E. Fienberg and Arnold Zellner, Amsterdam: North-Holland, pp. 195-208.
11. Sin, Chor-Yiu and C.W.J. Granger (1995) Estimating and forecasting quantiles with asymmetric least squares, Working paper, Economics Department, University of California, San Diego.
12. Zellner, A. (1986) Bayesian estimation and prediction using asymmetric loss functions, *Journal of the American Statistical Association*, 81, 446-451.

LEARNING AND FORECASTING WITH STOCHASTIC NEURAL NETWORKS

TZE LEUNG LAI
Department of Statistics, Stanford University,
Stanford, CA 94305-465, USA
E-mail: lait@stat.stanford.edu

SAMUEL PO-SHING WONG
Department of Information & Systems Management,
Hong Kong University of Science & Technology,
Clear Water Bay, Hong Kong
E-mail: imsam@ust.hk

Although the neural networks have been reported to be successful in different areas such as engineering, finance, computer science, applied mathematics and statistics, the commonly used "backpropagation" algorithm to estimate the network parameters is still difficult to apply directly without fine tuning and subjective tinkering, especially when the number of parameters is large. To circumvent the estimation difficulty, we propose a new model, namely, the stochastic neural network (SNN) by using neurons with stochastic firing mechanism. SNN shares the universal approximation property with neural networks and provides a parallel estimation procedure via the EM algorithm. We also suggest a stepwise model selection procedure for SNN to avoid overfitting. Applications to regression analysis and time series forecasting are also discussed.

1 Introduction

Recently, many researchers have been applying the neural networks methodology to signal processing, developing financial trading strategies, pricing financial derivatives, pattern recognition, nonparametric function estimation and non-linear time series forecasting. However, it is very difficult to estimate the network parameters by backpropagation algorithm without subjective tinkering. Assuming the neuron firing mechanism to be stochastic, we propose a new model, namely, the stochastic neural network (SNN). Since the expectation of SNN is simply the corresponding neural network with deterministic firing mechanism, the universal approximation property of neural networks must hold in SNN. Most importantly, SNN can be estimated by EM algorithm of Dempster, Rubin and Laird (1977) in a parallel manner because maximizing the expected complete log-likelihood function can be done via independent weighted least squares and logistic regression procedures. In fact, the parallel estimation procedure can be applied to a general type of SNN which corresponds to piecewise polynomial models. The universal approximation property also holds for general SNN. Moreover, we provide a stepwise model selection procedure for SNN to avoid overfitting. The methodology can also apply to non-linear time series forecasting.

The article is organized as follows. The neural networks and other related statistical tools are compared and summarized in Section 2. Section 3 is devoted to the theory and implementation of SNN. Some examples in the regression context are shown in Section 4. Section 5 studies the application of SNN to time series forecasting. Concluding remarks and future research directions are listed in Section 6.

2 Neural Networks and Related Tools

A single-layered feedforward neural network can be presented graphically as in Figure 1 and mathematically as:

$$f_K(\mathbf{x}) = h(\beta_0 + \sum_{j=1}^{K} \beta_j \varphi(\alpha_{0j} + \boldsymbol{\alpha}_j^T \mathbf{x})). \tag{1}$$

where $\varphi(t) = 1/(1+\exp(-t))$ is the logistic function. $\mathbf{x} \in \mathbb{R}^d$ is the vector of input variables. $f_K(\mathbf{x})$ is the output of the network. "Feedforward" means that the output of the neural network is not one of the input variables. Otherwise, the neural network is called "recurrent". The β's are called the output weights and the α's are called the input weights. The weights are the parameters of the model. For classification and pattern recognition problems, a natural choice for $h(\cdot)$ is the logistic function. However, for regression and time series forecasting applications, it is usually taken to be the identity function.

The most elementary unit of neural networks is neuron and it is of the form:

$$f(\mathbf{x}) = \varphi(\alpha_0 + \boldsymbol{\alpha}^T \mathbf{x}).$$

Since $\varphi(t)$ is an increasing function from 0 to 1, one can interpret that $\alpha_0 + \boldsymbol{\alpha}^T\mathbf{x}$ is the stimulus and the neuron fires if the stimulus reaches a sufficiently high level. A neuron with parameter vector $\boldsymbol{\theta} = (\alpha_0, \boldsymbol{\alpha})$ fires at a lower level of stimulus than the one with $c\boldsymbol{\theta}$ for $c > 1$, i.e. the magnitude of the parameter vector measures the sensitivity of the neuron to the inputs. Also, if c is increased without bound and $\boldsymbol{\theta}$ is fixed, the logistic neuron will converge to the indicator function which partitions $\mathbb{R}^d$ into two half-spaces by the hyperplane $\alpha_0 + \boldsymbol{\alpha}^T\mathbf{x}$. The model of the indicator or threshold neuron for $d = 1$ and $d = 2$ are shown in Figure 2. Therefore, the direction of $\boldsymbol{\theta}$ of a logistic neuron gives the orientation of the "separating" hyperplane.

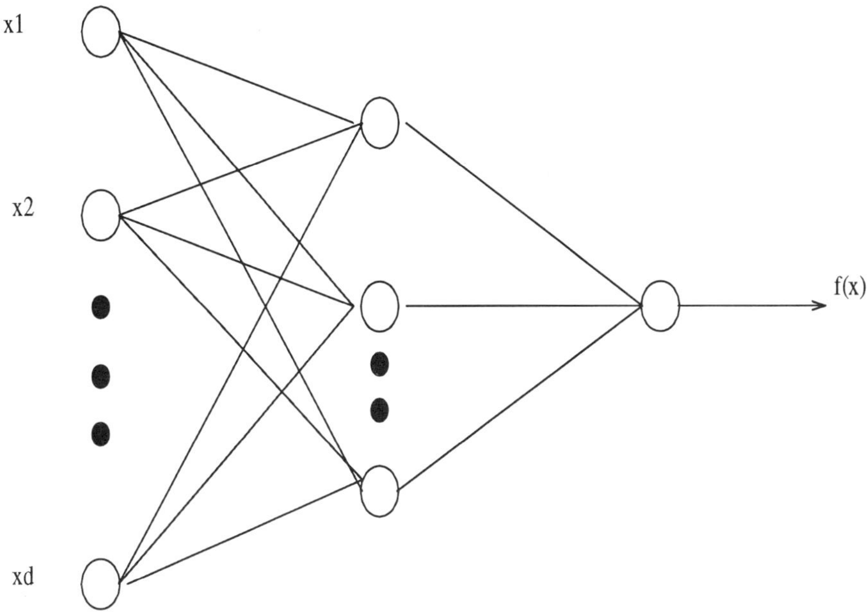

Figure 1: Single-hidden-layered perceptron.

Actually, the indicator neuron model, also known as the perceptron, was first suggested by McCulloch and Pitts (1943). Its on-line estimation procedure was proposed by Rosenblatt (1962) who successfully enabled the perceptron to reconstruct some simple logical functions by presenting examples.

The most attractive feature of the neural networks is the universal approximation property which was proved by Barron (1993). The main theorem in that paper says that any given "smooth" function defined on $\mathbb{R}^d$ can be approximated (in L^2 sense) by a neural network with sufficiently high number of neurons.

In Statistics, many nonparametric function estimation techniques for high dimension have emerged since 1980's. Some of them, actually, are very similar to neural networks in the form of (1). Classification and Regression Trees (CART) developed by Breiman, Friedman, Olshen and Stone (1984) is basically a linear combination of indicator function of hyper-reactangles in the input space. Therefore, it is equivalent to fixing the vector $\boldsymbol{\alpha}_j$ to have only one non-zero entry in the neural network with indicator neurons. Geometrically, that means constraining the separating hyperplanes to be perpendicular to the

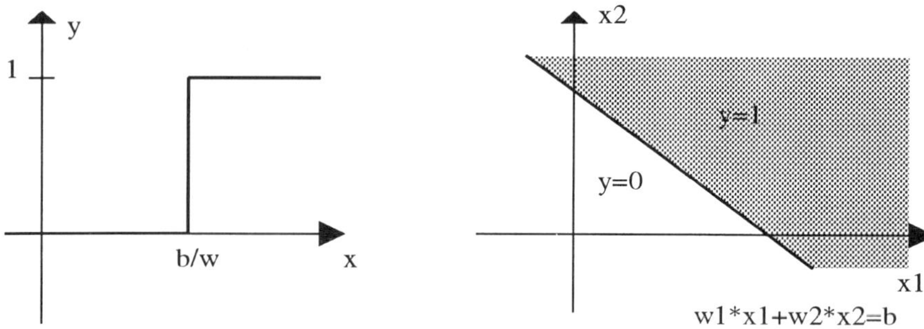

Figure 2: Perceptrons of d=1 and d=2.

co-ordinate axes. Projection Pursuit Regression (PPR) proposed by Friedman and Stuetzle (1981) can be viewed as a neural network without any assumption on the neuron firing mechanism. Instead of using the logistic function, they estimated the activation function by using a nonparametric technique named super-smoother. Generalized Additive Models (GAMS) of Hastie and Tibshirani (1990) is a special case of PPR which again fixes the vector $\boldsymbol{\alpha}_j$ to have only one non-zero entry. Multivariate Adaptive Regression Splines (MARS) of Friedman (1991) uses neurons which are the tensor products of truncated splines.

All these statistical tools can be estimated by computationally efficient and stable algorithms. For example, MARS and CART employ different forms of recursive partitioning while PPR and GAMS apply the idea of backfitting. Their accompanied model selection procedures are also helpful in avoiding overfitting.

Neural networks also come with an estimation algorithm, namely, the back-propagation which is developed by Rumelhart, Hinton and Williams (1986). It can be described as follows. Given i.i.d. $\{(\mathbf{X}_i, Y_i) : i = 1, \ldots, n\}$ where $\mathbf{X}_i$ and Y_i take values in $\mathbb{R}^d$ and $\mathbb{R}$, the parameters of the approximating function (1) are estimated by least squares, i.e.

$$\hat{\boldsymbol{\theta}} = \text{argmin}_{\boldsymbol{\theta}} S(\boldsymbol{\theta}) = \text{argmin}_{\boldsymbol{\theta}} \sum_{i=1}^{n} (Y_i - f_K(\mathbf{X}_i; \boldsymbol{\theta}))^2.$$

The backpropagation tries to minimize the $S(\boldsymbol{\theta})$ by the iteration:

$$\hat{\boldsymbol{\theta}}_k = \hat{\boldsymbol{\theta}}_{k-1} - \eta \frac{\partial S}{\partial \boldsymbol{\theta}}\Big|_{\hat{\boldsymbol{\theta}}_{k-1}} \qquad k = 1, 2, \ldots \tag{2}$$

where η is a positive constant known as the learning rate of the algorithm. The choice of η is essential to the minimization procedure. If it is too large, it may miss the optimum point. But if it is too small, the convergence will be slow and the algorithm may easily be locked into a local minimum. (2) is usually called the "batch" mode of backpropagation. The "on-line" version of the algorithm has the recursive form

$$\hat{\boldsymbol{\theta}}_k = \hat{\boldsymbol{\theta}}_{k-1} - \eta_k \frac{\partial S_k}{\partial \boldsymbol{\theta}}\bigg|_{\hat{\boldsymbol{\theta}}_{k-1}} \tag{3}$$

where $S_k(\boldsymbol{\theta}) = (y_k - f(\mathbf{x}_k; \boldsymbol{\theta}))^2$, $k = 1, \ldots, n$. There are various suggestions in the choice of learning rate including taking η as a decreasing sequence, varying η according to the values of $S(\boldsymbol{\theta}_k)$ or $S_k(\boldsymbol{\theta}_k)$, and using the "momentum" term which requires the specification of another unknown constant. It seems that the choice is actually problem-dependent and there is no clear answer on this important issue.

To avoid overfitting, early stopping of the backpropagation is usually suggested. That is, the iteration is stopped if the performance of the current estimates is "good" in an out-of-sample data set. However, the determination of good performance is highly subjective. There are researchers using the technique of shrinkage, i.e., instead of minimizing $S(\boldsymbol{\theta})$, they minimize $S(\boldsymbol{\theta}) + \lambda C(\boldsymbol{\theta})$ where $\lambda > 0$. Again, the choice of λ is critical and the computer intensive way of choosing λ among a grid of positive values is seldom employed given the prohibitive size of the problem. Weigend, Huberman and Rumelhart (1991) propose an updating rule on λ. They, however, provide no theoretical justification of the rule.

3 Stochastic Neural Networks

In the previous section, we highlighted several problems in the estimation of the neural network parameters: (a) the learning rate η in the gradient-type backpropagation method is hard to determine, and (b) it is very difficult to choose a suitable penalty factor λ to avoid overfitting. Stochastic Neural Networks (SNN) provide an alternative methodology to circumvent the difficulties in estimation without losing the universal approximation property of neural networks.

3.1 Definition and properties

Consider the application of neural networks to regression data. Given i.i.d. $\{\mathbf{X}_i, Y_i\}_{i=1}^n$ sampled from $(\mathbf{X}, Y)$ with $\mathbf{X} \in \mathbb{R}^d$ and $Y \in \mathbb{R}$, and assuming the function $\mathrm{E}[Y|\mathbf{X}=\mathbf{x}]$ is "smooth" (in the sense of Barron (1993)), the parameter estimates are obtained by minimizing the residual sum of squares. The methodology can be viewed as using maximum likelihood estimation to fit the data to the stochastic model

$$Y_i = f_K(\mathbf{x}_i;\boldsymbol{\theta}) + \epsilon_i, \quad i = 1,\ldots,n, \tag{4}$$

where the ϵ_i are i.i.d. normal with mean zero and variance σ^2. Since the logistic units can be viewed as expectations of Bernoulli random variables, an alternative stochastic model that makes use of the universal approximation property of neural networks is:

$$Y_i = \beta_0 + \textstyle\sum_{j=1}^K \beta_j I_{ij} + \epsilon_i, \tag{5}$$
$$I_{ij} \sim \mathrm{Bernoulli}(\pi_{ij}) \quad \text{and} \quad \pi_{ij} = \varphi(\alpha_{0j} + \boldsymbol{\alpha}_j^T \mathbf{x}_i); \quad i = 1,\ldots,n; j = 1,\ldots,K,$$

where I_{ij} are mutually independent, ϵ_i are i.i.d. normal with mean zero and variance σ^2 and are independent of the I_{ij}'s. The universal approximation property still holds for (5) because

$$\begin{aligned}\mathrm{E}[Y_i] &= \mathrm{E}[\beta_0 + \sum_{j=1}^K \beta_j I_{ij}] \\ &= \beta_0 + \sum_{j=1}^K \beta_j \varphi(\alpha_{0j} + \boldsymbol{\alpha}_j^T \mathbf{x}_i) \\ &= f_K(\mathbf{x}_i;\boldsymbol{\theta}).\end{aligned}$$

The I_{ij} in (5) can be interpreted as stochastic neuron that fires with probability π_{ij}. Therefore, (5) is a generalization of the perceptron that fires deterministically. Moreover, it should be noted that a neural network essentially smoothes a piecewise constant regression function via the logistic transform. We can refine the piecewise constant function to piecewise linear or polynomial functions, leading to the q-th order stochastic neural network (q-SNN), which takes the form

$$Y_i = \boldsymbol{\beta}_0^T \mathbf{v}_i + \sum_{j=1}^K \boldsymbol{\beta}_j^T \mathbf{v}_i I_{ij} + \epsilon_i, \quad i = 1,\ldots,n, \tag{6}$$

where $\mathbf{v}_i^T = (1, x_{i1}, x_{i1}^2, \ldots, x_{i1}^q, \ldots, x_{id}, \ldots, x_{id}^q)$ and $\boldsymbol{\beta}_j^T = (\beta_{00}^{(j)}, \beta_{11}^{(j)}, \cdots, \beta_{dq}^{(j)})$ are vectors in $\mathbb{R}^{dq+1}$.

It should be noted that the degenerate case of a stochastic neuron is the same as that of a logistic neuron, namely, the indicator function. The degenerate case of 1-SNN, under certain parameter configuration, is equivalent to the Hinging Hyperplanes methodology developed by Breiman (1993). Also, since the set of all q-SNN is a superset of the colection of all neural networks. The universal approximation property follows directly from Barron (1993) result. The property is explicitly stated as follows:

Theorem 3.1 *Let μ be a probability measure which is supported on the ball centered at the origin with radius R in $\mathbb{R}^d$ and $f(\mathbf{x})$ be any real-valued function on $\mathbb{R}^d$ such that*

$$\int \| \boldsymbol{\omega} \| |\tilde{f}(\boldsymbol{\omega})| d\boldsymbol{\omega} = c < \infty, \tag{7}$$

where $\tilde{f}(\boldsymbol{\omega})$ is the Fourier transform of $f(\mathbf{x})$ and $\|\cdot\|$ is the square norm in $\mathbb{R}^d$. There exist $\boldsymbol{\beta}_j$, $j = 0, \ldots, K$ and $\alpha_{0j}, \boldsymbol{\alpha}_j$, $j = 1, \ldots, K$ such that

$$\int (f(\mathbf{x}) - \boldsymbol{\beta}_0^T \mathbf{v} - \sum_{j=1}^{K} \boldsymbol{\beta}_j^T \mathbf{v} \varphi(\alpha_{0j} + \boldsymbol{\alpha}_j^T \mathbf{x}))^2 \mu(d\mathbf{x}) \leq \frac{(2Rc)^2}{K},$$

where $\mathbf{v} = (1, \mathbf{x}^T)^T$.

Another close relative of SNN is the mixture of experts (ME) which was proposed by Jordan and Jacobs (1994). ME can be stated as

$$Y_i = I\{E_i = 1\}\boldsymbol{\beta}_1^T \mathbf{X}_i + \cdots + I\{E_i = K\}\boldsymbol{\beta}_K^T \mathbf{X}_i + \epsilon_i,$$

where ϵ_i are i.i.d. normal with mean 0 and variance σ^2; E_i is a multinomial random variable independent of ϵ_i with $\Pr(E_i = j) = \exp(\boldsymbol{\alpha}_j^T X_i) / \sum_{k=1}^{K} \exp(\boldsymbol{\alpha}_k^T X_i)$ and $I\{\mathrm{A}\}$ is the indicator of the event A. Since the experts E_i are unobservable, the EM algorithm can be applied to get the maximum likelihood estimates of the parameters. Note that ME with $K = 2$ coincides with a 1-SNN with one neuron.

3.2 Estimation

Since the outcomes of the neurons are unobservable, the EM algorithm can be employed in evaluating the maximum likelihood estimates of the parameters. To describe the estimation procedure, we have the following notations. Let

$$\begin{aligned} \mathbf{I}_i^T &= (I_{i1}, \ldots, I_{iK}), \\ \boldsymbol{\psi}_i^T &= (\mathbf{v}_i^T, \mathbf{v}_i^T I_{i1}, \ldots, \mathbf{v}_i^T I_{iK}), \quad i = 1, \ldots, n; \end{aligned}$$

$$\begin{aligned}
\boldsymbol{\alpha}^T &= (\alpha_{01}, \boldsymbol{\alpha}_1^T, \ldots, \alpha_{0K}, \boldsymbol{\alpha}_K^T),\\
\boldsymbol{\beta}^T &= (\boldsymbol{\beta}_0^T, \ldots, \boldsymbol{\beta}_K^T),\\
\boldsymbol{\theta}^T &= (\boldsymbol{\alpha}^T, \boldsymbol{\beta}^T, \sigma),\\
\boldsymbol{\Psi}^T &= (\boldsymbol{\psi}_1^T, \ldots, \boldsymbol{\psi}_n^T) \quad \text{and}\\
\mathbf{Y}^T &= (Y_1, \ldots, Y_n).
\end{aligned}$$

Then the complete data log-likelihood is given by

$$l_c(\boldsymbol{\theta}) = H(\boldsymbol{\alpha}) - \frac{S(\boldsymbol{\beta})}{2\sigma^2} - \frac{n}{2}\log(2\pi\sigma^2),$$

where

$$H(\boldsymbol{\alpha}) = \sum_{j=1}^{K} H_j(\alpha_{0j}, \boldsymbol{\alpha}_j),$$

$$H_j(\alpha_{0j}, \boldsymbol{\alpha}_j) = \sum_{i=1}^{n} I_{ij}\log(\pi_{ij}) + (1 - I_{ij})\log(1 - \pi_{ij}) \quad j = 1, \cdots, K; \tag{8}$$

$$S(\boldsymbol{\beta}) = (\mathbf{Y} - \boldsymbol{\Psi}\boldsymbol{\beta})^T(\mathbf{Y} - \boldsymbol{\Psi}\boldsymbol{\beta}). \tag{9}$$

Thus, the E-step requires $\mathrm{E}[I_{ij}|Y_i]$ and $\mathrm{E}[I_{ij}I_{ik}|Y_i]$ which can be calculated by

$$\mathrm{E}(I_{ij}|Y_i) = \sum_{\mathbf{I}_i : I_{ij}=1} f(Y_i, \mathbf{I}_i)/f(Y_i); \quad \mathrm{E}(I_{ij}I_{ik}|Y_i) = \sum_{\mathbf{I}_i : I_{ij}I_{ik}=1} f(Y_i, \mathbf{I}_i)/f(Y_i),$$

where

$$f(Y_i, \mathbf{I}_i) = \frac{1}{\sigma}\phi(\frac{Y_i - \boldsymbol{\beta}^T\boldsymbol{\psi}_i}{\sigma}) \prod_{j=1}^{K} \pi_{ij}^{I_{ij}}(1 - \pi_{ij})^{1 - I_{ij}}; \tag{10}$$

$$f(Y_i) = \sum_{\mathbf{I}_i} f(Y_i, \mathbf{I}_i) \tag{11}$$

are the joint density of $(Y_i, \mathbf{I}_i)$ and the marginal density of Y_i, respectively; and $\phi(t)$ is the density function of the standard normal.

It is clear that maximizing the conditional expectation of (8) is equivalent to fitting logistic regression with $\mathrm{E}[I_{ij}|Y_i]$ as the dependent variable and $(1, \mathbf{x}_i)^T$ as the covariates. For any $i = 1, \ldots, n$ and $j = 1, \ldots, K$, let

$$\begin{aligned}
\mathbf{V}_j^T &= (\mathrm{E}[I_{1j}|Y_1], \ldots, \mathrm{E}[I_{nj}|Y_n]),\\
\mathbf{u}_i^T &= (1, \mathbf{x}_i),\\
\mathbf{U}^T &= (\mathbf{u}_1, \ldots \mathbf{u}_n), \quad \text{and}\\
\boldsymbol{\omega}_j^T &= (\alpha_{0j}, \boldsymbol{\alpha}_j^T).
\end{aligned}$$

By Newton's method, $\tilde{\boldsymbol{\omega}}_j = \text{argmax}_{\boldsymbol{\omega}_j} E[H(\boldsymbol{\omega}_j)|\mathbf{Y}]$ can be obtained by iterating

$$\boldsymbol{\omega}^{(r)} = (\mathbf{U}^T\mathbf{W}\mathbf{U})^{-1}\mathbf{U}^T\mathbf{W}\mathbf{Z}$$

until convergence, where $\mathbf{W} = \text{diag}\{p_i(1-p_i)\}$ with $p_i = \varphi(\mathbf{u}_i^T\boldsymbol{\omega}^{(r-1)})$ and $\mathbf{Z}$ has components

$$z_i = \mathbf{u}_i^T\boldsymbol{\omega}^{(r-1)} + \frac{\text{E}[I_{ij}|Y_i] - p_i}{p_i(1-p_i)}.$$

It is well known that the likelihood surface of logistic regression is logarithmically concave. Therefore, within each M-step, convergence of the logistic regression iterations is guaranteed.

Within the M-step, in addition to maximizing $\text{E}[H(\boldsymbol{\alpha})|\mathbf{Y}]$ over all $\boldsymbol{\alpha}$, $\text{E}[S(\beta)|\mathbf{Y}]$ has to be minimized. The minimization procedure is equivalent to weighted least squares because

$$\text{E}[S(\beta)|\mathbf{Y}] = \sum_{i=1}^{n}\sum_{\mathbf{I}_i}(Y_i - \psi_i^T\beta)^2\,\text{Pr}(\mathbf{I}_i|Y_i).$$

The solution $\hat{\beta}$ is unique if $\text{E}(\boldsymbol{\Psi}^T\boldsymbol{\Psi}|\mathbf{Y})$ is of full rank. The full rank condition simply means that there is no redundant hidden unit.

Therefore, the M-step is decoupled into K independent logistic regressions and a weighted least squares regression. This independent structure enables us to apply parallel computing to increase the speed of convergence. Each of the above logistic regressions can be interpreted as the training of the corresponding hidden unit, while the weighted regression corresponds to the training of the output unit. Besides, the sequence of the observed log-likelihoods of the EM steps is non-decreasing. This shows that EM convergence is quite insensitive to initial conditions, and we need not worry about other parameters such as the learning rate in backpropagation.

It should be noted that the model (6) can be relaxed so that each hidden unit carries two sets of input variables (not necessarily disjoint) : one for the logistic part which defines the hyperplane ($\alpha_{0j} + \boldsymbol{\alpha}_j^T x = 0$) as in the discussion of the perceptrons, and the other is for the output part which captures the variation of the underlying function on the half-space ($\alpha_{0j} + \boldsymbol{\alpha}_j^T x \geq 0$). Furthermore, the degree of the polynomial for each input variable can differ for different hidden units. It is clear that the local properties of the unknown function can be explored more efficiently given this flexibility by using a suitable model selection procedure.

3.3 Model Selection

The primary goal of model selection is to choose the optimal model. However, since it is computationally expensive to estimate all possible models, we propose a stepwise procedure to achieve this goal in a greedy manner. The procedure consists of two steps: forward selection and backward elimination. The forward selection determines the number of neurons needed and the input variables associated with each neuron, while the backward elimination removes the redundant parameters from the model selected by the forward step. All model selection procedures depend on their model selection criteria. In this paper, we use Schwarz's (1978) Bayesian Information Criterion (BIC). Other selection criteria, such as Akaike's Information Criterion (Akaike, 1974), may provide similar results.

For the backward elimination, we need to test if each parameter in the model is significant. In general, the Wald statistics are normally distributed by large sample theory. However, if the parameter corresponds to a reduction of the number of neurons, the hypotheses contain nuisance parameters only under the alternative hypothesis and the statistics is known to be non-normally distributed as reported in Davies (1977, 1987). Therefore, if the absence of a parameter can eliminate a neuron, that parameter will be kept in the model no matter what its Wald statistics is. That is, the number of neurons is fixed in the forward selection.

The details of the model selection procedure are quite tedious and are listed in the Appendix.

4 Regression Examples

In this section, we compare the SNN outlined in the previous section with some other commonly used nonparametric regression methods including Multivariate Adaptive Regression Splines (MARS), Projection Pursuit Regression (PPR) and Generalized Additive Models (GAMS). The first subsection is devoted to a study of simulated bivariate regression problem with correlated input variables. A housing cost data set from the *Places Rated Almanac* (Boyer and Savageau 1986) is used to demonstrate the performance of SNN in the second subsection.

4.1 Bivariate additive regression with multicollinearity

In this example, the data $(y_i, \mathbf{x}_i)$ are generated by

$$y_i = \frac{2}{3}\sin(1.3x_{i1}) - \frac{9}{20}x_{i2}^2 + \epsilon_i, \quad i = 1, \ldots, 100,$$

where ϵ_i are i.i.d. normal with mean zero and standard deviation 0.1, and $\mathbf{x}_i = (x_{i1}, x_{i2})^T$ are i.i.d. normal vectors with zero means, unit variances and correlation 0.4. To study performance, 50 replications are generated from the model. Figure 3 shows the fitted curves by GAMS with smoothing splines and by 2-SNN with $K_{\max} = 3$. The SNN is constrained to be additive. It is clear that the variance of SNN is higher in the first component but the bias of SNN is much smaller in the second component.

To compare the performance of different methods, we define the ASE to be the average squared error of the fitted regression function from the true regression function for each of the 50 samples. A box plot of the ratio of the ASE of the method to that of SNN is shown in figure 4. Since most of the relative ASE's are above 1, it shows that SNN outperforms its competitors. It should also be noted that Projection Pursuit Regression performs poorly in this example mainly because it does not involve the additivity constraint of the underlying model.

4.2 A data set of American cities

Boyer and Savageau (1986) rated 329 American cities on the nine criteria listed in Table 4.1. We attempt to model the housing cost as a function of the other eight criteria. In this example, 50 cities are randomly chosen to be the out-of-sample data and 279 cities are used to train the models.

Table 4.1

Y	housing costs
X_1	climate
X_2	health care and environment
X_3	crime rate
X_4	transportation
X_5	education
X_6	access to the arts
X_7	recreational opportunities
X_8	economics

Taking $K_{\max}$ to be 8 and constraining the first order SNN to be additive, the relative in-sample error (RIE) and the relative prediction error (RPE) are calculated based on the formulae

$$\text{RIE} = \frac{\sum_{i=1}^{279} (\hat{Y}_i - Y_i)^2}{\sum_{i=1}^{279} (Y_i - \bar{Y})^2},$$

$$\text{RPE} = \frac{\sum_{i=1}^{50} (\hat{Y_i^o} - Y_i^o)^2}{\sum_{i=1}^{50} (Y_i^o - \bar{Y^o})^2},$$

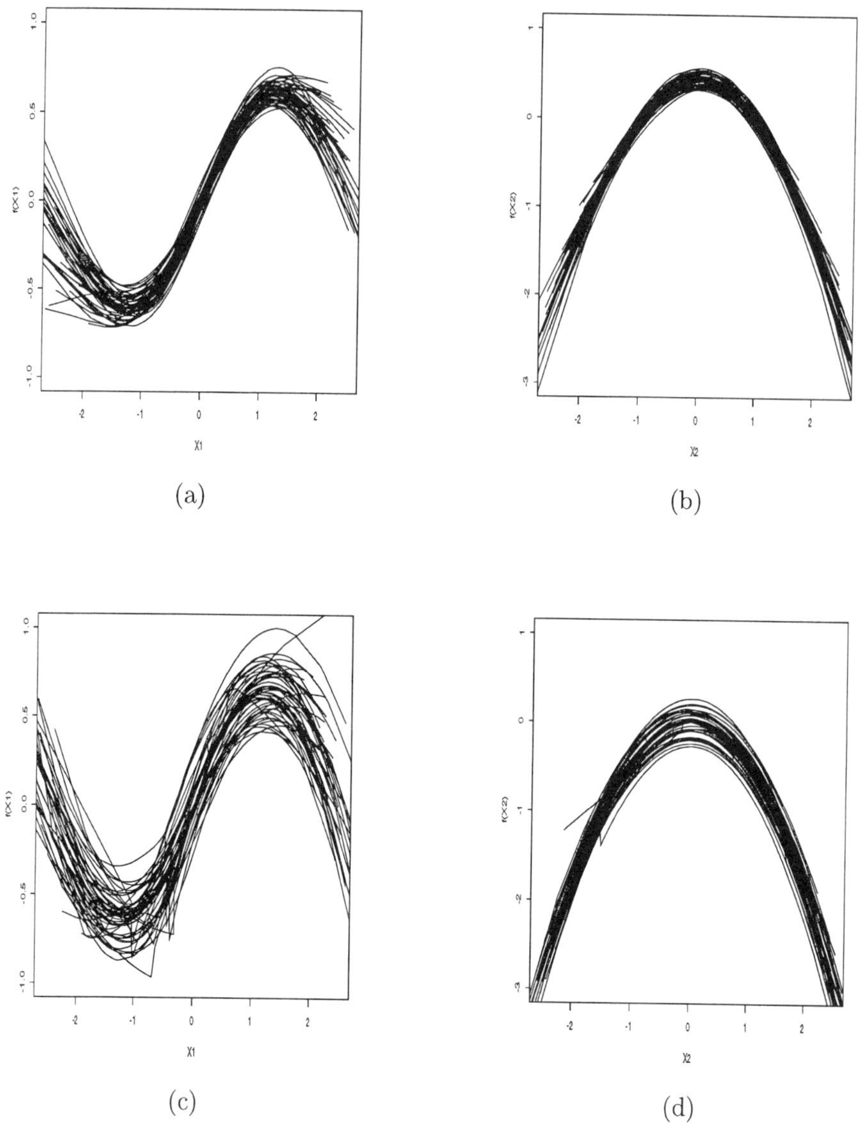

Figure 3: Fitted values by GAMS ((a) and (b)) and by SNN ((c) and (d)).

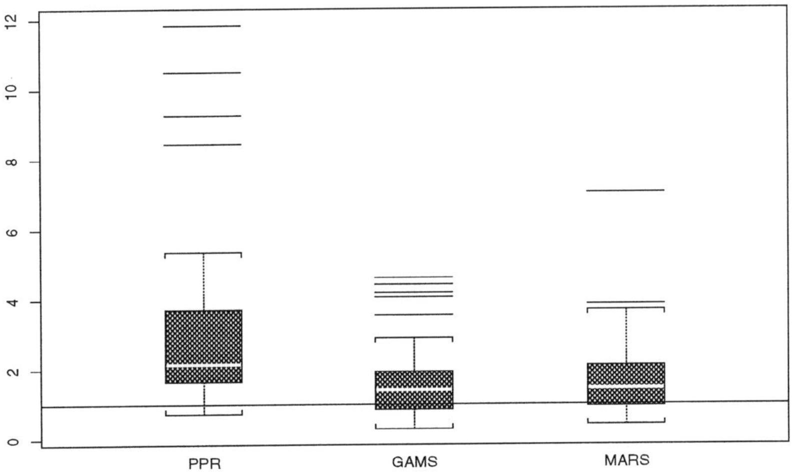

Figure 4: Comparison of the ratio of the ASE of different methods to that of SNN.

where Y_i and Y_i^o are the in-sample and out-sample data respectively.

The competitors include Multivariate Adaptive Regression Splines with additive constraint (MARS), Generalized Additive Models with spline components (GAMS), Projection Pursuit Regression (PPR) and Neural Network (NN). The variable metric algorithm of Venables and Ripley (1994) is employed to train the neural networks instead of the backpropagation algorithm because backpropagation requires fine tuning and subjective tinkering. Since we constrain the maximum number of hidden units of the neural networks to be only 8, the variable metric algorithm is able to obtain the least-squares estimates without encountering ill-conditioned matrices. Moreover, even though it is a batch algorithm, it is reasonably fast because the sample size is only 279 and the maximum network size is small. Using BIC as a measure of lack of fit, we choose the neural network with 3 hidden units (NN3) after fitting neural networks whose hidden unit numbers range from 1 to 8. For the other models, their standard estimations and the model selection procedures are used. The performance of all the competitors are listed in the following table.

Table 4.2

Model	RIE	RPE
MARS	0.444	0.567
GAMS	0.424	0.615
PPR	0.409	0.621
NN3	0.810	0.828
SNN	0.564	0.451

It is easy to see that SNN gives the best prediction but not the in-sample error. Also, those methods which are constrained to be additive perform reasonably well. It suggests that the underlying relationship may be nearly additive. Since the model selection of NN3 focuses only on the number of hidden units but not the input variables involved in a hidden unit, NN3 performs poorly both for in-sample and out-sample data. The model selection of PPR also chooses only the number of hidden units. However, the flexibility in taking the ridge functions helps PPR to give a small in-sample error.

Using the model selection procedure developed in Section 3, SNN chooses only 4 hidden units and the fitted values are shown in Figure 5. The result is different from that in Friedman (1991) where all 329 observations are used and only 3 variables are chosen in modeling the housing cost. It is easy to see that the housing cost is an increasing function of 4 components, namely, the climate, the health care and environment, the recreational opportunities and the economics. The housing cost is affected most heavily by the health care and environment. It is also interesting to see that the slope of the housing cost decreases marginally when the index of the health care and environment is greater than 1000. The climate factor plays a major role in housing cost if it goes beyond the level of 600. The recreational opportunities and the economics increase with the housing cost in a uniform way. The marginal effect of the recreational opportunities is higher than that of the economics.

5 SNN in Nonlinear Time Series Forecasting

Neural networks are popular not only in regression analysis but also in time series forecasting, such as forecasting the number of sunspots, predicting stock prices and foreign exchange rates. As SNN is a stochastic version of neural network, we would like to present how to apply SNN to time series data and compare its performance with other nonlinear time series models.

The SNN model is in the following form: Let $\mathcal{Y}_t$ be the σ-algebra generated by $\{Y_i : i \leq t\}$ and $\mathcal{I}_t$ be the σ-algebra generated by $\{\mathbf{I}_i = (I_{i1}, \ldots, I_{iK}) : i \leq$

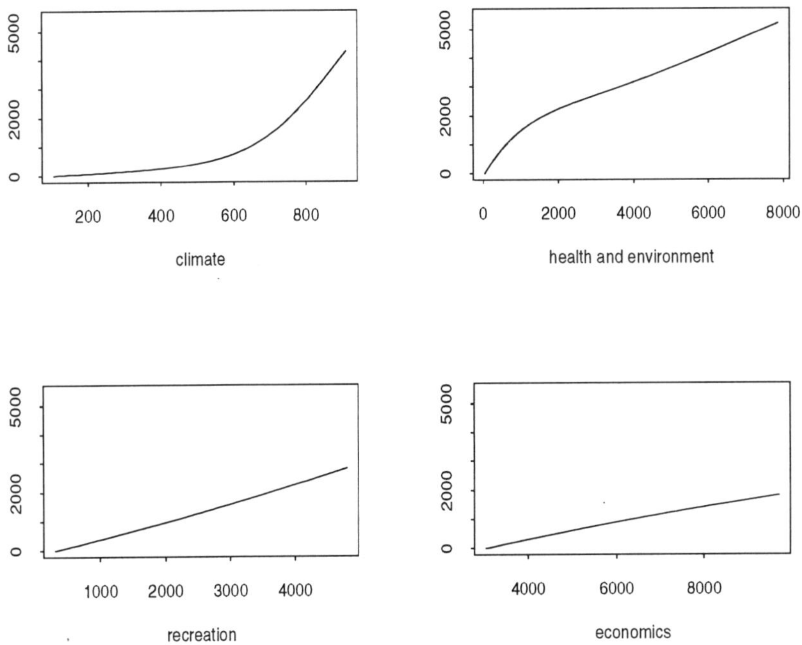

Figure 5: Fitted values of SNN for housing cost data.

$t\}$. Let

$$Y_t = \beta_0^T \mathbf{X}_{t-1} + \sum_{j=1}^{K} \beta_j^T \mathbf{X}_{t-1} I_{tj} + \epsilon_t, \tag{12}$$

with $\mathbf{X}_{t-1} = (1, Y_{t-1}, \ldots, Y_{t-p})^T$, and given $\mathcal{Y}_{t-1} \vee \mathcal{I}_{t-1}$,

$$I_{tj} \sim \text{Bernoulli}(\pi_{tj}(\mathbf{X}_{t-1})) \text{ with } \pi_{tj}(\mathbf{X}_{t-1}) = \phi(\boldsymbol{\alpha}_j^T \mathbf{X}_{t-1}).$$

It is also assumed that the I_{tj} are mutually independent and independent of the i.i.d. normal ϵ_t, with mean zero and variance σ^2.

It is easy to see that (12) is a local autoregressive model with partitions of the predictor space formed by the intersection of half-spaces with boundaries $\boldsymbol{\alpha}_j^T \mathbf{X}_{t-1} = 0$, $j = 1, \ldots, K$. For $K = 1$, one can also view the model as a generalization of the Threshold Autoregression (TAR) of Tong (1990) because TAR uses indicator neuron and it fixes the boundary to be perpendicular to one of the axes in the predictor space. Moreover, if $K = 1$ and all π_{tj} are constant, then SNN becomes the Mixture Autoregression (MAR) proposed by Wong and Li (1999). Furthermore, if the neuron does not depend on the lagged input values but is associated to the past value of the neuron, then SNN is equivalent to the Regime Switching models of Hamilton (1994).

For estimation of the model parameters, the conditional likelihood of the model is identical to the likelihood in (8) and (9). Therefore, the EM steps derived in Section 3 can be applied directly to get the maximum conditional likelihood estimator of the time series SNN. Details of the implementation and probabilistic properties of the time series model (12), particularly in connection with multistep-ahead forecasts, are given in Lai and Wong (1999).

6 Conclusion

SNN provides a way to use the universal approximation property with efficient estimation algorithm and systematic model selection. The extensions of SNN to accommodate heteroscedastic noise can be easily implemented by allowing each hidden unit to carry its own noise, i.e.

$$Y_t = \beta_0^T \mathbf{X}_t + \epsilon_{0t} + \sum_{j=1}^{K} I_{jt}(\beta_j^T \mathbf{X}_t + \epsilon_{jt}),$$

where for each j, the ϵ_{jt} are i.i.d. normal with mean zero and variance σ_j^2 and are all independent of the I_{jt}. The extended model can still be estimated by EM algorithm.

Another concern about noise is the robustness to outliers. In fact, one can assume the SNN with t-distribution in the sense of Lange, Little and Taylor (1989) and this modified SNN again can be estimated by EM algorithm. The only difference is an extra non-linear procedure for the degree of freedom parameter in the M-step. This model is found to be resistant to outliers.

There is a belief in the neural network community that any monotonically increasing function from zero to one can play the role of describing the firing mechanism of neuron and the choice of the activation function should not cause any significant difference in performance. One can explore the effect of various monotonic increasing function by incorporating an extra shape parameter to each neuron similar to Taylor (1988). The estimation procedure can be done via EM algorithm and the logistic regression will be affected but it is still computationally affordable.

Extension to time series is problem-specific. In particular, SNN can be modified in order to capture the key characteristics of the financial return series, namely, (a) fat-tailed marginal distribution, and (b) volatilities clustering. Duan and Wong (1999) proposed to use the following model of Regime Switching with Feedback for the return series.

$$\begin{aligned} X_{t+1} &= (\mu_0 + \sigma_0 \varepsilon_{t+1})(1 - I_{t+1}) + (\mu_1 + \sigma_1 \varepsilon_{t+1}) I_{t+1} \\ I_{t+1} &= 1 \quad \text{if} \quad q_1 \varepsilon_t^+ + q_2 \varepsilon_t^- + |\xi_t| > c(I_t) \\ \varepsilon_{t+1} | \mathcal{F}_t &\sim N(0,1) \\ \xi_{t+1} | \mathcal{F}_t &\sim N(0,1) \\ E(\varepsilon_{t+1} \xi_{t+1} | \mathcal{F}_t) &= 0 \end{aligned}$$

where q_1 and q_2 are both non-negative and for any real number a, $a^+ = \max(a, 0)$ and $a^- = a^+ - a$. If q_1 and q_2 are both zero, the model reduces to Hamilton's (1990) Regime Switching model. On the other hand, if $c(0) = c(1)$, the neuron reduces to SNN with $-\varepsilon_t^+$ and $-\varepsilon_t^-$ as input variables.

Another direction which has not been fully explored is the classification and pattern recognition application. One should expect that SNN can perform reasonably well at least compared with CART because SNN permits the separating hyperplanes to be oblique with respect to the axes of the covariates.

Appendix: Model Selection Procedure

The forward model selection procedure involves specification of the following:

1. Specify $K_{\max}$, the maximum number of hidden units; $K_{\max}$ should grow with the number of observations.

2. Specify the minimum number ($d_{\min}$) and maximum number ($d_{\max}$) of input variables in any hidden unit for the forward step.

3. Set q to be the maximum degree of the input variables. In general, $q \leq 3$.

It is obvious that $d \geq d_{\max} \geq d_{\min} \geq 1$. Only additive models will be fitted if $d_{\max} = d_{\min} = 1$.

Let $\mathcal{V}(K)$ be the set of input variables chosen for the K-th hidden unit. Any chosen input variable will be involved in both the logistic part and the weighted least squares part with all q terms of the hidden unit as well as the baseline polynomial. The forward selection procedure can be described as :

for $K := 1$ to $K_{\max}$;

```
        V(K) := ∅; lack-of-fit[K]:=∞;
        crit:=∞; (*)
        if #(V(K)) = d_max then goto (**);
        for j := 1 to d;
            if x_j ∉ V(K) then do;
                call SNN(V(1), ..., V(K − 1), V(K) ∪ {x_j}; BIC);
                if BIC<crit then do;
                    crit:=BIC; label:=j;
                endif;
            endif;
        endfor;
        if (#(V(K)) < d_min) or (lack-of-fit[K]>crit) then do;
            V(K):=V(K) ∪ {x_label}; lack-of-fit[K]:=crit; goto (*);
        endif;
    endfor; (**)
    K*:=argmin(lack-of-fit); BIC*:=lack-of-fit[K*];
```

The subroutine $\text{SNN}(\mathcal{V}(1), \ldots, \mathcal{V}(K-1), \mathcal{V}(K); \text{BIC})$ fits q-SNN with K hidden units, using input variables in $\mathcal{V}(j)$ in the j-th hidden unit, $j = 1, \ldots, K$, by the EM algorithm and gives the BIC value of the fitted model. Therefore, the inner loop selects the input variables with smallest BIC for the hidden unit. The chosen variable is added to the hidden unit if the number of variables assigned to the hidden unit is less than the $d_{\min}$ or if the BIC of the model is lowered by including the variable. The outer loop controls the increment of the number of the hidden units. The selected model is the q-SNN with K^* hidden units and the input variables in $\mathcal{V}(1), \ldots, \mathcal{V}(K^*)$.

With K^* fixed after the forward step, the backward elimination procedure proceeds as follows:

1. Calculate the Wald statistic for each parameter of the model.

2. Choose the one with the smallest absolute value of Wald's statistic from the set of estimated parameters. It corresponds to the most insignificant one in this set.

3. Check if elimination of the chosen parameter results in a reduction of the number of hidden units. If that is the case, keep this parameter away from elimination and go back to Step 2. Otherwise, go to the next step.

4. If the absolute value of the chosen Wald statistic is greater than some pre-specified threshold z, then stop; otherwise, go to the next step.

5. Re-estimate the model with the chosen parameter set to zero. Record the corresponding BIC. Go to Step 1 with the current model.

The output of the backward step is the model with the smallest BIC in the above elimination sequence. Step 2 above prevents the procedure from reducing the number of hidden units. The z in Step 4 is a percentile of the standard normal distribution. It controls the extensiveness of the elimination. If z is too large, it will result in a very long list of candidate models and will cause a lot of computing effort in estimating each of them. On the other hand, if z is too small, the model may not be trimmed adequately. We suggest $z = 1.96$, which corresponds at least approximately to a test with size 5% level. A more extensive elimination step can be obtained by setting $z = 2.58$ which corresponds to size 1%.

References

1. Akaike, H. (1974) A new look at the statistical model identification. *IEEE Transactions on Automatic Control* **19** 716-723.
2. Barron, A.R. (1993) Universal approximation bounds for superpositions of a sigmod function. *IEEE Transactions on Information Theory* **39** 930-945.
3. Boyer and Savageau (1986) *Places Rated Almanac.* Rand McNally.
4. Breiman, L. (1993) Hinging hyperplanes for regression, classification and function approximation. *IEEE Transactions on Information Theory* **39** 999-1013.
5. Breiman, L., Friedman, J.H., Olshen, R.A. and Stone C.J. (1984) *Classification and Regression Trees.* Monterey, CA: Wadsworth and Brooks/Cole.
6. Chan, K.S. and Tong, H. (1985) On the use of the deterministic Lyapunov function for the ergodicity of stochastic difference equations. *Advanced Applied Probability* **17** 667-678.
7. Davies, R.B. (1977) Hypothesis testing when a nuisance parameter is present only under the alternative. *Biometrika* **64** 247-254.
8. Davies, R.B. (1987) Hypothesis testing when a nuisance parameter is present only under the alternative. *Biometrika* **74** 33-43.
9. Dempster, A.P., Laird, N.M. and Rubin, D.B. (1977) Maximum likelihood from incomplete data via the EM algorithm. *Journal of Royal Statistical Society series B* **39** 1-38.

10. Duan, J.C. and Wong, S.P. (1999) Regime Switching with Feedback. *Working Paper, Hong Kong University of Science and Technology.*
11. Friedman, J.H. (1991) Multivariate adaptive regression splines (with discussion). *Annals of Statistics* **19** 1-141.
12. Friedman, J.H. and Stuetzle, W. (1981) Projection pursuit regression. *Journal of American Statistical Association* **76** 817-823.
13. Hamilton, J.D. (1995) *Time Series Analysis.* New Jersey: Princeton University Press.
14. Hastie, T.J. and Tibshirani, R.J. (1990) *Generalized Additive Models.* London: Chapman and Hall.
15. Jordan, M.I. and Jacobs, R.A. (1994) Hierarchical mixtures of experts and the EM algorithm. *Neural Computation* **6** 181-214.
16. Lai, T.L. and Wong, S.P. (1999) Stochastic neural networks with applications to nonlinear time series. *Working Paper.*
17. Lange, K.L., Little, R.J. and Taylor, J.M.G. (1989) Robust statistical modeling using the t distribution. *Journal of American Statistical Association* **84** 881-896.
18. Lewis, P.A.W. and Stevens, J.G. (1991) Nonlinear modeling of time series using multivariate adaptive regression splines (MARS). *Journal of American Statistical Association* **86** 864-877.
19. McCulloch, W.S. and Pitts, W. (1943) A logical calculus of ideas immanent in neural activity. *Bulletin of Mathematical Biophysics* **5** 115-133.
20. Rosenblatt, F. (1962) *Principles of Neurodynamics: Perceptron and Theory of Brain Mechanisms.* Spartan Books, Washington D.C.
21. Rumelhart, D.E., Hinton, G.E. and Williams, R.J. (1986) Learning representations by backpropagation errors, *Nature* **323** 533-536.
22. Schwarz, G. (1978) Estimating the dimension of a model. *Annals of Statistics* **6** 461-464.
23. Taylor, J.M.G. (1988) The cost of Generalizing Logistic Regression. *Journal of American Statistical Association* **83** 1078-1083.
24. Tjøstheim, D. (1990) Nonlinear time series and Markov chains. *Advanced Applied Probability* **22** 587-611.
25. Tong, H. (1990) *Non-linear Time Series: A Dynamical System Approach.* London:Oxford University Press.
26. Tweddie, R.L. (1975) Sufficient conditions for ergodicity and recurrence of Markov chain on a general state-space. *Stochastic Processes and Their Applications* **3** 385-403.
27. Venables, W.N. and Ripley, B.D. (1994) *Modern Applied Statistics with S-Plus.* Springer-Verlag.
28. Weigend, A., Rumelhart, D. and Huberman, B. (1991) Generalization by

weight-elimination with application to forecasting. *Advances in Neural Information Processing 3* San Mateo CA: Morgan Kaufmann. 875-882.

29. Wong, C.S. and Li, W.K. (1999) On a mixture autoregressive model. *Journal of Royal Statistical Society series B* forthcoming.

Part V: Applications

THE OVERREACTING BEHAVIOR OF REAL EXCHANGE RATE DYNAMICS

YIN-WONG CHEUNG
Department of Economics, University of California,
Santa Cruz, CA 95064, USA
E-mail: cheung@cats.ucsc.edu

KON S. LAI
Department of Economics and Statistics, California State University,
Los Angeles, CA 90032, USA
E-mail: klai@calstatela.edu

This study reviews and discusses empirical evidence corroborating the existence of overreaction in the short-term responses of real exchange rates. The amplification of shock responses, albeit occurring over a short time period only, can delay and substantially prolong the time it takes for the real exchange rate to converge to parity. Interestingly, the findings of short-term amplified responses of the real exchange rate—with its subsequent reversal and gradual reversion toward the long-run equilibrium—appear compatible with the chartist-fundamentalist model of the foreign exchange market microstructure.

1 Introduction

The purchasing power parity (PPP) theory, which suggests that two countries' price levels are equal at equilibrium when expressed in a common currency unit, has served as a major building block for many models of exchange rate determination. Under the PPP theory, nominal disturbances have no permanent effects on the real exchange rate, as implied by long-run monetary neutrality. Although short-run departures from parity are commonly recognized, many economists continue to hold the view that PPP, as a long-run proposition, will prevail. The faith in PPP has been weakened by the recent floating-rate experience, nonetheless. PPP deviations, gauged by real exchange rates, are often observed to be large, volatile and highly persistent. Such dynamics appear for the most part unexplained by economic fundamentals.

It is difficult to reconcile the immense short-term volatility of the real exchange rate with its very slow rate of convergence to parity. Slowly evolving changes in economic fundamentals—such as changes in tastes and technology—may contribute to the slow convergence, but they are not volatile enough over the short term to account for the vast exchange rate volatility. Sticky-price models, à la Dornbusch's (1976) overshooting analysis, are often used to show how monetary shocks can bring about large, volatile deviations from PPP. The strong short-term correlation generally found between exchange rates and real exchange rates can be viewed as indirect evidence for price stickiness (Mussa, 1986). With sticky prices, an unexpected change in money supply will alter real cash balances and interest rates, thereby affecting the value of the domestic currency. As prices gradually adjust later in response to the monetary shock, it will lead to reverse movements in interest rates and hence the currency value. Along with the adjustment in the currency rate during this phase, PPP deviations will dwindle at a rate depending upon how fast prices can adjust. The process of reversion continues until a long-run equilibrium consistent with PPP is reached. Accordingly, the real exchange rate will be more persistent the more sluggish the price adjustment is.

Rogoff (1996) points out that if PPP deviations are really driven by sluggish price adjustment, the real exchange rate should be expected to converge at a much faster rate than what has typically been found. The empirical rate of convergence appears far too slow to be explained by price stickiness, however. This poses a serious challenge for the literature. No existing macroeconomic models can consistently explain both the vast short-term volatility and the "excessively" high persistence observed in the real exchange rate.

In looking beyond the influence of macroeconomic fundamentals, Taylor (1995) recognizes the possible role of microstructural factors—including the behavior of foreign exchange market agents—in generating short-term PPP deviations. For example, the rising importance of chartists in currency trading can extend and magnify the short-term impact of market shocks on exchange rate movements. Based on survey expectations data for major currencies, Frankel and Froot (1990) report that, "at short horizons, [traders] tend to forecast by extrapolating recent trends, while at long horizons they tend to forecast a return to a long-run equilibrium such as purchasing power parity" (p.183). If short-term exchange rate responses can be magnified by trend-following currency trading, they will impart similar behavior into real exchange rates, especially when operating under sticky prices.

This study reviews and discusses empirical evidence, which corroborates the existence of amplified short-term responses of real exchange rates. These amplified shock responses—which tend to magnify PPP deviations initially—can not only contribute to the short-term volatility of the real exchange rate but also prolong substantially the time it takes for the real exchange rate to converge to parity.

2 Direct Evidence of Parity Reversion

In analyzing the mean-reverting property of real exchange rates, conventional unit root tests are known to be afflicted by low statistical power, leading to the widespread failure to find reversion toward PPP in early studies of the modern floating-rate period. The problem may be aggravated by the high volatility of floating exchange rates, making it difficult to detect parity reversion in the noisy data. Several approaches have been advanced to overcome the power problem, nevertheless. They include the use of long-horizon data to extend the sample period (Diebold, Husted and Rush, 1991; Lothian and Taylor, 1996). This method involves using data from the pre-float period. Another approach uses cointegration tests with good power to explore the long-run relationship between exchange rates and relative prices (Cheung and Lai, 1993; Edison, Gagnon and Melick, 1997). An alternative approach considers pooling data across real exchange rates in panel unit root tests (Frankel and Rose, 1996; Oh, 1996; Papell, 1997), but the robustness of panel test results has been called into question (Engel, Hendrickson and Rogers, 1997; O'Connell, 1998; Taylor and Sarno, 1998). Still another approach is to use efficient unit root tests with optimal power. Cheung and Lai (1998) employ this direct approach and unveil significant evidence of PPP reversion without using either long-horizon or panel data. The direct approach is adopted here.

The data under study are monthly real exchange rates constructed from nominal exchange rates and consumer price indices. Specifically, the real exchange rates of four European countries—France (FR), Germany (GE), Italy (IT), and the United Kingdom (UK)—vis-à-vis the United States (US) are investigated. Taken from the International Monetary Fund's *International Financial Statistics* data CD-ROM, the data cover the sample period from April 1973 through December 1996. All the series of real exchange rates are expressed in logarithms, following the common practice in previous PPP studies.

Before analyzing the intertemporal path of adjustment, we first establish evidence of mean reversion for the individual series of real exchange rates. The efficient unit root test devised by Elliott, Rothenberg and Stock (1996) is carried out. These authors establish the asymptotic power envelope for unit root tests by analyzing the sequence of Neyman-Pearson tests of the null hypothesis H_0: $\rho = 1$ against the local alternative H_a: $\rho = 1 + \bar{c}/T$, where ρ is the largest autoregressive (AR) root in the AR(k + 1) model, T is the sample size and $\bar{c} < 0$. Based on asymptotic power calculation, it is shown that a modified Dickey-Fuller test, called the DF-GLS test, can achieve significant power gains over standard unit root tests. The superior performance of the DF-GLS test is also supported by the Monte Carlo results reported by Stock (1994). Although the DF-GLS test shares very similar size properties as the ADF test, the former shows much better test power than the latter.

For a real exchange rate series, denoted by $\{y_t\}$, the DF-GLS test entails the following regression:

$$(1 - L)y_t^\mu = \phi_0 y_{t-1}^\mu + \Sigma_{j=2}^k \phi_j (1 - L) y_{t-j+1}^\mu + v_t \tag{1}$$

where L is the usual lag operator such that $Ly_t = y_{t-1}$; v_t is the random error term; and y_t^μ, the locally demeaned data process under the local alternative of $\bar{\rho} = 1 + \bar{c}/T$, is given by

$$y_t^\mu = y_t - \tilde{g}' z_t \tag{2}$$

Table 1: Results from the ADF and DF-GLS Unit Root Tests

Test	Series	k	Statistic	10% CV	5% CV
ADF	FR/US	4	-2.148	-2.864	-2.571
	GE/US	2	-1.868	-2.859	-2.566
	IT/US	2	-2.021	-2.859	-2.566
	UK/US	4	-2.337	-2.864	-2.571
DF-GLS	FR/US	4	-2.152**	-1.684	-2.000
	GE/US	2	-1.872*	-1.689	-2.006
	IT/US	2	-2.009**	-1.689	-2.006
	UK/US	4	-1.874*	-1.684	-2.000

Notes: The column beneath "k" gives the lag parameter chosen using the Akaike information criterion. Finite-sample critical values (CV) for the ADF test are obtained from Cheung and Lai (1995a) based on response surface estimation for $T = 285$. Finite-sample CVs for the DF-GLS test are from Cheung and Lai (1995b) for $T = 285$, as described by Eq. (4). Asymptotic CVs for the DF-GLS test are given, respectively, by -1.62 and -1.95 for the 10% and 5% significance levels. Statistical significance is indicated by a single asterisk (*) for the 10% level and a double asterisk (**) for the 5% level.

with $\hat{g}$ being the least squares coefficient estimated from regressing $\tilde{y}_t$ on $\tilde{z}_t$:

$$\tilde{y}_t = g'\tilde{z}_t + \epsilon_t \tag{3}$$

for which $\tilde{y}_t = (y_1, (1 - \bar{\rho}L)y_2, ..., (1 - \bar{\rho}L)y_T)'$ and $\tilde{z}_t = (z_1, (1 - \bar{\rho}L)z_2, ..., (1 - \bar{\rho}L)z_T)'$. In general, $z_t = (1, t)$, allowing for a linear trend. No time trend is considered in our case here, so $z_t = 1$. The DF-GLS statistic is given by the conventional t-ratio, testing H_0: $\phi_0 = 0$ against H_a: $\phi_0 < 0$. The parameter, $\bar{c}$, which defines the local alternative through $\rho = 1 + \bar{c}/T$, is recommended to be set equal to -7 for the no-trend case.

Finite-sample size properties of the DF-GLS test have been explored by Cheung and Lai (1995b). Approximate finite-sample critical values (CV) can be computed from a response surface equation of a polynomial form:

$$CV_{T,k} = \eta_0 + \Sigma_{i=1}^{2}\eta_i(1/T)^i + \Sigma_{j=1}^{3}\xi_j(k/T)^j \tag{4}$$

where $CV_{T,k}$ is the critical value estimate for a sample size T and lag k; and the relevant parameter values for $\{\eta_0, \eta_1, \eta_2, \xi_1, \xi_2$ and $\xi_3\}$ are tabulated by Cheung and Lai (1995).

Table 1 contains the statistical results obtained from the DF-GLS test. To facilitate comparison, results from the ADF test are reported as well. When a time trend was included, it was statistically insignificant in all the four cases. Accordingly, the results for the no-trend case are reported. For the choice of the lag parameter, k, data-dependent lag selection is implemented using the Akaike information criterion. In contrast to the standard ADF test, which consistently fails to identify stationarity, the test results from the efficient DF-GLS test indicate significant evidence in favor of PPP reversion in all the cases under examination. More specifically, the hypothesis of a unit root can be rejected in favor of stationary alternatives at either the 10% significance level in the GE/US and UK/US cases or the 5% significance level in the FR/US and IT/US cases. The results are consistent with those reported by Cheung and Lai (1998), who examined a shorter sample of real exchange rate data for FR/US, GE/US, and UK/US (but not for IT/US) and uncovered parity reversion in the data series using the DF-GLS test.

In analyzing long historical data, Culver and Papell (1995) and Perron and Vogelsang (1992) report that the behavior of real exchange rates can be characterized by trend-break models. Hegwood and Papell (1998) illustrate that the presence of structural breaks can cause a significant upward bias in the estimation of half-life persistence of PPP deviations in long-horizon data. As part of the preliminary data analysis, different trend-break unit root tests devised by Banerjee, Lumsdaine and Stock (1992)—henceforth BLS—were performed on the recent float data. The BLS tests involve the following regression:

$$(1 - L)y_t = \mu_0 + \mu_1 t + \mu_2 d_t(n) + \beta_0 y_{t-1} + \Sigma_{j=1}^{p}\beta_j(1 - L)y_{t-j} + \zeta_t \tag{5}$$

where $d_t(n)$ is a dummy variable and ζ_t is the random error term. When a trend shift is allowed for at time n, $d_t(n) = (t - n)I(t > n)$, with $I(\cdot)$ being the indicator function. Alternatively, when a mean shift (or a break in the trend) is allowed for at time n, $d_t(n) = I(t > n)$. For the usual Dickey-Fuller test, $d_t(n) = 0$. A sequence of t-statistics for testing $\beta_0 = 0$, denoted by $\tau_{DF}(n)$, can be generated by varying n over the sample. BLS discuss different versions of the mean-shift or trend-shift sequential test. The minimal sequential test is applied in this analysis, and its test statistic is defined by

$$\tau_{DF}^{min} \equiv \min_{r \le n \le T-r} \tau_{DF}(n) \tag{6}$$

for the sample size, T, and a trimming parameter, r. Following BLS, r is set equal to the integer part of $.15T$. According to the BLS test results (not reported here), in no case could significant evidence be found to support the relevance of trend-break models in explaining the real exchange rate dynamics over the recent float, regardless of whether mean-shift or trend-shift models were entertained.

3 Analyzing the Adjustment Process Toward Parity

Although the findings of no unit root in the real exchange rate confirm the long-run convergence to parity, they offer no specific information on the dynamic process of adjustment itself. The question is, How do deviations from PPP behave over the short or medium run? Such information may bear upon the issue concerning the slow rate of parity convergence. To obtain the relevant information, impulse response analysis can be used.

Given that the real exchange rate is shown to be stationary, its dynamics can be captured in general by an autoregressive moving-average (ARMA) model as follows:

$$B(L)y_t = D(L)u_t \tag{7}$$

where $B(L) = 1 - b_1L - ... - b_pL^p$; $D(L) = 1 + d_1L + ... + d_qL^q$; all roots of $B(L)$ and $D(L)$ are stable; and u_t is the white-noise innovation term. The persistence of the process over different time horizons can be analyzed by studying the moving-average representation for y_t: $y_t = C(L)u_t$ with

$$C(L) = B^{-1}(L)D(L) \tag{8}$$

where $C(L) = 1 + c(1)L + c(2)L^2 + ... + c(j)L^j + ...$ Consider a unit shock to the process. The impact of a unit innovation at time t on the level of y at time $t+j$ is given by $c(j)$. This $c(j)$ measure—referred to as the impulse response—summarizes the basic information concerning persistence over all time spans up to infinite after the initial shock. For a stationary process, which contains no unit root, the infinite impulse response is $c(\infty) = 0$. A stationary process thus has zero long-run persistence. Over horizons much shorter than infinity, on the other hand, $c(j) \neq 0$ and sizable persistence can still exist over the short or medium run.

Instead of studying the entire sequence of $c(j)$, $j = 1, 2, ...$, a simple summary measure of persistence typically employed in the PPP literature is the half-life, which indicates how long it takes for the impact of a unit shock on the real exchange rate to dissipate by half. By definition, the half-life, denoted by ℓ_h, is given by $c(\ell_h) = ½$. Since discrete-time data are analyzed, and the half-life does not have to be exactly an integer number, the approximate value of ℓ_h will be calculated using a simple interpolation method when $c(j) < c(\ell_h) = ½ < c(j+1)$ for some j.

4 Empirical Findings of Overreaction in Initial Responses

The adjustment dynamics of real exchange rates in response to a shock to parity are examined through the sequence of cumulative impulse responses, $c(j)$. The DF-GLS test applied earlier is based on approximating AR models. Using the GLS estimates of the fitted AR models, impulse response functions are constructed for the individual series of real exchange rates.

Table 2 reports estimates of up to the first 120 cumulative impulse responses, which cover a time span of 10 years for monthly data. The results reveal the existence

Table 2: Time-Profiles of $c(j)$ Estimates for Real Exchange Rates

j	FR/US	GE/US	IT/US	UK/US
0	1.000	1.000	1.000	1.000
1	1.263	1.277	1.337	1.339
2	1.282	**1.336**	1.431	1.358
3	1.319	1.329	**1.437**	1.365
4	**1.364**	1.304	1.412	**1.391**
5	1.333	1.273	1.376	1.371
6	1.307	1.241	1.338	1.334
7	1.282	1.209	1.298	1.301
8	1.245	1.178	1.260	1.267
9	1.206	1.148	1.222	1.232
10	1.168	1.119	1.185	1.196
11	1.131	1.090	1.150	1.162
12	1.093	1.062	1.116	1.129
13	1.056	1.035	1.082	1.096
14	1.021	1.008	1.050	1.065
15	0.987	0.982	1.018	1.034
20	0.830	0.862	0.874	0.893
25	0.699	0.757	0.751	0.771
30	0.588	0.665	0.645	0.666
35	0.495	0.584	0.554	0.575
40	0.416	0.513	0.476	0.497
50	0.295	0.395	0.351	0.370
60	0.209	0.305	0.259	0.276
70	0.148	0.235	0.191	0.206
80	0.105	0.181	0.141	0.154
90	0.074	0.140	0.104	0.115
100	0.052	0.108	0.077	0.085
110	0.037	0.083	0.056	0.064
120	0.026	0.064	0.042	0.047

Notes: Columns 2-5 give the cumulative impulse responses of individual real exchange rate series at different time horizons, *j*, following a unit shock to parity. For each real exchange rate series, the number in boldface shows where the value of $c(j)$ peaks after the shock and the corresponding size of the shock amplification.

of non-monotonicity in the process of convergence to parity. The cumulative impulse responses are not a monotonic function of the adjustment horizon. Although the eventual decay of $c(j)$ toward zero affirms the existence of parity reversion, the real exchange rate initially tends to overreact to the shock in the sense that it continues its momentum to move farther apart from its long-run equilibrium level. Consequently, the PPP deviation tends to magnify first before diminishing (see Figure 1). Such non-monotonic responses of the real exchange rate—albeit they occur over merely a short time period—can delay and substantially prolong the process of convergence. In the cases studies here, it takes at least 15 months to offset the impact of the amplified responses and for $c(j)$ to simply return back to the unity level. The results here are similar to those obtained by Cheung and Lai (1999) from ARMA models, supporting the robustness of the impulse response results with respect to model specifications.

The presence of short-term overreaction in shock responses leaves the real exchange rate to adjust to, in effect, a more-than-unit shock during the subsequent reversion phase. For a given rate of decay of PPP deviations, the half-life persistence for the real exchange rate can vary widely, depending on the size of the short-term overreaction. If the size of the overreaction is considerable, the calculated half-life will show up to be long even when PPP deviations can die out at a relatively fast speed. This can be illustrated using a numerical example. Suppose that the real exchange rate reverts at the same speed as price adjustment, say, at a speed of $\ell_h = 2$ years. This value of ℓ_h implies that PPP deviations will dampen out at a rate of about 29.3 percent per year. Suppose also that there is shock amplification by a factor of 1.5 during the first month in response to a monetary shock. Under this situation, it can be shown that it will take about 1.2 year just to reverse the impact of the initial overreaction. Moreover, the situation will produce a ℓ_h estimate of roughly 3.3 years, much higher than the corresponding ℓ_h value of 2 years in the absence of the initial shock amplification. Table 3 illustrates a positive relationship between the size of the shock amplification and its contribution (in percentage) to the observed half-life persistence. In the case in which the amplification factor equals 1.5, the short-term overreaction can explain in excess of one-third of the observed persistence.

The half-life estimates are computed from our actual data on real exchange rates. Confidence intervals of the estimates are presented as well. For a given horizon, j, $c(j)$ is a nonlinear function of the ARMA parameter vector, $\zeta = (b_1, \ldots, b_p, d_1, \ldots, d_q)$. Using the delta method, standard errors for $c(j)$ can be estimated from

$$\mathrm{Var}(c(j)) = \nabla c'(j)\Omega\nabla c(j) \tag{9}$$

where $\nabla c(j)$ is $\partial c(j)/\partial\zeta$ and Ω is the variance-covariance matrix of ζ (Campbell and Mankiw, 1987). The sample standard errors will be used to construct confidence intervals for $c(j)$, from which confidence intervals for ℓ_h estimates are derived. It should be noted that these asymptotic estimates of standard errors do not correct for any finite-sample bias and can understate the level of potential imprecision associated

Table 3: The Impact of Overreacting Responses on Half-Life Persistence

Amplification Factor	Proportion of ℓ_h Explained
1.0	0.0%
1.1	15.3%
1.2	23.4%
1.3	29.6%
1.4	34.4%
1.5	38.5%
1.6	41.9%
1.7	44.6%
1.8	47.1%
1.9	49.2%
2.0	51.0%
2.1	52.6%
2.2	54.1%

Notes: In the benchmark case of no short-term overreaction, the amplification factor equals 1.0. The second column gives the computed proportion of the observed half-life attributable to the overreaction. The computation assumes that the overreaction occurs during the first month only.

Table 4: Half-Life Estimates

Variable	FR/US	GE/US	IT/US	UK/US
ℓ_h	2.89	3.41	3.20	3.31
L_{95}	1.36	1.44	1.40	1.41
L_{90}	1.50	1.61	1.56	1.57
U_{90}	5.38	6.85	6.09	6.56
U_{95}	5.79	7.41	6.56	7.07

Notes: The column "ℓ_h" provides the point estimates of the half-life adjustment speed (in years). [L_{90}, U_{90}] represents the 90% confidence interval for ℓ_h; whereas, [L_{95}, U_{95}] gives the 95% confidence interval for ℓ_h.

with half-life estimation. Berkowitz and Kilian (1998) and Kilian (1998) recently advocate the use of bootstrapping methods to evaluate sampling uncertainty. For example, the distribution of the innovation term can be approximated by the empirical distribution of the estimated residual using resampling (with replacement) techniques.

Table 4 gives the half-life persistence estimates for the individual real exchange rate series. The point estimates of ℓ_h range from 2.9 to 3.4 years and yield an average of about 3.2 years. As noted by Rogoff (1996), these half-life estimates seem too long to be explained by price stickiness because they suggest—according to Eq. (10) below—a very slow rate of convergence of about 19 percent per year on average.

Pesaran and Shin (1996) investigate the speed of convergence to PPP under a multivariate cointegration framework. Unlike the univariate time series method considered here, these authors analyze the equilibrium relations between exchange rates, prices and interest rates in the case of the UK. Specifically, the persistence profiles of both the PPP relation and the UIP (uncovered interest-rate parity) relation are estimated simultaneously. Their point persistence estimates show that the estimated rate of convergence to PPP is very slow, while the convergence to UIP seems rather fast. Interestingly, these authors also observe that the persistence profile for the PPP relation is hump-shaped such that a shock to PPP tends to magnify first before reverting.

Reversion speeds have often been computed indirectly from half-life estimates under an implicit assumption of monotonic convergence at a constant speed (AS): the speed is computed from ℓ_h as

$$AS = 1 - \exp[\ln(\tfrac{1}{2})/\ell_h]. \tag{10}$$

Such an assumption is not valid in general situations, nonetheless. In particular, the convergence can be not monotonic. Non-monotonic convergence may arise when undershooting occurs before reverting back to parity. The convergence may also come in the form of oscillating dynamics. In short, the simple half-life measure unsatisfactorily ignores the actual structure of the adjustment process. In our case, shock responses are found to amplify before dissipating, leading to non-monotonic dynamics. Such short-term amplification of shock responses can delay and protract the adjustment process, leading to the slow convergence to parity. The non-monotonicity confounds the half-life measure to give distorting estimates of the adjustment speed.

5 A Direct Measure of Adjustment Speed

A direct measure of the adjustment speed, which is more informative than the half-life measure, can be obtained from the cumulative impulse response function. Unlike the half-life measure, which is single-valued, the alternative measure is a function of time, giving the actual rate at which the impact of a shock die out along the entire path of adjustment. More specifically, the adjustment speed of the real exchange rate at time

Table 5: Time-Profiles of the Adjustment Speed (Per Month) of Real Exchange Rates

j	FR/US	GE/US	IT/US	UK/US
0	-26.29%	-27.71%	-33.65%	-33.94%
1	-1.49%	-4.61%	-7.05%	-1.39%
2	-2.91%	**0.49%**	-0.45%	-0.52%
3	-3.40%	1.93%	**1.75%**	-1.86%
4	**2.30%**	2.38%	2.53%	**1.42%**
5	1.94%	2.51%	2.82%	2.67%
6	1.93%	2.55%	2.93%	2.51%
7	2.89%	2.56%	2.97%	2.58%
8	3.13%	2.56%	2.99%	2.80%
9	3.09%	2.57%	3.00%	2.86%
10	3.24%	2.57%	2.99%	2.86%
11	3.33%	2.57%	2.99%	2.87%
12	3.32%	2.57%	3.00%	2.88%
13	3.35%	2.57%	2.99%	2.89%
14	3.38%	2.56%	3.00%	2.88%
15	3.39%	2.58%	3.00%	2.89%
20	3.40%	2.57%	3.00%	2.89%
25	3.39%	2.58%	3.00%	2.89%
30	3.40%	2.56%	3.01%	2.90%
35	3.40%	2.57%	3.00%	2.89%
40	3.41%	2.58%	2.99%	2.90%
45	3.40%	2.56%	3.01%	2.89%
50	3.40%	2.58%	2.99%	2.89%
60	3.41%	2.59%	2.98%	2.90%
70	3.39%	2.56%	2.99%	2.87%
80	3.44%	2.54%	2.98%	2.93%
90	3.38%	2.57%	3.08%	2.88%
100	3.35%	2.60%	3.00%	2.93%
110	3.50%	2.53%	3.01%	2.98%
120	3.44%	2.50%	3.02%	2.95%

Notes: Columns 2-5 present the adjustment speed (AS_j) estimates for individual real exchange rate series at different time horizons, *j*, subsequent to a shock to parity. For each real exchange rate series, the number in boldface indicates when the shock amplification ends and when the PPP deviation begins to die out.

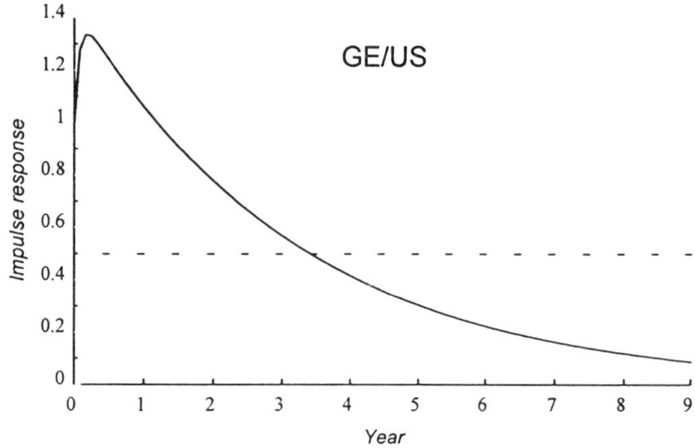

Figure 1. A sample plot (the GE/US case) of the dynamic responses of the real exchange rate to a unit shock.

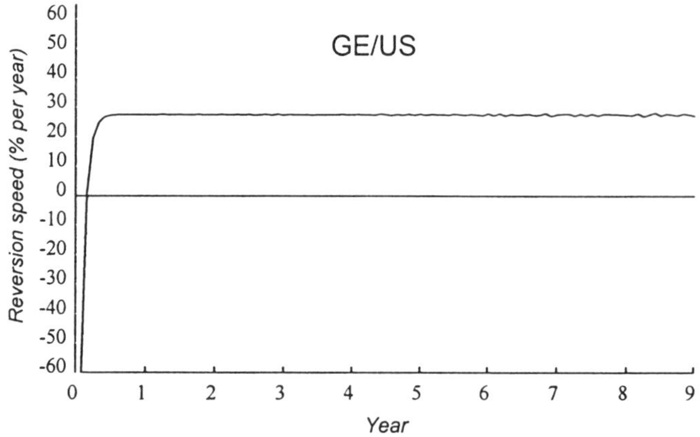

Figure 2. A sample plot (the GE/US case) of the different speeds of real exchange rate adjustment over time.

t is given by

$$AS_t = -[dC(t)/dt]/C(t) \tag{11}$$

which corresponds to the instantaneous (percentage) rate of decrease in the cumulative impulse response at time t. The AS_t measure is easy to interpret. When $AS_t > 0$, the real exchange rate is reverting toward parity at time t, and the magnitude of AS_t indicates the relevant speed. When $AS_t < 0$, on the other hand, the real exchange rate is moving further away from parity at time t at a speed of $|AS_t|$ per unit time. In this way, a researcher can compute and gauge both the direction and the speed at which the adjustment takes place at any time horizons after the initial shock.

Table 5 reports the time profile of adjustment speeds at various time horizons following a shock to parity. In every case, the adjustment speed starts from a negative value, showing the impact of the initial shock amplification (see also Figure 2). Reverting dynamics then take over quickly and then attain a steady positive speed toward parity. The AS_t estimates show that subsequent to the short-term amplified responses, real exchange rates converge at a rate of between 2.6 to 3.4 percent per month—an equivalent rate of between 31 to 41 percent per year—which is much faster than what half-life estimates have implied. Accordingly, the short-term amplified responses may create the appearance of slow reversion when measuring in terms of the half-life.

6 Concluding Remarks

The short-term adjustment of the real exchange rate has been found to be characterized by overreaction and amplified shock responses. Such dynamics can contribute to the large, volatile short-term PPP deviations. They can also delay and prolong the process of convergence to parity. Although the short-term overreacting dynamics may be viewed as overshooting behavior in the broad sense that reversion occurs only after persistent movements away from the long-run equilibrium, the dynamic adjustment pattern identified for the real exchange rate seems not compatible with the Dornbusch-type rational expectations models of overshooting. Specifically, short-term exchange rate overshooting under Dornbusch's (1976) model happens initially at the time of the shock only such that the maximal impact of the shock occurs contemporaneously. Following the shock, the real exchange rate reverts to its long-run value monotonically. This contrasts with our findings, in which the full impact of the shock is not felt immediately but until a few periods after the initial shock. It follows that the amplified responses observed in the real exchange rate cannot be explained by the conventional models of exchange rate overshooting.

On the other hand, the findings of short-term amplified responses of the real exchange rate—with its subsequent reversal and gradual reversion toward parity —seem consistent with the chartist-fundamentalist view of exchange rate dynamics, as identified in survey data on expectations of foreign exchange market participants (Allen and Taylor, 1990; Frankel and Froot, 1990, 1993). Chartists are market agents who like to follow recent trends and tend to have bandwagon expectations. Fundamentalists, in contrast, are market agents who base their forecasts on economic fundamentals and such forecasts tend to be regressive. Empirical findings from survey data generally suggest that exchange rate forecasts over short horizons are dominated by chartist analysis; whereas, exchange rate forecasts over long horizons are governed by fundamental analysis. Cheung and Wong (1999) explore the market practitioners' views on exchange rate dynamics and confirm the presence of bandwagon effects and short-term overreaction to news. To the extent that short-term currency trading is in large part spurred by bandwagon expectations, the effects of market shocks on real exchange rates will tend to amplify before dissipating.

The chartist-fundamentalist model also offers a possible explanation for persistent deviations from UIP, as noted by Eichenbaum and Evans (1995). In response to an expansionary monetary shock, for example, the domestic interest rate falls and the exchange rate (the dollar price of foreign currency) rises. The rise in the exchange rate will continue for a while after the initial shock because there are chartist traders who jump on the bandwagon, buying the foreign currency and causing further appreciation. When the rise in the exchange rate extends beyond the initial shock, the relative decrease in the U.S. interest rate will not be offset by an expected appreciation of the dollar, thereby giving rise to persistent expected excess returns and sustained deviations from UIP.

Acknowledgments

The authors would like to thank an anonymous referee, the editors, Kaushik Chaudhuri, Casper de Vries, Charles Engel, Clive Granger, Nils-Petter Lagerlof, Pushkar Maitra, Ulrich Muller, Hashem Pesaran, Tony Phipps, Jeff Sheen, Graham White, as well as participants of the 1999 Hong Kong International Workshop on Statistics in Finance and of the seminar at the University of Sydney for comments and suggestions. Any remaining errors are certainly ours.

References

1. H. Allen and M.P. Taylor, Charts, Noise and Fundamentals in the London Foreign Exchange Market, *Economic Journal*, 100, 49-59 (1990).
2. A. Banerjee, R.L. Lumsdaine and J.H. Stock, Recursive and Sequential Tests of the Unit-Root and Trend-Break Hypotheses: Theory and International Evidence, *Journal of Business and Economic Statistics*, 10, 271-287 (1992).
3. J. Berkowitz and L. Kilian, Recent Developments in Bootstrapping Time Series, Discussion Paper, Department of Economics, University of Michigan (1998).
4. J.Y. Campbell and N.G. Mankiw, Are Output Fluctuations Transitory?, *Quarterly*

Journal of Economics, 102, 857-880 (1987).
5. Y.W. Cheung and K.S. Lai, Long-Run Purchasing Power Parity During the Recent Float, *Journal of International Economics*, 34, 181-192 (1993).
6. Y.W. Cheung and K.S. Lai, Lag Order and Critical Values of the Augmented Dickey-Fuller Test, *Journal of Business and Economic Statistics*, 13, 277-280 (1995a).
7. Y.W. Cheung and K.S. Lai, Lag Order and Critical Values of a Modified Dickey-Fuller Test, *Oxford Bulletin of Economics and Statistics*, 57, 411-419 (1995b).
8. Y.W. Cheung and K.S. Lai, Parity Reversion in Real Exchange Rates During the Post-Bretton Woods Period, *Journal of International Money and Finance*, 17, 597-614 (1998).
9. Y.W. Cheung and K.S. Lai, On the Purchasing Power Parity Puzzle, *Journal of International Economics*, forthcoming (1999).
10. Y.W. Cheung and C.Y.P. Wong, A Survey of Market Practitioners' Views on Exchange Rate Dynamics, *Journal of International Economics*, forthcoming.
11. S.E. Culver and D.H. Papell, Real Exchange Rates Under the Gold Standard: Can They be Explained by the Trend Break Model?, *Journal of International Money and Finance*, 14, 539-548 (1995).
12. F.X. Diebold, S. Husted and M. Rush, Real Exchange Rates Under the Gold Standard, *Journal of Political Economy*, 99, 1252-1271 (1991).
13. R. Dornbusch, Expectations and Exchange Rate Dynamics, *Journal of Political Economy*, 84, 1161-1176 (1976).
14. H. Edison, J.E. Gagnon and W.R. Melick, Understanding the Empirical Literature on Purchasing Power Parity: The Post-Bretton Woods Era, *Journal of International Money and Finance*, 61, 1-17 (1997).
15. C. Engel, M.K. Hendrickson and J.H. Rogers, Intranational, Intracontinental, and Intraplanetary PPP, *Journal of the Japanese and International Economies*, 11, 480-501 (1997).
16. G. Elliott, T.J. Rothenberg and J.H. Stock, Efficient Tests for an Autoregressive Unit Root, *Econometrica*, 64, 813-836 (1996).
17. J.A. Frankel and K.A. Froot, Chartists, Fundamentalists, and Trading in the Foreign Exchange Market, *American Economic Review*, 80, 181-185 (1990).
18. J.A. Frankel and K.A. Froot, Understanding the U.S. Dollar in the Eighties: The Expectations of Chartists and Fundamentalists, in *On Exchange Rates*, ed. J.A. Frankel (The MIT Press, Cambridge, MA, 1993).
19. J.A. Frankel and A.K. Rose, A Panel Project on Purchasing Power Parity: Mean Reversion Within and Between Countries, *Journal of International Economics*, 40, 209-224 (1996).
20. N.D. Hegwood and D.H. Papell, Quasi Purchasing Power Parity, *International Journal of Finance and Economics*, 3, 279-289 (1998).
21. L. Kilian, Confidence Intervals for Impulse Responses under Departures from Normality, *Econometric Reviews*, 17, 1-29 (1998).

22. J.R. Lothian and M.P. Taylor, Real Exchange Rate Behavior: The Recent Float from the Perspective of the Past Two Centuries, *Journal of Political Economy*, 104, 488-509 (1996).
23. M. Mussa, Nominal Exchange Rate Dynamics, *Carnegie Rochester Conference on Public Policy*, 25, 117-214 (1986).
24. P.G.J. O'Connell, The Overvaluation of Purchasing Power Parity, *Journal of International Economics*, 44, 1-19 (1998).
25. K.-Y. Oh, Purchasing Power Parity and Unit Root Tests Using Panel Data, *Journal of International Money and Finance*, 15, 405-418 (1996).
26. D.H. Papell, Searching for Stationarity: Purchasing Power Parity Under the Current Float, *Journal of International Economics*, 43, 313-332 (1997).
27. P. Perron and T.J. Vogelsang, Nonstationarity and Level Shifts with an Application to Purchasing Power Parity, *Journal of Business and Economic Statistics*, 10, 301-320 (1992).
28. M.H. Pesaran and Y. Shin, Cointegration and Speed of Convergence to Equilibrium, *Journal of Econometrics*, 71, 117-143 (1996).
29. K. Rogoff, The Purchasing Power Parity Puzzle, *Journal of Economic Literature*, 34, 647-668 (1996).
30. J.H. Stock, Unit Roots, Structural Breaks and Trends, in *Handbook of Econometrics*, Vol. 4, eds. R.F. Engle and D.L. McFadden (North-Holland, New York, 1994).
31. M.P. Taylor, The Economics of Exchange Rates, *Journal of Economic Literature*, 33, 13-47 (1995).
32. M.P. Taylor and L. Sarno, The Behavior of Real Exchange Rates During the Post-Bretton Woods Period, *Journal of International Economics*, 46, 281-312 (1998).

PORTFOLIO MANAGEMENT AND MARKET RISK QUANTIFICATION USING NEURAL NETWORKS

JÜRGEN FRANKE
Department of Mathematics
Universität Kaiserslautern
Erwin-Schrödinger-Str.
67663 Kaiserslautern
Germany

We discuss how neural networks may be used to estimate conditional means, variances and quantiles of financial time series nonparametrically. These estimates may be used to forecast, to derive trading rules and to measure market risk.

1 Introduction

Neural networks are now a well-established tool in financial engineering. The main applications, considered up to now, are to classification, forecasting and portfolio management, but also to option pricing (compare, e.g., Anders[1], Bol et al.[2] and Refenes et al.[11]). In this paper, we first introduce the basic concepts, relating them to nonlinear time series models. Then, we give a short review of asymptotic theory, including a study of an appropriate resampling method. To illustrate the potential of neural network based procedures in practice, we also discuss too realistic case studies from stock and FX markets.

In the last two sections, we propose procedures which allow to estimate conditional variances and quantiles of nonlinear time series using neural networks. These nonparametric approaches may be used to quantify the risk of financial assets either by estimating the conditional volatility or the conditional value-at-risk. The kind of information conditioned upon may be rather arbitrary and of a high-dimensional structure.

2 Nonlinear time series models based on neural networks

One of the well-known stylized facts about financial time series is their serial uncorrelatedness, i.e. the univariate data appear to be white noise. Hence, we expect only nonlinear predictors to show any reasonable performance, and, additionally, we should use in forecasting not only past observations of the time series of interest, but also other economic information from the past. For forecasting the time series S_t, we therefore consider as basic model a nonlinear $AR(\tau)$ - process with exogeneous components $X_t \in R^d$

$$S_{t+1} = m(S_t, S_{t-1}, ..., S_{t-\tau}, X_t) + \varepsilon_{t+1} \tag{1}$$

The conditional expectation of the ε_t given information up to time t is 0. More specific assumptions on these innovations will be made later on. The d-variate exogeneous component X_t consists of values of other financial and economic time series up to time t. We do not assume a particular parametric form of the predictor function m which is the conditional expectation of S_{t+1} given $S_t, S_{t-1}, ..., S_{t-\tau}, X_t$. Therefore, we have to estimate it nonparametrically if we want to use it in forecasting S_{t+1}. As we have situations in mind where the autoregressive order $\tau + 1$ and the dimension d are large, familiar smoothing methods like kernel estimators, discussed e.g. by Kreiss[9], are not applicable without assuming a particular, e.g. additive, structure of the function m on $R^{\tau+1+d}$. Neural networks offer an alternative class of estimators which are flexible and computationally feasible.

To keep the notation simple, we first give a short review of neural network function estimators in the context of a heteroscedastic regression model similar to the time series model (1):

$$Z_t = m(X_t) + \varepsilon_t \tag{2}$$

where $X_1, X_2, \ldots$ are independent identically distributed with density $p(x), x \in R^d$, and the residuals $\varepsilon_1, \varepsilon_2, \ldots$ are independent with

$$\mathcal{E}\{\varepsilon_t | X_t = x\} = 0\,,\; \mathcal{E}\{\varepsilon_t^2 | X_t = x\} = \sigma_\varepsilon^2(x) < \infty.$$

We assume that the conditional mean $m(x)$ and the conditional variance $\sigma_\varepsilon^2(x)$ of Z_t given $X_t = x$ are continuous and bounded functions.

We want to estimate the function m on R^d using feedforward neural networks with one hidden layer. As the basic building block we consider the so-called *neuron* as a nonlinear transformation of a linear combination of the inputs $x = (x_1, ..., x_d)'$:

$$x \mapsto \psi(b + u_1 x_1 + ... u_d x_d)$$

ψ is a fixed *activation function*; in the following we always choose the centered sigmoid function

$$\psi(s) = \frac{2}{1 + e^{-s}} - 1.$$

Combining H neurons, we get the *network function*

$$f_H(x, \vartheta) = v_0 + \sum_{h=1}^{H} v_h \psi(b_h + w_h' x)$$

where $\vartheta = (b_1, \ldots, b_H, w_1', \ldots, w_H', v_0, \ldots, v_H)'$ denotes the parameter vector consisting of the *network weights* with $w_h' = (w_{1h}, \ldots, w_{dh}),\ h = 1, \ldots, H$.

$f_H(x, \vartheta)$ specifies a mapping from the input space R^d to the output space which, in our case, is one-dimensional. Such network functions are universal approximators (Hornik et al.[7]), i.e. any regression function $m(x)$ may be approximated arbitrarily well using a large enough number H of neurons and appropriate parameters ϑ. In practice, feedforward networks with more than one hidden layer of neurons may provide a more parsimonious fit to m. As the theory and numerical practice is essentially the same for this more general case, we restrict our considerations here mainly to networks with only one hidden layer.

To estimate the conditional expectation $m(x) = \mathcal{E}\{Z_t | X_t = x\}$ from a sample $(X_1, Z_1), \ldots, (X_N, Z_N)$, we fix the number H of neurons and calculate the nonlinear least squares estimate $\widehat{\vartheta}_N$ of the parameter ϑ by solving

$$\widehat{D}_N(\vartheta) \equiv \frac{1}{N} \sum_{t=1}^{N} (Z_t - f_H(X_t, \vartheta))^2 = \min_{\vartheta \in \Theta_H} !$$

$\widehat{\vartheta}_N$ is consistent in the sense that $\widehat{\vartheta}_N \longrightarrow \vartheta_0$ for $N \to \infty$, where ϑ_0 is the parameter for which the given network provides the best approximation of m, i.e.

$$\mathcal{E}(m(X_t) - f_H(X_t, \vartheta))^2 = \min_{\vartheta \in \Theta_H} !$$

Under the above conditions, $\widehat{\vartheta}_N$ is asymptotically Gaussian:

Theorem: For $N \to \infty$,

$$\sqrt{N}(\hat{\vartheta}_N - \vartheta_0) \longrightarrow \mathcal{N}(0, \Sigma_1 + \Sigma_2).$$

with covariance matrices
$\Sigma_i = A(\vartheta_0)^{-1} B_i(\vartheta_0) A(\vartheta_0)^{-1}, i = 1, 2$, where $A(\vartheta) = \nabla^2 D_0(\vartheta)$,
$B_1(\vartheta) = 4 \cdot \int \sigma_\varepsilon^2(x) \nabla f_H(x, \vartheta) \nabla' f_H(x, \vartheta) p(x) dx$,
$B_2(\vartheta) = 4 \cdot \int (m(x) - f_H(x, \vartheta))^2 \nabla f_H(x, \vartheta) \nabla' f_H(x, \vartheta) p(x) dx$.

The second part Σ_2 of the asymptotic covariance matrix represents the effect of misspecification due to fitting a network function with given H to an arbitrary regression function m. In the correctly specified case, where

$m(x) = f_H(x, \vartheta_0)$, we have $\Sigma_2 = 0$.

A simple proof of the theorem is given by Franke and Neumann[6]. A much more general result, which, under appropriate assumptions, also covers the time series model (1), has been given by White[13]. An immediate consequence of the theorem is

$$f_H(x, \widehat{\vartheta}_N) \longrightarrow f_H(x, \vartheta_0) \qquad \text{for } N \to \infty.$$

By the universal approximation property of neural networks, $f_H(x, \vartheta_0)$ converges to $m(x)$ for $H \to \infty$. Therefore, $f_H(x, \widehat{\vartheta}_N)$ should become a consistent nonparametric estimate of $m(x)$ if H increases with N with an appropriate rate. White[14] has proven a corresponding result. In practice, H is chosen by comparing the performance of the function estimators $f_H(x, \widehat{\vartheta}_N)$ for various H on a *validation set* of data, which has not been used in calculating the estimate $\widehat{\vartheta}_N$. Alternatively, one could use the *neural information criterion* of Murata et al.[10] which is a version of Akaike's AIC adapted to neural network based regression and autoregression models.

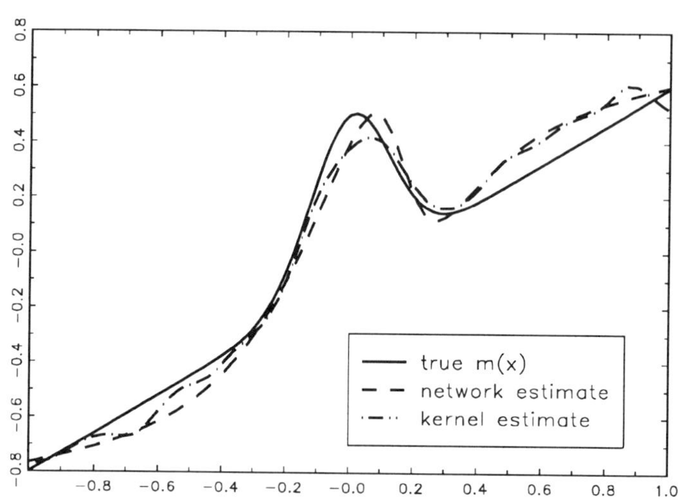

Figure 1: autoregressive function and estimates

Resampling may be used to improve the asymptotic normal approximation for the law of $f_H(x, \widehat{\vartheta}_N)$, where, for practical purposes, the covariance matrices Σ_1 and Σ_2 would have to be estimated anyhow. We present a residual-based bootstrap for the simple nonlinear autoregression of order 1 or NLAR(1)

$$S_{t+1} = m(S_t) + \varepsilon_t, \tag{3}$$

but the generalization to higher order models is straightforward. We start the procedure with some initial estimate $\widehat{m}_N$ which may be a neural network function estimate itself or some other consistent estimate for m. It allows for calculating sample versions of the innovations ε_t by

$$\widehat{\varepsilon}_t = Y_t - \widehat{m}_N(X_t)\,, \quad t = 1, \ldots, N,$$

which have to be centered around 0:

$$\widetilde{\varepsilon}_t = \widehat{\varepsilon}_t - \frac{1}{N}\sum_{k=1}^{N} \widehat{\varepsilon}_k\,, \quad t = 1, \ldots, N.$$

Let $\widetilde{F}_N$ denote the empirical distribution given by $\widetilde{\varepsilon}_1, \ldots, \widetilde{\varepsilon}_N$.

To generate the bootstrap resamples of the original time series, we first draw independent bootstrap innovations $\varepsilon_1^*, \ldots, \varepsilon_N^*$ from $\widetilde{F}_N$, i.e.

$$\varepsilon_t^* = \widetilde{\varepsilon}_k \text{ with probability } \frac{1}{N}, \quad k = 1, \ldots, N.$$

Then, we generate the bootstrap data as

$$S_{t+1}^* = \widehat{m}_N(S_t^*) + \varepsilon_t^*, \; t = 1, \ldots, N.$$

Using standard Monte Carlo techniques, we may mimic the behaviour of any quantity of interest based on a whole family of independent bootstrap resamples $S_0^*(i), \ldots, S_N^*(i),\; i = 1, \ldots, B$. The mean-squared error of the function estimate at x ,

$$\mathrm{mse}(x) = \mathcal{E}(m(x) - f_H(x, \widehat{\vartheta}_N))^2,$$

may, e.g., be approximated by its bootstrap analogue

$$\mathrm{mse}^*(x) = \frac{1}{B}\sum_{i=1}^{B} (\widehat{m}_N(x) - f_H(x, \widehat{\vartheta}_{N,i}^*))^2$$

where $\widehat{\vartheta}_{N,i}^*$ is the weight vector estimated from fitting the network function to the i-th bootstrap resample. The validity of this bootstrap approach has been shown for the regression model (2) by Franke and Neumann[6]. The proof can be generalized to the autoregressive case, too. However, the innovations ε_t have to be independent and identically distributed as, otherwise, the first step of drawing independent, identically distributed bootstrap innovations would make no sense. In the heteroscedastic case, other bootstrap procedures have to be considered.

We illustrate the performance of neural network estimates for nonlinear autoregressive functions and of the bootstrap approximations for their distribution with a small Monte Carlo study. The data $S_0, \ldots, S_N$, where $N = 200$, were generated by the NLAR(1)-scheme (3) with independent Gaussian innovations ε_t with mean 0 and standard deviation $\sigma_\varepsilon = 0.3$. The autoregressive function is a bump function

$$m(x) = 0.7x - 0.1 + 1.5\phi(x), \tag{4}$$

where ϕ denotes the standard normal density. On the interval $[-1, +1]$, where the stationary law of S_t is mainly concentrated, m is quite well approximated by a neural network function $f_3(x, \vartheta_0)$ with $H = 3$ hidden neurons and, therefore, 10-dimensional parameter vector ϑ_0. Figure 1 shows $m(x)$, the network function estimate $f_3(x, \widehat{\vartheta}_N)$ and, for sake of comparison, a Nadaraya-Watson-type kernel estimate $\widehat{m}(x, b)$ with bandwidth $b = 0.7$. The latter also served as initial estimate of the bootstrap procedure.

To investigate the performance of the bootstrap, we approximated the distribution of $d(x) = f_3(x, \widehat{\vartheta}_N) - m(x)$ by the distribution of $d^*(x) = f_3(x, \widehat{\vartheta}^*_N) - \widehat{m}(x, b)$. The quantities of interest were calculated from $M = 500$ independent Monte Carlo copies of the true sample and from $B = 500$ bootstrap resamples from the original data set $S_0, ..., S_N, N = 200$ resp. Figure 2a shows the function m together with the "true" 90 % - confidence band for m based on 500 Monte Carlo runs, where the band is not a uniform one, but formed by interpolating confidence intervals for $m(x)$ for various x. The neural network provides a good estimate of the autoregression function m, in particular around the origin where most of the observations are concentrated. Figure 2b compares this "true" confidence band with the corresponding 90 % - bootstrap confidence band. Remembering that the bootstrap is based on only one medium-sized time series sample, both bands agree remarkably well. Finally, for 4 different x Figure 3a-d show kernel density estimates, each with Gaussian kernel and bandwidth $b = 0.02$, of the estimation error $d(x)$ and its bootstrap approximation $d^*(x)$. Again, the performance of the bootstrap is quite satisfactory.

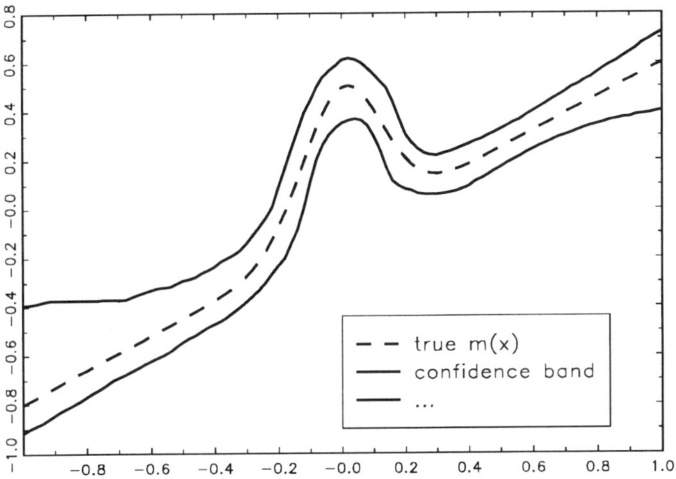

Figure 2a: Monte Carlo 90%-confidence band for m(x)

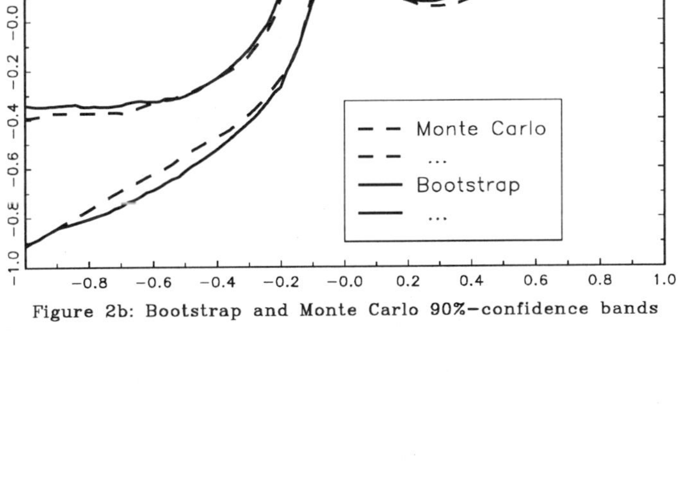

Figure 2b: Bootstrap and Monte Carlo 90%-confidence bands

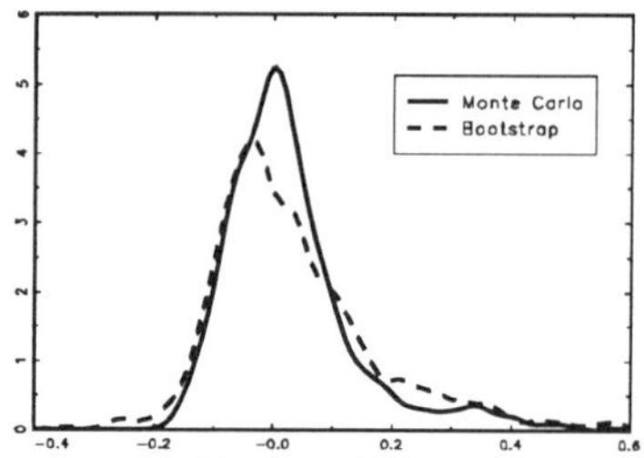

Figure 3a: error density at x = -0.8

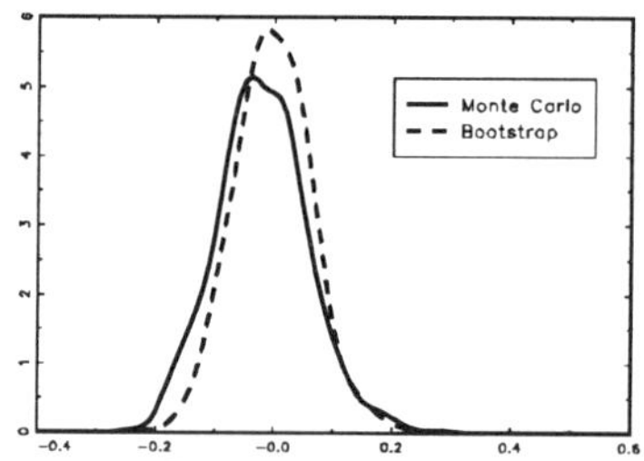

Figure 3b: error density at x = 0.0

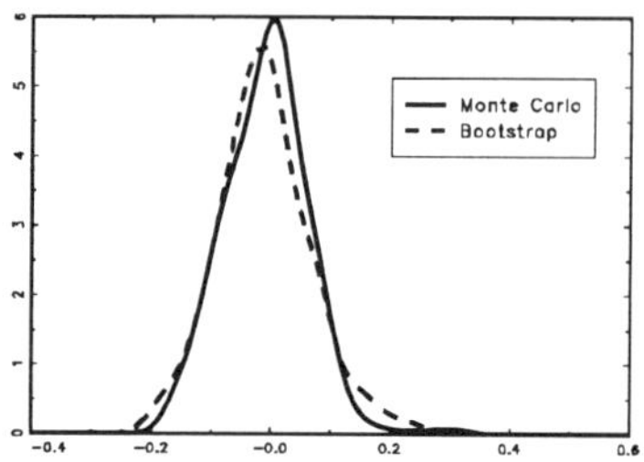

Figure 3c: error density at x = 0.2

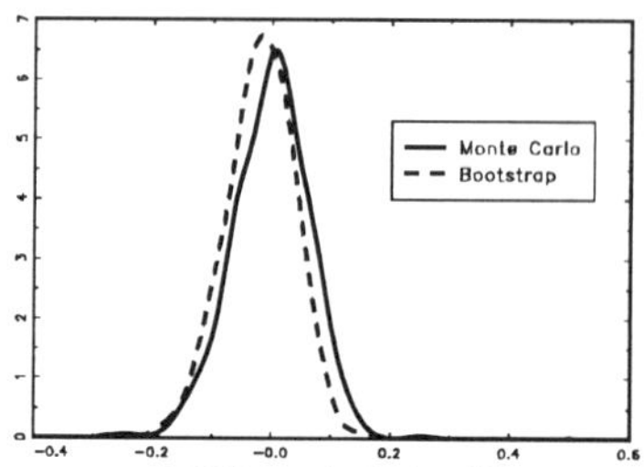

Figure 3d: error density at x = 0.8

3 Managing portfolios using neural networks

To illustrate the performance of neural networks in real applications which are of considerable complexity we give a short sketch of two case studies. In the first example, the task was to predict stock prices three months (60 trading days) ahead where the main goal was to generate trading signals for managing a portfolio of those stocks. The candidates for inclusion in the portfolio were 28 Dutch stocks dominating the CBS index. The available data were daily closing prices of all those stocks from 1993 to 1996. For model building and network parameter estimation, the data up to the end of 1995 were used. The data of 1996 were put aside for model validation.

As potential arguments for the forecasting function $f_H(x, \widehat{\vartheta}_N)$ several linear and nonlinear transformations of past stock prices $S_{t-\tau}, ..., S_t$, were considered, e.g. moving averages, envelopes, average directional movement indicators and other familiar tools of technical market analysis. Additionally, as exogeneous variables X_t in (1), the CBS index itself, foreign exchange rates, international interest rates, the MG base metal price and other intermarket data were taken into account. More than 60 candidates were investigated as potential coordinates of the input vector x. The final inputs were selected using experience of expert traders and statistical model selection procedures. More details are given by Franke[4]. The best network consisted of only $H = 3$ hidden neurons, but used 25-dimensional input vector x. The total number of parameters, therefore, was $dim(\widehat{\vartheta}_N) = 82$.

The point forecasts of stock prices varied considerably which is not surprising in view of the long forecasting period of 60 lags. However, they were condensed to a mere trend forecast, i.e. the information used in trading was solely if the stock price will
- increase significantly (by more than 5 %)
- decrease significantly (by more than 5 %)
- stay at approximately the same level.

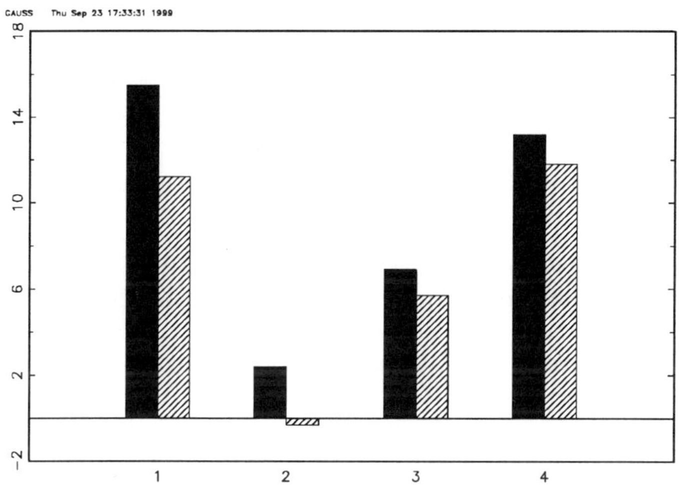

Figure 4: accumulated returns of stock portfolio

Using these forecasts, capital was allocated to the 28 stocks at the beginning of each quarter in the validation year 1996, and the resulting portfolio was held for 3 months unchanged. Only those stocks were included in the portfolio for which the prices were predicted to increase significantly up to the end of the holding period. This buy-and-hold strategy relying on neural network forecasts of stock prices was compared with the simple strategy of just buying the CBS index. Figure 4 shows the returns in percent for the network portfolio (solid bars) and the index portfolio (shaded bars). In each quarter, the network portfolio outperformed the index portfolio considerably which is even more remarkable as stock prices generally increased during the whole year of 1996, a situation in which it is not easy to beat the index.

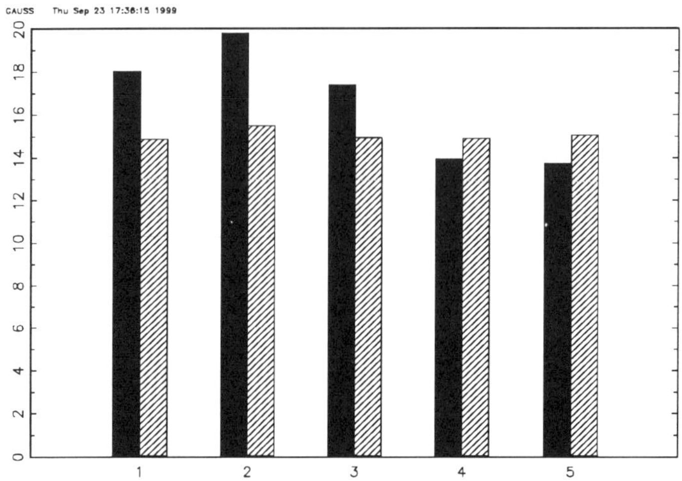

Figure 5: accumulated returns of currency portfolio

In the second example, the task was to construct a rule for allocating capital in a portfolio of three major currencies (US-Dollar, British Pound and Japanese Yen). A weekly buy-and-hold strategy was considered, i.e. at a particular day of the week, e.g. Tuesday, the portfolio composition was decided upon, based on the output of a neural network, and then the portfolio was held unchanged for one week. As inputs for the network, technical indicators calculated from past foreign exchange rates and intermarket data as in the above example were considered. Data from 1989 - 1995 were used for model building and parameter estimation, and the performance of the resulting allocation rules were evaluated using data from 1996 - September 1997. In this case, feedforward neural networks with more than one hidden layer proved to be more efficient than networks with only one layer of hidden neurons considered elsewhere in this paper. A typical network showing a good performance had two hidden layers with $H_1 = 9$ and $H_2 = 5$ neurons, respectively, and a 17-dimensional input vector x, resulting in $dim(\widehat{\vartheta}_N) = 230$ parameters to be estimated from the data. The details are given by Franke and Klein[5].

The network allocation rules were compared with various other portfolios, those consisting of one currency only, an equilibrium allocation of one third of the capital to each of the currencies and a well-established portfolio from real trading. For the validation period 1996 - September 1997, Figure 5 shows the annualized accumulated return in percent of one particular network allocation (solid bars) compared to the best of the competitors (shaded bars) which, during that period, always happened to be the portfolio containing only the, then, strong British pound. The performance is given for alle 5 possible weekly holding periods *1: Monday-Monday, 2: Tuesday-Tuesday, ... , 5: Friday-Friday.* That particular network outperformed all other allocations for the first three periods, but did not do so well for Thursdays and Fridays. This observation is not so surprising as differences in general trading behaviour between the start and the end of a week are well known. Therefore, in practice, one neural network did not suffice, but a system of networks, one for each day of the week, had to be developped.

4 Neural network estimates of volatility

The last two sections have illustrated that neural networks provide good estimates for the conditional mean of a financial time series even given a rather complex information set. In this section, we show how estimates of the conditional variance and volatility may be constructed following the same kind of approach. We now consider the following nonlinear heteroscedastic time series model:

$$S_{t+1} = m(S_t, S_{t-1}, ..., S_{t-\tau}, X_t) + \sigma_t \eta_{t+1} \tag{5}$$

where $\eta_1, \eta_2, \ldots$ are independent identically distributed with mean 0 and variance 1. We assume that the stochastic volatility σ_t is of a similar functional form as the conditional mean

$$\sigma_t = \sigma(S_t, S_{t-1}, ..., S_{t-\tau}, X_t) \tag{6}$$

Time series satisfying (5) and (6) are nonlinear AR-ARCH-processes with exogeneous components $X_t \in R^d$. The familiar parametric AR-ARCH-models are just a special case of this general type of stochastic process.

We construct a nonparametric estimate of the volatility function σ using neural networks as in section 2. As σ^2 is the conditional variance of S_{t+1} given the past we could fit a neural network function with inputs $S_t, S_{t-1}, ..., S_{t-\tau}, X_t$ as before and with outputs S_{t+1}^2 instead of S_{t+1} to the data. We would get an estimate of the conditional second moment and, subtracting the squared neural network estimate $f_H(x, \widehat{\vartheta}_N)$ for the conditional mean, an estimate of the conditional variance, too. For kernel estimates, however, Fan and Yao[3] have shown that it is more efficient to use $f_H(x, \widehat{\vartheta}_N)$ instead to calculate squared sample residuals and to smooth them instead of S_{t+1}^2 to get a nonparametric estimate of the conditional variance. We follow their approach in the neural network setting. To simplify notation, we describe the procedure for the nonlinear AR(1)-ARCH(1)-model

$$S_{t+1} = m(S_t) + \sigma(S_t)\eta_{t+1} \tag{7}$$

only. The generalization to time series models given by (5) and (6) is straightforward. In a first step, we calculate estimates of the innovations $\varepsilon_t = \sigma(S_t)\eta_{t+1}$ using the estimate $f_H(x, \widehat{\vartheta}_N)$ for $m(x)$ from section 2:

$$\widehat{\varepsilon}_{t+1} = S_{t+1} - f_H(S_t, \widehat{\vartheta}_N), t = 1, \ldots, N.$$

As $\sigma^2(x) = \mathcal{E}\{\varepsilon_t^2 | S_t = x\}$, we can estimate this function by fitting a neural network function $f_G(x, \gamma)$ with G neurons in the hidden layer to the data $S_0, ..., S_N$, where the parameter estimate $\widehat{\gamma}_N$ is given by

$$\frac{1}{N} \sum_{t=1}^{N} (\widehat{\varepsilon}_t{}^2 - f_G(S_{t-1}, \gamma))^2 = \min_{\gamma \in \Theta_G} !$$

The square root of $f_G(x, \widehat{\gamma}_N)$ is, the, a neural network based estimate of the volatility function $\sigma(x)$, i.e. of the conditional standard deviation of S_{t+1} given $S_t = x$. The consistency of this estimate for increasing sample size N and suitably increasing number G of hidden neurons again follows essentially from the work of White (1989, 1990) on neural network estimates for conditional expectations involving time series.

We study the performance of the neural network volatility estimates in a simulation study where we generate $M = 500$ Monte Carlo samples $S_0, ..., S_N$ with $N = 500$ from the nonlinear AR(1)-ARCH(1)-model (7). The sample size has to be larger as in section 2 as variances are harder to estimate than means in general. The η_t are standard normal random variables, the autoregressive function $m(x)$ is the same bump function (4) as in section 2, and the conditional variance function is chosen as in a common ARCH(1)-model as

$$\sigma^2(x) = 0.1 + 0.7x^2.$$

Figure 6a shows the true function m and a 90 % - confidence band based on the neural network function estimates $f_H(x, \widehat{\vartheta}_N)$ for the interval $[-2, +2]$, which contains the majority of the data. Comparing it to Figure 2a, we remark that the neural network estimates of the conditional expectations perform still reasonably well in the heteroscedastic case, in particular, if one recalls the heavy-tailedness of the stationary distribution of the S_t introduced by the ARCH(1)-innovations $\varepsilon_t = \sigma(S_t)\eta_{t+1}$. Even the mean standard deviation $\mathcal{E}\{\sigma(S_t\}$ is about 0.95 and, therefore, more than three times as large as in the simulation study of section 2.

Figure 6b shows the true squared volatility function σ^2 together with a 90 % - confidence band from the Monte Carlo study. Considering the heavy-tailed law of the data S_t and the general difficulty of estimating variances the neural network estimates does reasonably well. Additionally, the simulation still suffers from numerical problems. In contrast to the homoscedastic model considered in section 2, the numerical procedure (a quasi-gradient method) for calculating the nonlinear least-squares parameters $\widehat{\vartheta}_N$ and $\widehat{\gamma}_N$ was prone to end up in local extrema with quite a bad performance of the corresponding function estimates. We solved this problem by starting the minimization routine with lots of different randomly selected initial values. Using an appropriate numerical algorithm like simulated annealing would be an alternative.

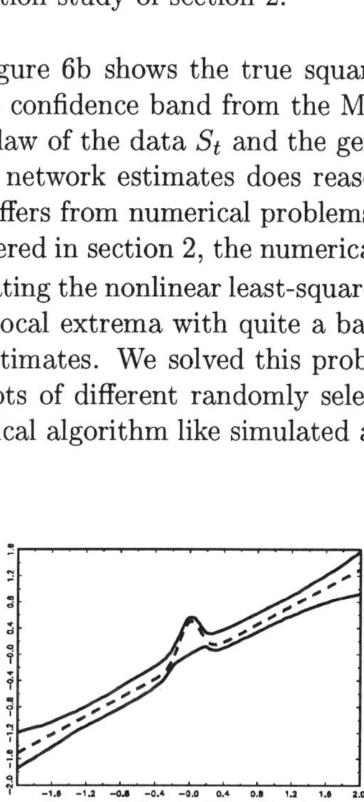

Figure 6a: m(x) (– – –) and 90%–confidence band

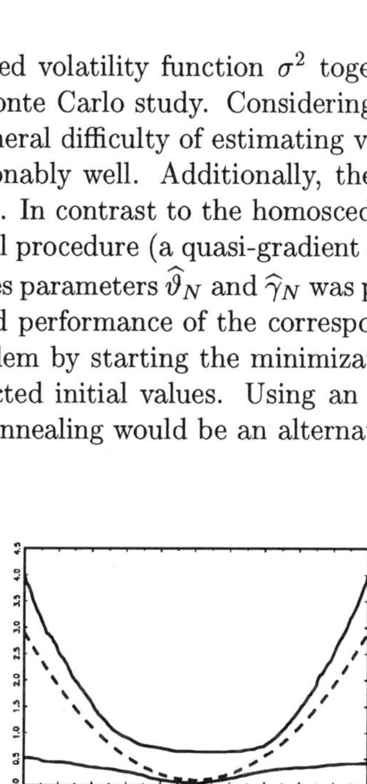

Figure 6b: σ^2(x) (– – –) and 90%–confidence band

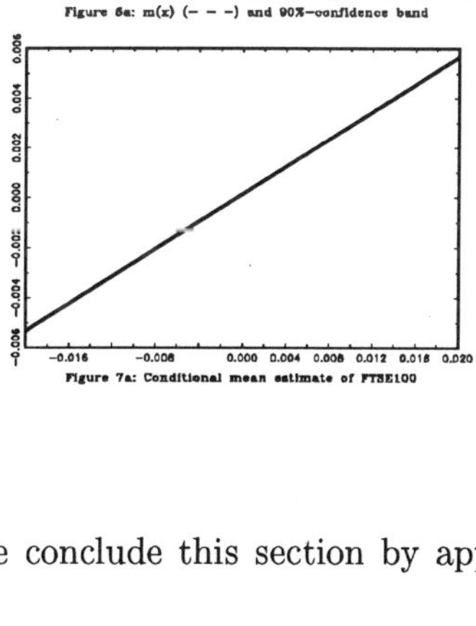

Figure 7a: Conditional mean estimate of FTSE100

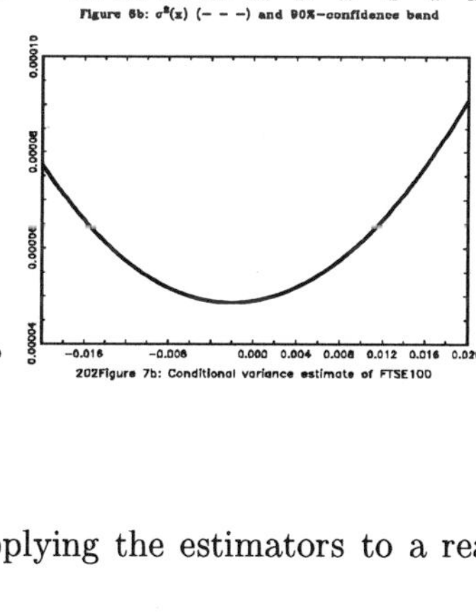

202Figure 7b: Conditional variance estimate of FTSE100

We conclude this section by applying the estimators to a real data set.

We selected the British FTSE100 index from January 4, 1993 to November 4, 1994, totalling 480 observations Z_t. Then, we fitted the model (7) to the daily returns $S_t = (Z_t - Z_{t-1})/Z_{t-1}$ estimating the conditional mean m and the conditional variance σ^2 by neural networks with $H = G = 3$ hidden neurons, corresponding to 10 parameters each. We also tried networks with up to 7 hidden neurons, but the estimates essentially did not change. Figure 7a and 7b show the estimates of conditional mean and variance of S_t given S_{t-1}. The mean is almost, but not exactly linear whereas the variance resembles an ARCH(1)-term apart from the asymmetry.

5 Estimating conditional value-at-risk with neural networks

Apart from volatility, another popular measure for financial hazards is the *value at risk* (VaR) as a bound which is exceeded by losses with small probability α only. There are various definitions of VaR (compare, e.g., Jorion[8]), but the crucial quantity is always the α-quantile of the return distribution of the financial asset. We consider here conditional quantiles given the information up to the present time t, and we discuss how to estimate them using neural networks. For our exposition, we concentrate on the simple nonlinear autoregression of order 1 given by (3). Generalizations to more complicated models are again straightforward. The conditional α-quantile function $q_\alpha(x)$ is given as solution of $F(q_\alpha(x)/x) = \alpha$, where $F(s/x)$ denotes the conditional distribution function of S_{t+1} given $S_t = x$

$$F(s/x) = pr\{S_{t+1} \leq s | S_t = x\}$$

Nonparametric conditional quantile estimates based on common smoothing methods are closely related to kernel density estimates. Following, e.g., Samanta[12] we could estimate the joint density of S_{t+1} and S_t and the marginal density of S_t by kernel smoothing, getting the conditional density as a ratio. By integration, we get an estimate $F_N(s/x)$ for $F(s/x)$. Then, an estimate $q_{\alpha,N}(x)$ for the conditional quantile function $q_\alpha(x)$ is derived by solving $F_N(q_{\alpha,N}(x)/x) = \alpha$.

We could mimick this approach using neural networks. $F(s/x)$ is a conditional expectation of the indicator function $1_{(-\infty,s]}$ and could be approximated by neural networks as the conditional mean and variance in previous sections. However, for solving $F_N(q_{\alpha,N}(x)/x) = \alpha$ numerically, we would have to train neural networks frequently to get $F_N(s/x)$ for various values of s. If we are interested in estimating $q_\alpha(x)$ for only a few α, this approach is too cumbersome from a numerical point of view. We, therefore, follow a different approach which is based on the observation that the conditional quantile function $q_\alpha(x)$ solves

$$\mathcal{E}\{\,|S_t - q|\,\big(\alpha 1_{[0,\infty]}(S_t - q) + (1-\alpha)1_{(-\infty,0]}(S_t - q)\big)\,\big|S_{t-1} = x\} = \min_{q\in R}!$$

We get a neural network estimate $f_Q(S_{t-1}, \widehat{\chi}_N)$ for $q_\alpha(x)$ by minimizing a sample version of this conditional expectation:

$$\frac{1}{N}\sum_{t=1}^{N} \big|S_t - f_Q(S_{t-1},\chi)\big|\big(\alpha 1_{[0,\infty]}(S_t - f_Q(S_{t-1},\chi))$$
$$+(1-\alpha)1_{(-\infty,0]}(S_t - f_Q(S_{t-1},\chi))\big) = \min_{\chi\in\Theta_G}!$$

$f_Q(x,\chi)$ denotes a network function as in section 2 with Q hidden neurons. This approach has been studied by White[15] who proved the consistency of the conditional quantile estimate f_Q if N and Q increase with appropriate rates to ∞.

We illustrate the performance of this quantile estimator with a simulation study where the generated data follow exactly the same nonlinear autoregression and specifications as in the Monte Carlo study of section 2. In particular, the sample size is $N = 201$ and the number of Monte Carlo runs is $M = 500$. Figure 8 shows the true conditional 5 % - quantile function $q_{.05}(x)$ for this time series together with a 90 % - confidence band based on the neural network quantile estimates $f_Q(x, \widehat{\chi}_N)$ with $Q = 10$. As for estimating the conditional mean, the performance is quite good in this homoscedastic situation.

Finally, we estimate the conditional 5 % - quantile function for the next return of the FTSE100-index series given the present return, where we used the same data as in section 4. Figure 9 shows the resulting estimate.

Acknowledgement:

section 3 is based on joint work with Commerzbank AG, Frankfurt, in particular with D. Oppermann and U. Kern. The data, we used, were provided by DATASTREAM.

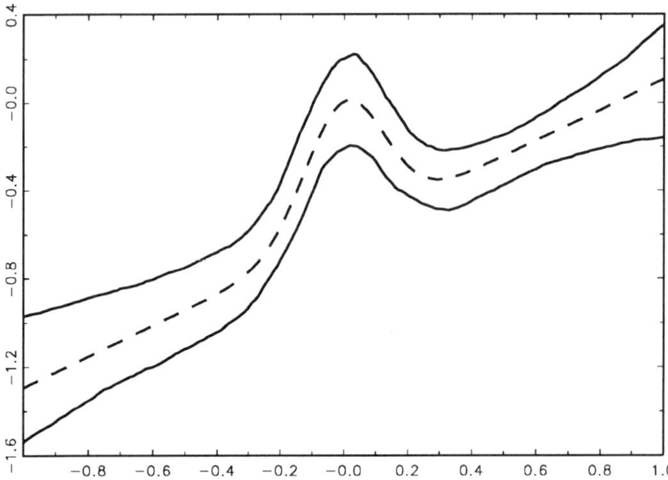

Figure 8: q(x) (– – –) and 90%-confidence band

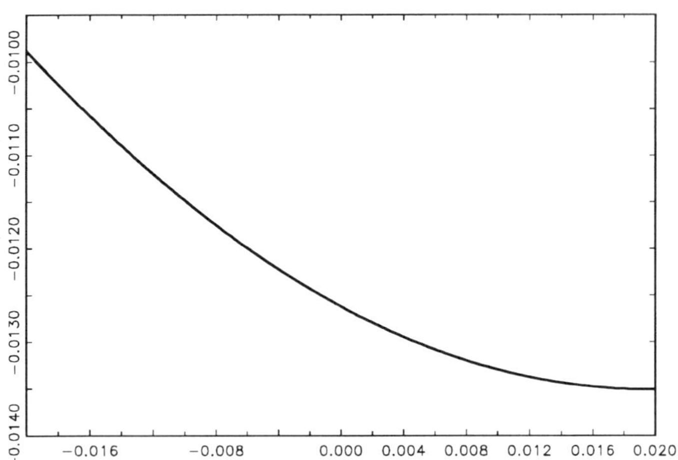

Figure 9: Conditional 5%-quantile estimate of FTSE100

References

1. U. Anders, *Statistische neuronale Netze.* (Vahlen, München, 1997)
2. G. Bol, G. Nakhaeizadeh and K.-H. Vollmer eds., *Finanzmarktanalyse und -prognose mit innovativen quantitativen Verfahren.* (Physica-Verlag, Heidelberg, 1996)
3. J. Fan and Q. Yao, *Efficient estimation of conditional variance functions in stochastic regression. Biometrika.* **85**, 645-660 (1998).
4. J. Franke, *Nonlinear and Nonparametric Methods for Analyzing Financial Time Series.* In: *Operation Research Proceedings 98,* P. Kall und H.-J. Luethi eds. (Springer-Verlag, Berlin, 1999).
5. J. Franke and M. Klein, *Optimal portfolio management using neural networks - a case study. Report in Wirtschaftsmathematik* (University of Kaiserslautern, 1999).
6. J. Franke and M. Neumann, *Bootstrapping neural networks.* Tentatively accepted for publication in *Neural Computation.*
7. K. Hornik, M. Stinchcombe and H. White. *Multilayer feedforward networks are universal approximators, Neural Networks* **2**, 359-366 (1989)
8. Ph. Jorion *Value at Risk: The New Benchmark for Controlling Market Risk.* (Irwin, Chicago, 1996).
9. J. P. Kreiss, *Nonparametric estimation and bootstrap for financial time series.* In: this volume.
10. N. Murata, S. Yoshizawa and S. Amari, *Network information criterion - Determining the number of hidden units for an artificial neural network model. IEEE Trans. Neural Networks* **5**, 865-872 (1994).
11. A.-P.N. Refenes, A.D. Zapranis and J. Utans, *Neural model identification, variable selection and model adequacy.* In: *Neural Networks in Financial Engineering,* A. Weigend et al. eds. (World Scientific, Singapore, 1996)
12. M. Samanta, *Nonparametric estimation of conditional quantiles. Statistics & Probability Letters* **7**, 407-412 (1989).
13. H. White, *Some asymptotic results for learning in single hidden-layer feedforward network models. J. Amer. Statist. Assoc.* **84**, 1008-1013 (1989).
14. H. White *Connectionist nonparametric regression: multilayer feedforward networks can learn arbitrary mappings. Neural Networks* **3**, 535-550 (1990).
15. H. White, *Nonparametric estimation of conditional quantiles using neural networks.* In: *Computing Science and Statistics,* C. Page and R. Le Page eds. (Springer-Verlag, Berlin, 1992).

OPTIMAL ASSET ALLOCATION UNDER GARCH MODEL

W. C. HUI, H. YANG AND K. C. YUEN
Department of Statistics and Actuarial Science,
University of Hong Kong

We use a discrete time model to investigate the optimal asset allocation strategy of a risk averse investor whose wealth consists of a single risky asset and a riskless asset. The objective is to maximize the expected utility of wealth over a planning horizon. We assume that the return of the risky asset follows the generalized autoregressive conditional heteroscedastic (GARCH) process. We illustrate the approach through numerical examples.

Keywords: Optimal allocation, GARCH process, heteroscedasticity, power utility, risk aversion, risk premium.

1 Introduction

Asset allocation problem is one of the key topics in investment finance. It also plays a major role in actuarial science. The investment of a pension fund among different assets is an obvious application. Since the real financial market is extremely complex, it is almost impossible to find the optimal allocation. In spite of this, we believe that models with simplifying assumptions can provide useful insights. The asset allocation problem is sometimes called the portfolio selection problem. In this paper, both names are used interchangeably.

Merton (1969) considered the lifetime portfolio selection problem from a continuous time perspective and obtained closed-form solutions under certain assumptions in which the return of the risky asset is governed by a geometric Brownian motion and the investor's utility function yields constant relative risk aversion. On the other hand, Samuelson (1969) considered a similar model in discrete time. He advocated a dynamic stochastic programming approach and succeeded in obtaining the optimal decision for a consumption-investment model. Grauer and Hakansson (1982) handled the portfolio selection problem by updating the joint return distribution for the assets every period. They were able to incorporate time variation in the distribution of the returns of the assets. Their results show that the gains from active reallocation among the major asset categories are substantial.

The problem becomes very complex when transaction costs are taken into account. In the case that an investor has a power utility and an infinite horizon and that the transaction costs are proportional to the amount of risky asset traded, Constantinides (1986) obtained approximate solutions for the boundaries of the no-transaction region. For real applications, it may be better to assume a finite terminal date. With a finite-time horizon, Gennotte and Jung (1994) developed a numerical method to obtain the approximate values of the boundaries. Boyle and Lin (1997) used a discrete time approach to tackle the same problem, and developed analytical expressions for the investor's indirect utility function as well as the boundaries.

Volatility plays an important role in much of the modern finance theory. Many financial and econometric models were developed under the assumption of constant volatility. However, empirical evidence shows that the volatilities of most risky assets do change through time. Recently, many researchers in financial economics have put their efforts in modeling time variation of volatility. One of the most famous tools that has emerged for characterizing such change in volatility is the generalized autoregressive conditional heteroscedastic (GARCH) model of Bollerslev (1986). In the GARCH model, the current conditional variance is allowed to change over time and is specified as a linear function of past squared errors and past conditional variances. Using this model, Engle and Mustafa (1992) estimated the implied stochastic process of the volatility of an asset from option prices written on the asset. The GARCH model has been applied to the pricing of contingent claims intensively. Duan (1995) developed a GARCH option pricing model which captures the changes in the conditional volatility of the underlying asset.

Due to the inappropriateness of the constant volatility assumption in practice, a dynamic optimization of an investment model capturing the phenomenon of changing variance should be of great interest to investors. This paper uses the GARCH model to model the volatility of the underlying return process. We formulate the asset allocation problem by a discrete time model. The set up of the problem is presented in Section 2. The optimal portfolio policy is derived in Section 3. Numerical examples are given to illustrate our method in Section 4. Section 5 briefly discusses the problem with proportional transaction costs. Finally, some concluding remarks are given in Section 6.

2 Formulation of the Problem

In our problem, an investor has to decide how to allocate his wealth among two assets. The first one is a risky asset for which the rate of return is assumed to be conditionally lognormally distributed. The second one is a riskless asset which earns a constant rate of return periodically. The investor aims to maximize the expected utility of his wealth over a finite-time horizon. We further assume that the investor does not consume his wealth in the planning horizon.

In this paper we distinguish the physical probability measure from the subjective (risk preference) probability measure. The risk preference of the investor is represented by a so-called power utility function which is under the subjective probability measure and takes the form

$$U(W) = \frac{W^\gamma}{\gamma}\,, \quad \gamma \le 1\,, \quad \gamma \neq 0\,, \tag{2.1}$$

or

$$U(W) = \ln W\,, \quad \gamma = 0\,, \tag{2.2}$$

where W denotes the wealth of the investor. The logarithm of W in (2.2) represents the limiting case for $\gamma = 0$. This utility function yields a constant relative risk aversion of $1-\gamma$. The parameter γ is an index of risk preference (or subjective view). The risk aversion is the lowest when $\gamma = 1$; and it increases as γ decreases. Hence, an investor chooses a value of γ in accordance with his own risk preference.

Given the power utility function, the objective of the asset allocation problem is to

$$\max_{\{\pi_t\}} E_t \sum_{i=0}^{t+1} (1+\rho)^{-i} U(W_i), \tag{2.3}$$

where $t = 0, 1, \ldots, T-1$,

E_t = expectation operator conditional on the information up to time t,

W_t = total wealth at time t, and

ρ = discount factor.

The investor's investment opportunities occur at discrete, equally spaced points in time. These points divide the time horizon of the investor into T periods. The state of the system at the beginning of each period, $t = 0, 1, \ldots, T-1$, is denoted by W_t, and W_T represents the terminal wealth at the end of the time horizon. At each period, the investor chooses to invest a proportion, π_t, of his wealth in the risky asset and thus $1 - \pi_t$ of his wealth in the riskless asset. The investor makes his investment decisions in the way that the expected utility of his wealth over the planning horizon is maximized.

Denote S_t and B_t as the price of the risky asset and the price of the riskless asset at time t respectively. Following the work of Duan (1995), we assume that the rate of return of the risky asset is conditionally lognormally distributed under the physical probability measure $\mathcal{P}$. That is,

$$\ln Z_t = r + \lambda\sqrt{h_t} - \frac{1}{2}h_t + \epsilon_t\,, \quad Z_t = \frac{S_t}{S_{t-1}}\,, \quad r = \frac{B_t}{B_{t-1}} - 1, \tag{2.4}$$

where ϵ_t has a zero mean and conditional variance h_t under $\mathcal{P}$, r is the fixed one-period interest rate, and λ is the excess return of the risky asset over the riskless asset. Duan (1995) interpreted the parameter λ as the unit risk premium. Using his interpretation, λ and γ are correlated and λ can be expressed in terms of γ. Since Duan (1995) considered the option pricing problem, it does not matter how to interpret the measure $\mathcal{P}$ in his case. However, we assume that measure $\mathcal{P}$ is the physical measure for the movement of stock price. Equation (2.4) describes the movement of the return of stock price under measure $\mathcal{P}$ which is independent of the investor's risk preference. Hence, the parameter γ of the power utility function does not relate to λ.

3 Optimal Portfolio Policy under GARCH Model

To describe the varying variances over time, we consider the GARCH(p, q) process of Bollerslev (1986) and link it to our portfolio selection model. Assume that the discrete-time stochastic process ϵ_t in (2.4) follows a GARCH(p, q) process under measure $\mathcal{P}$. The formal expression of the process is given by

$$\epsilon_t \,\Big|\, \mathcal{F}_{t-1} \sim N(0, h_t) \quad \text{under measure } \mathcal{P} \,, \tag{3.1}$$

$$h_t = \alpha_0 + \sum_{i=1}^{q} \alpha_i \epsilon_{t-i}^2 + \sum_{j=1}^{p} \beta_j h_{t-j} \,, \tag{3.2}$$

where $\alpha_0 > 0$, $p \geq 0$, $q > 0$, $\alpha_i \geq 0$ $(i = 1, \ldots, q)$, $\beta_j \geq 0$ $(j = 1, \ldots, p)$, and $\mathcal{F}_t$ is the information obtained from observing the asset prices up to and including time t. The sum $\sum \alpha_i + \sum \beta_j$ is assumed to be less than one in order to ensure the wide-sense stationarity of the GARCH(p, q) process. For $p = 0$, ϵ_t of (3.1) reduces to the ARCH(q) process, and for $p = q = 0$, it is simply white noise.

The portfolio selection problem is to maximize (2.3) with respect to a set of decision variables $\pi_0, \ldots, \pi_{T-1}$. Equivalently, we have the problem

$$\max_{\{\pi_t\}} E_t \sum_{i=0}^{t+1} (1 + \rho)^{-i} U(W_i), \tag{3.3}$$

subject to the constraint

$$W_{t+1} = W_t \cdot \big((1 - \pi_t)(1 + r) + \pi_t Z_{t+1} \big) \,, \tag{3.4}$$

with a fixed initial wealth W_0.

At time 0, the investor chooses to invest π_0 of his wealth W_0 in the risky asset and put the rest in the riskless asset. One period later, that is at time

1, the return of the risky asset Z_1 is observed and thus W_1 is known. The investor then uses this information to decide the value of π_1. The same step is repeated until the final choice of π_{T-1}.

To derive a forward recursion formula for the optimal solution, we define a set of functions

$$J_t(W_t, \pi_t) = E_t \sum_{i=0}^{t+1} (1+\rho)^{-i} U(W_i), \tag{3.5}$$

for $t = 0, 1, \ldots, T-1$. There are totally T allocation dates. From (3.5), we have

$$J_0(W_0, \pi_0) = \left(U(W_0) + E_0(1+\rho)^{-1} U(W_1)\right) , \tag{3.6}$$

at date $t = 0$. Using (2.1) and (3.4), we rewrite (3.6) as

$$J_0(W_0, \pi_0) = \frac{W_0^\gamma}{\gamma} + \frac{(1+\rho)^{-1}}{\gamma} W_0^\gamma \left\{ E_0 \left[(1-\pi_0)(1+r) + \pi_0 Z_1\right]^\gamma \right\} . \tag{3.7}$$

Differentiating (3.7) with respect to π_0, we get

$$E_0 \left\{ \left[(1-\pi_0)(1+r) + \pi_0 Z_1\right]^{\gamma-1} (Z_1 - r - 1) \right\} = 0 . \tag{3.8}$$

The optimal portfolio selection at time 0, π_0', is the solution of (3.8). Substituting π_0' into (3.7), we obtain the maximum value of J_0

$$J_0(W_0, \pi_0') = (1 + b_0) \frac{W_0^\gamma}{\gamma} ,$$

where $b_0 = (1+\rho)^{-1} E_0 \{[(1-\pi_0')(1+r) + \pi_0' Z_1]^\gamma\}$.

At allocation date $t = 1$, we observe Z_1 and have another function

$$J_1(W_1, \pi_1) = \left(U(W_0) + E_1(1+\rho)^{-1} U(W_1) + E_1(1+\rho)^{-2} U(W_2)\right) .$$

Again, we differentiate J_1 with respect to π_1 and obtain

$$E_1 \{ [(1-\pi_1)(1+r) + \pi_1 Z_2]^{\gamma-1} (Z_2 - r - 1) \} = 0 . \tag{3.9}$$

Let π_1' be the solution of (3.9). The maximum of J_1 is given by

$$J_1(W_1, \pi_1') = (1 + b_0 + b_1 c_1) \frac{W_0^\gamma}{\gamma} ,$$

where $b_1 = (1+\rho)^{-2}E_1\{[(1-\pi_1')(1+r)+\pi_1' Z_2]^\gamma\}$ and $c_1 = [(1-\pi_0')(1+r)+\pi_0' Z_1]^\gamma$. By similar arguments, we derive the following forward recursion formula

$$J_t(W_t, \pi_t') = \left[1 + \sum_{i=0}^{t} b_i \Big(\prod_{j=0}^{i} c_j\Big)\right] \frac{W_0^\gamma}{\gamma} ,$$

for $t = 0, 1, \ldots, T-1$, where

$$\begin{aligned} b_i &= (1+\rho)^{-(i+1)} E_i\{[(1-\pi_i')(1+r)+\pi_i' Z_{i+1}]^\gamma\} , \\ c_j &= [(1-\pi_{j-1}')(1+r)+\pi_{j-1}' Z_j]^\gamma , \\ c_0 &= 1 . \end{aligned}$$

Hence, the set $(\pi_0', \ldots, \pi_{T-1}')$ forms the optimal allocation decisions associated with (3.3).

4 Numerical Examples

To illustrate our method, we simulate the data from the GARCH(1, 1) process. We consider a four-period problem in our numerical study. Before implementing the results in the previous section, we need to estimate the parameters of the GARCH(1, 1) model. From (2.4), (3.1), and (3.2), we have

$$\begin{aligned} \epsilon_t &= \ln\ Z_t - r - \lambda\sqrt{h_t} + \frac{1}{2}h_t , \\ \epsilon_t \,\big|\, \mathcal{F}_{t-1} &\sim N(0, h_t) , \\ h_t &= \alpha_0 + \alpha_1 \epsilon_{t-1}^2 + \beta_1 h_{t-1} . \end{aligned}$$

As usual, we estimate the parameters by the maximum likelihood method. Apart from some constants, the likelihood, for a sample of size n, can be expressed as

$$L = -\frac{1}{2}\sum_{i=1}^{n} \ (\ln h_i + \frac{\epsilon_i^2}{h_i}) .$$

Let the length of each period be 0.25. For each period, we set $r = 0.01$, $\gamma = -1$, and $\lambda = 0.11754$. From the GARCH model of (3.10), we generate 1000 stock prices with $h_0 = 0.015625$, $\alpha_0 = 3 \times 10^{-5}$, $\alpha_1 = 0.2$, and $\beta_1 = 0.5$. The GARCH parameter estimates are $\hat{\alpha}_0 = 3.2703 \times 10^{-5}(4.584 \times 10^{-6})$, $\hat{\alpha}_1 = 0.288156(0.0319)$, and $\hat{\beta}_1 = 0.440531(0.0472)$. The standard errors are given in the parentheses.

Different values of γ are used to illustrate the effect of risk attitude on the optimal fraction of wealth invested in the risky asset. Table 1 summarizes the results of 9 different portfolio strategies corresponding to the values of γ ranging from 0.3 (low risk aversion) to -5 (high risk aversion). From Table 1, we see that as the investor becomes more risk averse, less is invested in the risky asset. When the investor has a low risk aversion, say $\gamma = 0.3$, he almost limits his portfolio to the risky asset throughout the four-period time. For other cases, the investor keeps adjusting the fraction from period to period.

Table 1. Optimal fraction for different values of γ.

γ	π_0	π_1	π_2	π_3
-5	0.13	0.11	0.12	0.13
-3	0.20	0.17	0.18	0.19
-2	0.27	0.22	0.24	0.25
-1	0.40	0.33	0.36	0.38
-0.5	0.53	0.45	0.48	0.50
0	0.80	0.67	0.72	0.76
0.1	0.89	0.74	0.80	0.84
0.2	1.00	0.84	0.90	0.94
0.3	1.00	0.96	1.00	1.00

Furthermore, we adjust the values of h_0 and λ to examine the effects of the initial variance and the excess return on the optimal fraction respectively. The results displayed in Tables 2 and 3 are based on $\gamma = -1$. An increase in variance implies that an additional risk is imposed on the risky asset. As shown in Table 2, when the asset becomes more risky and the excess return remains constant, the investor tends to reduce the fraction for the risky asset and to invest more in the riskless asset. The increase of λ makes the risky asset more attractive to the investor. Table 3 indicates that the optimal fraction invested in the risky asset increases with λ in the four-period time.

Note that if the GARCH(1, 1) process is mistreated as a constant variance process, the optimal fraction will not change at all.

Table 2. Optimal fraction for different values of h_0.

h_0	π_0	π_1	π_2	π_3
3×10^{-3}	0.87	0.63	0.59	0.56
5×10^{-3}	0.50	0.40	0.44	0.45
1.5625×10^{-2}	0.40	0.33	0.36	0.38
2×10^{-2}	0.33	0.29	0.32	0.34
3×10^{-2}	0.27	0.25	0.27	0.30
5×10^{-2}	0.19	0.18	0.20	0.22

Table 3. Optimal fraction for different values of λ.

λ	π_0	π_1	π_2	π_3
0.05	0.16	0.13	0.14	0.15
0.08	0.32	0.27	0.29	0.30
0.11754	0.40	0.33	0.36	0.38
0.12	0.63	0.53	0.58	0.60
0.15	0.95	0.80	0.86	0.91

5 Optimal Portfolio Policy with Proportional Transaction Costs

In this section, we examine the portfolio selection problem with proportional transaction costs. When the investor adjusts the composition of the portfolio at the trading time, he needs to pay a transaction cost which is proportional to the size of the trade. As before, the portfolio of the investor consists of a risky asset and a riskless asset. The investor who has a power utility function maximizes the expected utility of his wealth through time. The general setting in this section is similar to that in Section 2. However, in the presence of transaction costs, it is better to express the asset holdings in the portfolio in terms of dollar amount.

Again we assume that there are T trading times, $t = 0, 1, \ldots, T-1$, available over a finite-time horizon. The rates of return on the risky and riskless assets are given in (2.4). We further assume that ε_t of (2.4) follows a GARCH (p, q) process under the physical measure $\mathcal{P}$. The investor holds a portfolio with a dollar amount of x_t^0 in the riskless asset and a dollar amount of x_t in the risky asset just before trading time t. After a trade of u_t dollars in the risky asset, the investor needs to pay a proportional transaction cost of $\theta|u_t|$, for $0 < \theta < 1$, which is charged to the riskless asset. If u_t is positive (negative), it means that the risky asset is bought (sold). After the transaction at time t, the dollar amounts of the risky and riskless assets in the portfolio become

$$y_t = x_t + u_t,$$

and

$$y_t^0 = x_t^0 - u_t - \theta|u_t|,$$

respectively.

Parallel to (3.3) and (3.4), the objective of the investor is to maximize

$$E_t \sum_{i=0}^{t+1} (1+\rho)^{-i} U(W_i),$$

with respect to $\{u_t\}$ and subject to the constraint

$$\begin{aligned} W_{t+1} &= x_{t+1}^0 + x_{t+1} \\ &= y_t^0(1+r) + y_t Z_{t+1} \\ &= (x_t^0 - u_t - \theta|u_t|)(1+r) + (x_t + u_t)Z_{t+1}, \end{aligned}$$

for $t = 0, 1, \ldots, T-1$. As before, the initial wealth $W_0 = x_0^0 + x_0$ is given.

To determine the maximum of the expected utility of wealth, we define

$$J_t(W_t, u_t) = E_t \sum_{i=0}^{t+1} (1+\rho)^{-i} U(W_i),$$

for $t = 0, 1, \ldots, T-1$. At the beginning of the trading time $t = 0$, we consider

$$\begin{aligned} J_0(W_0, u_0) &= U(W_0) + E_0(1+\rho)^{-1}U(W_1) \\ &= \frac{W_0^\gamma}{\gamma} \\ &\quad + \frac{(1+\rho)^{-1}}{\gamma} E_0 \left\{ [(x_0^0 - u_0 - \theta|u_0|)(1+r) + (x_0 + u_0)Z_1]^\gamma \right\}. \end{aligned} \tag{5.1}$$

Differentiating (5.1) with respect to u_0, we get

$$E_0 \left\{ [(x_0^0 - u_0 - \theta|u_0|)(1+r) + (x_0 + u_0)Z_1]^{\gamma-1} [Z_1 - (1+r)(1+sgn(u_0)\theta)] \right\} = 0, \tag{5.2}$$

where

$$sgn(u_0) = \begin{cases} +1, & u_0 \geq 0, \\ -1, & u_0 < 0. \end{cases}$$

The optimal portfolio selection at time $t = 0$, u_0', is the solution of (5.2). Using the similar approach to the problem without transaction costs, the optimal trading strategy at time t, u_t', is the solution of

$$E_t \left\{ [(x_t^0 - u_t - \theta|u_t|)(1+r) + (x_t + u_t)Z_{t+1}]^{\gamma-1} [Z_{t+1} - (1+r)(1+sgn(u_t)\theta)] \right\} = 0,$$

where

$$sgn(u_t) = \begin{cases} +1, & u_t \geq 0, \\ -1, & u_t < 0. \end{cases}$$

6 Conclusion

Using the GARCH model, we have examined the optimal asset allocation of an investor whose objective is to maximize the expected utility of his wealth. This model allows the conditional variance changing over time. The investor adjusts his portfolio through time in response to the changing variances and other market conditions. The numerical results suggest that our methodology should be of practical use. In real applications, it is important to look for a GARCH process which fits the data well before performing the proposed method.

In this paper, we only consider the GARCH model with conditionally normally distributed errors. However, empirical evidence shows that the distribution of stock returns has fatter tails than the normal distribution (see Mandelbrot (1963)). In order to capture the fat-tail phenomenon in financial data, we may use the GARCH model with conditionally t-distributed errors which was proposed by Bollerslev (1987).

The GARCH process is just one of the possible tools to model the changing variances. The ideas presented here may be extended to other stochastic variance time series models in the literature. For further research, one may formulate the problem through a consumption-investment model and may consider the problem with several risky assets.

Acknowledgments

This work was partially supported by grants from Research Grants Council of HKSAR (Project No. HKU 7168/98H and HKU 7202/99H).

References

1. T. Bollerslev. Generalized Autoregressive Conditional Heteroscedasticity, *Journal of Econometrics* **31**, 307–327 (1986).
2. T. Bollerslev. A Conditional Heteroscedastic Time Series Model for Speculative Prices and Rates of Return, *Review of Economics and Statistics* **69**, 542–547 (1987).
3. P.P. Boyle and X. Lin. Optimal Portfolio Selection with Transaction Costs, *North American Actuarial Journal* **1**, 27–39 (1997).
4. G.M. Constantinides. Capital Market Equilibrium with Transaction Costs, *Journal of Political Economy* **94**, 842–862 (1986).
5. J.C. Duan. The GARCH Option Pricing Model, *Mathematical Finance* **5**, 13–32 (1995).

6. R.F. Engle and C. Mustafa. Implied ARCH Models from Option Prices, *Journal of Econometrics* **52**, 289-311 (1992).
7. G. Gennotte and A. Jung. Investment Strategies under Transaction Costs: The Finite Horizon Case, *Management Science* **3**, 385–404 (1994).
8. R.R. Grauer and N.H. Hakansson. Higher Return, Lower Risk: Historical Returns on Long-Run, Actively Managed Portfolio of Stocks, Bonds and Bills, 1936-1978, *Financial Analysts Journal* **38**, 39–53 (1982).
9. B. Mandelbrot. The Variation of Certain Speculative Prices, *Journal of Business* **36**, 394-419 (1963).
10. R.C. Merton. Lifetime Portfolio Selection under Uncertainty: The Continuous-Time Case, *Review of Economics and Statistics* **51**, 247–257 (1969).
11. P.A. Samuelson. Lifetime Portfolio Selection by Dynamic Stochastic Programming, *Review of Economics and Statistics* **51**, 239–246 (1969).

STATISTICAL MODELLING OF THE J-CURVE EFFECT IN TRADE BALANCE: A CASE STUDY

W. C. IP, H. WONG
Department of Applied Mathematics, The Hong Kong Polytechnic University, Hung Hom, Kowloon, Hong Kong
E-mail: mathipwc@polyu.edu.hk

Z. J. XIE and Y.L. LIU
Department of Probability and Statistics, Peking University Beijing 100871, China

The sparse coefficient regression has been applied satisfactorily to model the effect of Mexican Pesos exchange rate on the nation's trade balance. The full effect is found to take fourteen quarters to pass through and the J-curve effect in the trade balance lasts for twenty quarters.

1. Introduction

Devaluation of the domestic currency has been commonly used by a nation facing a hefty trade deficit problem. Devaluation causes foreign goods to become more expensive to domestic consumers and domestic goods to become cheaper to foreigners. With the Marshall-Lerner condition, these circumstances imply a drop in imports and a boost in exports, leading to trade balance improvements. Often, while a devaluation increases import prices quickly, import quantities adjust gradually. Depreciation may therefore increase the value of imports in the short-run, inducing greater trade balance deficits than before. Over time, the quantity of exports rises and the quantity of imports falls. Export values catch up with import values so that the initial deterioration in the trade balance is halted and then reversed. This phenomenon of trade balance to first deteriorates before improving as a result of a devaluation of the domestic currency is referred to as the J-curve effect. For example, the J-curve was a topic of discussion in Britain after a 1976 sterling devaluation was followed by a worsening trade balance.

Magee[7] gave a detailed account for the evolution of the J-curve in the trade balance. The empirical evidence of Bahmani-Oskooee[3] for Greece, India, Korea and Thailand supports the pattern of movement described by the J-curve. This paper intends to examine the statistical relationship between the exchange rate and trade balance of Mexico, using quarterly data on the relevant variables for the period 1983 Q1 - 1997 Q1. Specifically, it attempts to estimate of the duration of the J-curve effect, if any, in the Mexican trade balance as resulted from the continual depreciation of Pesos during this period. The paper is organized as follows. Section 2 gives a brief account of the trade balance data. Section 3 introduces a statistical model relating the trade balance to the exchange rate and its lagged values. This model is enhanced in Section 4 by including other economic fundamentals. Section 5 concludes.

2. Mexican Pesos Exchange Rate and Trade Balance Data

The Mexican Pesos exchange rate and trade balance data between the first quarter of 1983 and the first quarter of 1997 are depicted in figures 1 and 2 respectively.

Figure 1: US$/Pesos Exchange Rate

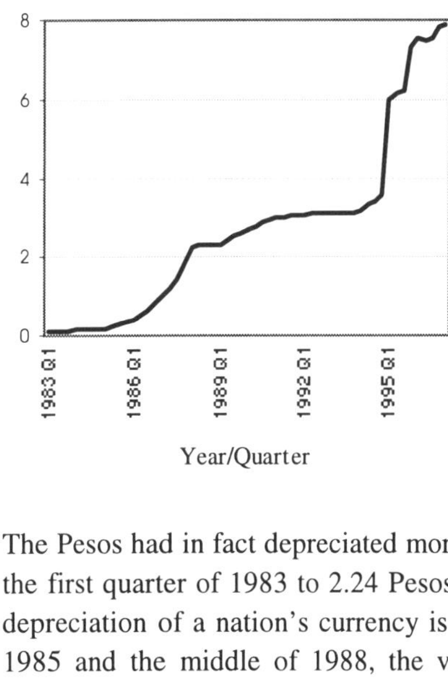

Figure 2: Trade Balance in US$M

The Pesos had in fact depreciated more than twenty times, from US$=0.10 Pesos in the first quarter of 1983 to 2.24 Pesos in the fourth quarter of 1988. A continuous depreciation of a nation's currency is not uncommon. For example, between early 1985 and the middle of 1988, the value of the US dollar measured against the currencies of the Other Ten countries declined more than 40 percent. Readers may refer to Meade[8] for more details. While the Pesos had experienced a hefty devaluation during this six-year period the trade balance continued to worsen from a substantial surplus position to a just balanced one in the same period. The devaluation of Pesos during 1983-1988 had not been seen to bring about any improvement to the nation's trade balance. The worsening trade balance had not reversed until the first quarter of 1995 when the Pesos was devaluated tremendously from 3.6 in the last quarter of 1994 to 6.0 in the first quarter of 1995, and to 8.0 in the first quarter of 1997.

3. Statistical Modelling of J-Curve

As evidence has suggested that the worsening trade balance situation of a nation usually improves after a certain lag of domestic currency devaluation, the movement of a nation's trade balance immediately before and after a currency devaluation is therefore largely influenced by the exchange rate. It is therefore of interest to study the effects of devaluation on the trade balance. In what follows we shall introduce the sparse coefficient modelling of time series data of An[1]. This model is particularly useful in studying the full effect on trade balance of changes in the exchange rate.

Let $\{x(t), y(t)\}$, $t = 0,1,2,..,n$, denote a realization of two time series. We wish to establish a regression equation of $y(t)$ on $x(t)$, possibly with a predetermined maximum lag p:

$$y(t) = \alpha_1 x(t-d_1) + \alpha_2 x(t-d_2) + .. + \alpha_m x(t-d_m) + \varepsilon(t)$$

where m and $\{d_k\}_1^m$ are unknown integers, $0 \le m$, $d_k \le p$, k=1,2,.., m and $\varepsilon(t)$ is the error term. An[1] and An and Gu[2] suggested a method to estimate the parameters m and the set of integers $\{d_k\}_1^m$ under iid Gaussian errors. They also proved that these estimates are consistent in probability sense. Unfortunately, their method is difficult to operationalize. Recently, Liu[5] suggested a computational algorithm for An's method, based on Gram-Schmidt orthogonalization procedure to choose m and $\{d_k\}_1^m$ which are optimal by a mean square error (MSE) criterion. The computational procedure is described briefly as follows:

Let

$$S = \{1, 2, \cdots, p\},$$
$$M_0 = \phi;\ M_s = \{d_1, d_2, \cdots, d_s\}, s = 1,2,\cdots, p_0.$$

Write $\mathbf{A}^{(t)} = (x(t-1), x(t-2), \cdots, x(t-p))$ and

$$\mathbf{A} = \begin{pmatrix} \mathbf{A}^{(p+1)} \\ \mathbf{A}^{(p+2)} \\ \vdots \\ \mathbf{A}^{(n)} \end{pmatrix},$$

where $p_0 \le p$, $\{d_1, \cdots, d_{p_0}\}$ are integers selected from S.

Step 1. Put $p_0 = 0; d_0 = 0; \mathbf{Q}_{d_0} = \mathbf{0};\ i = 1$

Step2. Denote the *j*-th column of $\mathbf{A}$ by $\mathbf{A}_j$.

Compute $\mathbf{P}_j = \mathbf{A}_j - \sum_{k=0}^{i-1} (\mathbf{A}'_j \mathbf{Q}_{d_k}) \mathbf{Q}_{d_k},\ j \in S$.

Replace S by $S \setminus \{j : \mathbf{P}_j = \mathbf{0}\}$. If $S = \phi$, or there exists $j \in S$, such that $\mathbf{P}'_j \mathbf{Y}_n = 0$, where $\mathbf{Y}_n = (y(p+1), \cdots, y(n))'$, then stop; otherwise go to the next step.

Step 3. Find

$$d_i = \arg \underset{j \in S}{Max} (\mathbf{P}'_j \mathbf{Y}_n)^2 / (\mathbf{P}'_j \mathbf{P}_j)$$

$$\mathbf{Q}_{d_i} = (\mathbf{P}'_{d_i} \mathbf{P}_{d_i})^{-1/2} \mathbf{P}_{d_i}.$$

Replace S by $S \setminus \{d_i\}$ and p_0 by $p_0 + 1$. If $i < p$, replace i by $i+1$ and return to Step 2.

The optimality properties of the sets M_s, $s = 1, 2, \cdots, m$ is given by the following theorem, proof of which may be found in Liu[5]:

THEOREM. Let the sets $M_s, s = 1, 2, \cdots, m$ be the integer sets selected by the above algorithm, then the following holds

$$E(M_{s+1}) = \underset{j \in M_s^c}{Min}\ E(M_s \cup \{j\}), s = 0, 1, .., p_0 - 1$$

where

$$M_s^c = \{1, 2, \cdots, p\} \setminus M_s,$$

and

$$E(M_s) = \mathbf{Y}'_n \mathbf{Y}_n - \mathbf{Y}'_n \mathbf{A}(M_s)(\mathbf{A}'(M_s)\mathbf{A}(M_s))^{-1} \mathbf{A}'_n (M_s) \mathbf{Y}_n$$

is the least *MSE* at the iterative step *s*.

The sparse coefficient model is applied to the Mexican data between the first quarter of 1983 and the first quarter of 1997. Here the data consist of observations on 57 quarters and we have chosen a maximum lag length *p*=15. Lags are necessary to capture the full effect on one economic variable of the change in another. Meade[8] showed that the full effect of the US dollar exchange rate change took on the average two decades to pass through to the prices of non-oil imports, though about fifty percent adjustment occurred within two quarters. He also showed that the full effect on export volume took eight quarters. Junz and

Rhomberg[4] presented empirical evidence to support lags of up to five years in the effects of exchange rate changes on market share of countries in world trade. For Mexican trade balance (*TB*) and exchange rate (*ER*), we obtain

$$M = 9, M_9 = \{1, 2, 4, 7, 9, 11, 12, 13, 15\}.$$

The resulting regression equation has an R^2=0.833 and p-value=$1.8\text{x}10^{-6}$. We have also used the stepwise regression to choose the best subset of current and lagged exchange rates. The procedure has included the current exchange rate and those of lags 1, 4, 5 and 9 with an R^2=0.815 and p-value=0.07. The Wald test favours the sparse coefficient regression with a test statistic W=24.1 and p-value=0.0002. The sparse regression is seen preferable to the stepwise. The results support that exchange rate with a lag structure alone will sufficiently explain the behaviour of a nation's trade balance shortly before and after a domestic currency devaluation. Similar study may be found, for example, in Salant[13].

It should be mentioned that the coefficient of determination and the Wald test are used as exploratory tools here. This is because, we understand that the data series considered in our work are nonstationary. Thus the condition of asymptotic normality of the relevant statistics may not hold. We believe, however, the approaches used form a reasonable basis for comparison. A good introduction to statistical inference with nonstationary time series is the book by Maddala and Kim[6]. Some of the original and detailed work are Park and Phillips[9,10], Phillips and Hansen[12], and Phillips[11].

4. Other Economic Fundamentals

Although exchange rate is a main attribute to a nation's trade balance, other economic fundamentals may also be crucial. For example, in addition to exchange rate, Bahmani-Oskooee[3] included domestic income, world income, domestic money supply, world money supply, etc, in his model. Here, we have included short run interest rate (*SIR*), export price index (*XPI*), import price index (*MPI*) and international reserve (*INR*). Only the interest rate is, however, found significant and the other variables are therefore dropped in subsequent analysis. The interest rate itself is nonetheless highly correlated with the exchange rate. Its correlation with the current exchange rate and included lagged exchange rates is 0.87. As strong collinearity exists only the orthogonal component (denoted by *SIR.ORG*) is used in the regression. Owing to the availability of data on the other economic fundamentals the final regression equation is fitted for observations from 1986 Q3 to 1997 Q1 (42 quarters) and the equation, after discarding insignificant coefficients, obtained is

$$TB(t) = -1.627 + 4.130ER(t) + 4.596ER(t-1) - 2.267ER(t-2) - 2.207ER(t-4)$$
$$-9.787ER(t-9) + 6.969ER(t-11) - 6.016ER(t-15) + 0.133SIR.ORG(t).$$

An R^2=0.94 and p-value=1.7x10^{-8} are recorded. The fitted values are plotted together with the actual data in Figure 3. The J-curve effect is obvious in the fitted model, from which the duration of the J-curve may be estimated to be around twenty quarters.

Figure 3: Fitted and Actual Trade Balance
(in US$ '000 M) 1987Q1 - 1997Q1

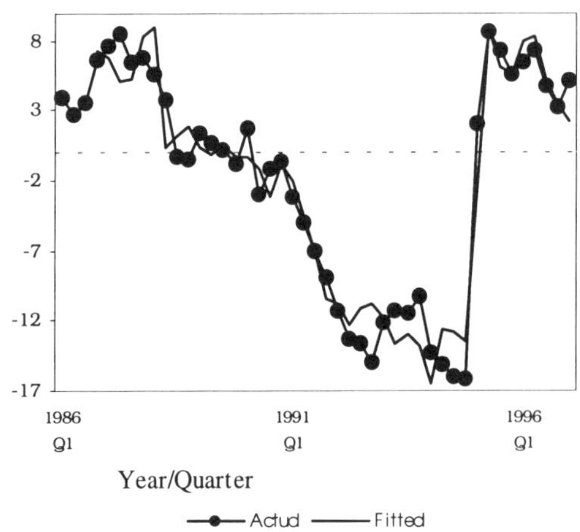

5. Conclusion

The J-curve effect in the trade balance of Mexico has been satisfactorily modelled by a sparse coefficient regression with a lag structure of the Pesos exchange rate. The existence of the well-known J-curve effect becomes obvious and its duration may also be readily determined. The inclusion of other relevant economic fundamentals often improves the fitting. An important use of modelling the J-curve that appeals most to policy makers is to determine the optimal amount of domestic currency devaluation which will lead to a reverse in the worsening trade balance as soon after as possible.

Acknowledgements

The first author's research was supported by a grant from The Hong Kong Polytechnic University Research Committee and the second author's research was supported by a grant from the University Research Council. The third author's research was partly supported by grants from the NNFC (No.79790130) and the PKU-Lianzheng Financial Laboratory.

The authors are also indebted to anonymous referees and editors for their comments and suggestions which lead to improvements of the paper.

References

1. H. An, The method of estimating parameters in regressive and autoregressive mixed models, *Statistics and Applied Probability* **2**, 19-27 (1987).
2. H. An and L. Gu, On the selection of regression variables, *ACTA Mathematics and Applications* **2**, 27-36 (1985).
3. M. Bahmani-Oskooee, Devaluation and the J-curve: Some evidence from LDCs, *The Review of Economics and Statistics* **67**, 500-504 (1985).
4. H.B. Junz and R.R. Rhomberg, Price competitiveness in export trade among industrial countries, *American Economic Review* **63**, 412-418 (1973).
5. Y.L. Liu, *Modelling of time-varying AR(p) and the forecasting of financial data by Neural Networks*, Unpublished Thesis, (Department of Probability and Statistics, Peking University, 1999).
6. G.S. Maddala and I. Kim, Unit Roots, Cointegration, and Structural Change, *Cambridge University Press*, (1998).
7. S.P. Magee, Currency contracts, pass-through, and devaluation, *Brooking Papers on Economic Activity*, 303-325 (1973).
8. E.E. Meade, Exchange rates, adjustment, and the J-curve, *Federal Reserve Bulletin* **74**(10), 633-644 (1988).
9. J.Y. Park and Phillips, P.C.B., Statistical Inference in Regressions with Integrated Processes : Part I, *Econometric Theory*, **4,** 468-497 (1988).
10. J.Y. Park and Phillips, P.C.B., Statistical Inference in Regressions with Integrated Processes : Part II, *Econometric Theory*, **5,** 95-131 (1989).
11. P.C.B. Phillips, Fully Modified Least Squares and Vector Autoregression, *Econometrica*, **63**, 1023-1078 (1995).
12. P.C.B. Phillips and B.E. Hansen, Statistical Inference in Instrumental Variables Regression with I(1) Processes, *Review of Economic Studies*, **57**, 99-125 (1990).

13. M. Salant, Devaluations improve the balance of payments even if not the trade balance in *Effects of exchange rates Adjustments*, (Washington Treasury Department, OASIA Res., 97-114 (1974).

RUIN THEORY WITH INTEREST INCOMES

H. YANG
Department of Statistics and Actuarial Science
The University of Hong Kong
Hong Kong
E-mail: hlyang@hkusua.hku.hk

L. ZHANG
Department of Applied Mathematics
Beijing Institute of Technology
Beijing, China

This paper summerizes the main results of our recent research on the ruin theory under compound Poisson model with constant interest force. We present some results on the distribution of the severity of ruin, the distribution of the surplus immediately prior to ruin, and the joint distribution of the surplus immediately before and after ruin. The probability of ruin for good is defined. By adapting the techniques of Sundt and Teugels (1995), integral equations satisfied by the above distributions and probability are obtained. The *Laplace Transforms* of the above distributions and probability are also obtained. Some asymptotic results and upper and lower bounds for the above distributions and probability are discussed. Some new results on the classical models are obtained as special cases of our model.

1 Introduction

For a long period of time ruin probabilities have been of a major interest in mathematical insurance, and have been investigated by many authors. The early work on this problem can be, at least, tracked back to Lundberg (1903). When we consider the ruin problems, the quantity of interest is the amount of surplus (by surplus, we mean the excess of some initial fund plus premiums collected over claims paid), we say ruin happens if the surplus becomes negative. In order to track surplus, we need to model the claim payments, premiums collected, investment incomes, and expenses, along with any other item that impacts the cash flow. For more detailed discussions on this subject, see Bühlmann (1970), Daykin, Pentikäinen and Pesonen (1994), Gerber (1979), Grandell (1991, 1997), Klugman, Panjer and Willmot (1998), Rolski, Schmidli, Schmidt and Teugels (1999) and the references therein.

For mathematical simplicity, up to now, the models described in risk theory are idealized. The most common used model in actuarial science is the compound Poisson model: Let $\{U(t); t \geq 0\}$ denote the surplus process which measures the surplus of the portfolio at time t, $U(0) = u$ be the initial surplus.

The surplus at time t can be written as:

$$U(t) = u + pt - X(t) \tag{1.1}$$

where $p > 0$ is a constant, it represents the premium rate, $X(t) = \sum_{i=1}^{N(t)} Y_i$ is the claim process, $\{N(t); t \geq 0\}$ is the number of claims up to time t, while the sequence $\{Y_1, Y_2, \cdots\}$ are independent and identically distributed (i.i.d.) variables with the same distribution $F(x)$ which has mean μ and $\{Y_1, Y_2, \cdots\}$ is independent of $\{N(t); t \geq 0\}$, $N(t)$ is a homogeneous Poisson process with intensity λ.

Define

$$\begin{aligned} \psi(u) &= P\{\bigcup_{t\geq 0} \{U(t) < 0\} | U(0) = u\} \\ &= P\{T < \infty | U(0) = u\} \end{aligned} \tag{1.2}$$

be the probability of ruin with initial surplus u, where $T = \inf\{t \geq 0 : U(t) < 0\}$ is called the ruin time.

The main classical results about ruin probability for the classical risk model are due to Lundberg (1926) and Cramér (1930), while the general ideas underlying collective risk theory go back as far as to Lundberg (1903). Assume the net-profit condition $p > \lambda\mu$ is hold, we list some of the main results here:

$$\psi(0) = \frac{\lambda\mu}{p}, \tag{1.3}$$

$$\psi(u) = \frac{\lambda}{p}\int_u^\infty (1 - F(x))\,dx + \frac{\lambda}{p}\int_0^u \psi(u - x)(1 - F(x))\,dx. \tag{1.4}$$

If we assume that the moment generating function of $F(x)$ exits, we have that

$$\lim_{u\to\infty} e^{Ru}\psi(u) = \frac{\rho\mu}{h'(R) - \frac{p}{\lambda}}, \tag{1.5}$$

where $h(r) = \int_0^\infty e^{rx} dF(x) - 1$, and $\rho = \frac{p-\lambda\mu}{\lambda\mu}$ is called safety loading. The result in (1.5) is called the "Cramér-Lundberg approximation". Here R satisfies

$$\frac{\lambda}{p}\int_0^\infty e^{Rx}(1 - F(x))\,dx = 1 \tag{1.6}$$

and R is called adjustment coefficient (or Lundberg exponent). Moreover,

$$\psi(u) \leq e^{-Ru} \tag{1.7}$$

(1.7) is referred to as the "Lundberg inequality"

When $F(x)$ is an exponential distribution, $\psi(u)$ has a closed form:

$$\psi(u) = \frac{1}{1+\rho} \exp(-\frac{\rho u}{\mu(1+\rho)}). \tag{1.8}$$

Recently, people in actuarial science have also started paying attention to the severity of ruin. Gerber, Goovaerts and Kass (1987) considered the probability that ruin occurs with initial surplus u and that the deficit at the time of ruin is less than y:

$$G(u,y) = P\{T < \infty, -y \leq U(T) < 0 | U(0) = u\} \tag{1.9}$$

which is a function of the variable $u \geq 0$ and $y \geq 0$. In their paper, an integral equation for $G(u,y)$ was obtained. In the case where the $Y_i' s$ have an exponential-mixture or Gamma-mixture distribution, closed form solutions of $G(u,y)$ were obtained.

Later Dufresne and Gerber (1988) introduced the distribution of the surplus immediately prior to ruin in the classical compound Poisson risk model, denote this distribution function by $F(u,y)$, then

$$F(u,x) = P\{T < \infty, 0 < U(T-) \leq x | U(0) = u\}. \tag{1.10}$$

Similar results to $G(u,y)$ were obtained in the paper. Dickson (1992) used a different way to deal with the function $F(u,y)$. Using the relationship of various events, he found the relationship among $G(u,y)$, $F(u,y)$ and $\psi(u)$. In that paper, Dickson used $G(u,y)$ and $\psi(u)$ to express $F(u,y)$, then the results of $G(u,y)$ and $\psi(u)$ are used to obtain the results for $F(u,y)$. Dickson and Reis (1994) extended the method of Dickson (1992) by using dual events to explain the relationship between the density of the surplus immediately prior to ruin, and the joint density of the surplus immediately prior to ruin and severity. Gerber and Shiu (1997,1998) examined the joint distribution of the time of ruin, the surplus immediately before ruin and the deficit at ruin. They showed that as a function of the initial surplus, the joint density of the surplus immediately before ruin and the deficit at ruin satisfies a renewal equation, but for the time of ruin, there is hardly any result.

The impact of investment risk on the ruin probability and other issues is of both theoretical interest and practical importance. Ruin theory with interest incomes should be examined carefully. Academic actuaries have for too long neglected this important (crucial) aspect of modelling. In recent years, we have seen an increasing interest in risk models with interest incomes. Sundt and Teugels (1995) considered a compound Poisson model with constant interest force, by using some similar techniques to the classical model, equation for the ruin probability as well as approximations and upper and lower bounds were discussed. Two special cases, zero reserve and exponential claim sizes, were treated in more detail. Yang (1999) considered a discrete time risk model with constant interest force. By using martingale inequalities, both Lundberg type inequality and non-exponential upper bounds for ruin probabilities were obtained. Paulsen and Gjessing (1997) considered a diffusion perturbed classical risk model. Under the assumption of stochastic investment income, a Lundberg type inequality was obtained. Paulsen (1998) provided a very good survey on this subject.

In this paper, we consider a continuous time compound Poisson model with a constant interest force. The first part of this paper summerizes the main results of Yang and Zhang (1999a)- (1999c), then the probability of ruin for good is discussed. Comparing our results with the corresponding results of Gerber, Goovaert and Kass (1987), Dufresene and Gerber (1988), Dickson (1992), Dickson and dos Reis (1994) and Gerber and Shiu (1997), we find that although most of the results for the models with interest income are analogue with the corresponding results for the models with no interest incomes, however there are some new properties for the models with interest incomes.

2 On the Distribution of Severity of Ruin

In this section, we first introduce the model, then summerize the main results on the distribution of severity of ruin which were obtained in Yang and Zhang (1999a). The model in this paper is same as in Sundt and Teugels (1995). Let $U_\delta(t)$ denote the value of the reserve at time t. $U_\delta(t)$ is governed by:

$$dU_\delta(t) = p\,dt + U_\delta(t)\cdot\delta\,dt - dX(t) \tag{2.1}$$

that is

$$U_\delta(t) = u\,e^{\delta t} + p\bar{s}_{\overline{t}|}^{(\delta)} - \int_0^t e^{\delta(t-v)}dX(v) \tag{2.2}$$

where $u = U_\delta(0)$, p is the premium rate that the insurance company receives, δ is the interest force, and $X(t)$ denotes the accumulated amount of the claims occurring in the time interval $(0,t]$, that is

$$X(t) = \sum_{j=1}^{N(t)} Y_j\ . \tag{2.3}$$

Here $N(t)$ denotes the number of claims occurring in the time interval $(0,t]$ and $\{N(t); t \geq 0\}$ is a homogeneous Poisson process with intensity λ, Y_i denote the amount of the ith claim and the Y_i are positive and mutually independent and identically distributed with common distribution F, where F satisfies $F(0) = 0$. Finally,

$$\bar{s}_{\overline{t}|}^{(\delta)} = \int_0^t e^{\delta v} dv = \begin{cases} t & \text{if} \quad \delta = 0 \\ \frac{e^{\delta t} - 1}{\delta} & \text{if} \quad \delta > 0 \end{cases} . \tag{2.4}$$

For convenience we will drop the index δ when the force of interest is zero.

The moments of the claim size distribution F will be denoted by $\mu_k = \int_0^\infty x^k dF(x)$ where in particular $\mu = \mu_1$. The quantity $\lambda\mu$ is the expected claim amount per time unit.

Let $\psi_\delta(u)$ denote the probability of ruin with initial reserve u, that is :

$$\psi_\delta(u) = P\left\{ \bigcup_{t \geq 0} (U_\delta(t) < 0) \,\Big|\, U_\delta(0) = u \right\}. \tag{2.5}$$

The non-ruin probability is denoted by $\bar{\psi}_\delta(u) = 1 - \psi_\delta(u)$, and it is the probability that ruin never occurs.

We are interested in the probability of ruin with an initial reserve u and the deficit (negative surplus) immediately after the claim causing ruin is at most y, denoted by $G_\delta(u,y)$. It is easy to see that $\psi_\delta(u) = \lim_{y \to +\infty} G_\delta(u,y)$

$$G_\delta(u,y) = P\big(-y \leq U_\delta(T) < 0 \,\big|\, U_\delta(0) = u \big) \tag{2.6}$$

where T is the ruin time.

We have the following theorem.

Theorem 2.1.

$$\begin{aligned} G_\delta(u,y) = {} & \frac{p}{p+\delta u} G_\delta(0,y) + \frac{1}{p+\delta u} \int_0^u G_\delta(u-z,y)[\delta + \lambda(1-F(z))]dz \\ & + \frac{\lambda}{p+\delta u} \int_0^y (F(z) - F(u+z))dz. \end{aligned} \tag{2.7}$$

Proof : See Yang and Zhang (1999a).

In the case of $\delta = 0$, let F_1 be the equilibrium distribution of F given by

$$F_1(x) = \frac{1}{\mu} \int_0^x (1 - F(v))dv \; . \tag{2.8}$$

From Sundt and Teugels (1995), we know that the relationship between the moments of the equilibrium distribution $\nu_k = \int_0^\infty x^k d\,F_1(x)$ and the moments of F are given by:

$$\nu_k = \frac{1}{k+1} \cdot \frac{\mu_{k+1}}{\mu} . \tag{2.9}$$

The Laplace transform of F_1 is defined as:

$$\phi(s) = \int_0^{+\infty} e^{-sz} dF_1(z) = \phi_0(s), \tag{2.10}$$

$$\phi_y(s) = \int_y^{+\infty} e^{-sz} dF_1(z). \tag{2.11}$$

The following theorem provide a asymptotic result for $G(u, y)$.

Theorem 2.2.

$$G(u,y) \sim \frac{1 - e^{-Ry} \int_y^{+\infty} \frac{\lambda}{p} e^{Rz}(1 - F(z))dz - \frac{\lambda\mu}{p} F_1(y)}{\frac{\lambda}{p}(-\phi'(-R) - \frac{p}{\lambda})} e^{-Ru}. \quad u \to +\infty . \tag{2.12}$$

where R is a positive solution of (1.6) and is called the adjustment coefficient.

Proof : See Yang and Zhang (1999a).

If we let $y \to +\infty$, then we have

$$\lim_{y\to\infty} G(u,y) = \psi(u) \sim \frac{1 - \frac{\lambda\mu}{p}}{\frac{\lambda}{p}(-\phi'(-R) - \frac{p}{\lambda})} e^{-Ru}. \tag{2.13}$$

We obtain the same result as in Grandell (1991).

The Laplace transform of $G_\delta(u, y)$ has been obtained in Yang and Zhang (1999a). Notice that in Yang and Zhang (1999a), when we solved the equation (2.7), we did not work on $G_\delta(u, y)$ directly. We made some transformation first.

When the initial surplus $u = 0$, we have the following result:

Theorem 2.3

$$G_\delta(0,y) = \lambda\mu\Big(p \int_0^{+\infty} e^{-\int_0^z (p-\lambda\mu\phi(\delta w))dw} dz\Big)^{-1}$$
$$\cdot\Big(\int_0^{+\infty} e^{-\int_0^z (p-\lambda\mu\phi(\delta w))dw} (\phi(\delta z) - e^{\delta zy}\phi_y(\delta z))\Big)dz.$$

Lundberg type bounds for $G_\delta(u, y)$ were obtained in Yang and Zhang (1999a). There, again, we did not give the bound for $G_\delta(u, y)$ directly.

3 On the Distribution of Surplus Immediately before Ruin

We now consider $F_\delta(u,y)$, the probability that the surplus immediately before ruin is less than y, i.e.

$$\begin{aligned} F_\delta(u,y) &= P\big(T < +\infty, 0 < U_\delta(T-) \le y | U_\delta(0) = u\big) \\ &= \psi_\delta(u) - P\big(T < +\infty, U_\delta(T-)) > y | U_\delta(0) = u\big) \end{aligned} \tag{3.1}$$

where T is the ruin time.

An integral equation for the distribution of surplus immediately before ruin is obtained in Yang and Zhang (1999b). Now we state the result in the following theorem.

Theorem 3.1.

$$\begin{aligned} F_\delta(u,y) = {} & \frac{p}{p+\delta u} F_\delta(0,y) + \frac{1}{p+\delta u} \int_0^u F_\delta(u-z,y)\big[\delta + \lambda(1-F(z))\big]\,dz \\ & - \frac{\lambda}{p+\delta u}\Big[I_{(u\le y)} \int_0^u (1-F(v))\,dv + (1 - I_{(u\le y)}) \int_0^y (1-F(v))\,dv\Big] \end{aligned} \tag{3.2}$$

where

$$I_{\{u\ge y\}} = \begin{cases} 1 \text{ if } & u \ge y \\ 0 \text{ otherwise} \end{cases}$$

is an indicator function.

Proof : See Yang and Zhang (1999b).

An asymptotic result for $F(u,y)$ is given in the following theorem:

Theorem 3.2. When $\delta = 0$ and the adjustment coefficient R exists, then we have when $u \le y$,

$$\lim_{u\to\infty} e^{Ru} F(u,y) = \frac{(1 - \frac{\lambda\mu}{p})}{\frac{\lambda}{p}(-\phi'(-R) - \frac{p}{\lambda})} \tag{3.3}$$

and when $u \ge y$,

$$\lim_{u\to\infty} e^{Ru} F(u,y) = \frac{F^*(y) \quad \frac{\lambda\mu}{p} F_1(y)}{\frac{\lambda}{p}\big(-\phi'(-R) - \frac{p}{\lambda}\big)} \tag{3.4}$$

where

$$F^*(x) = \int_0^x \frac{\lambda}{p} e^{Rz} (1 - F(z))\,dz$$

and $\phi(s)$ is given by (2.10).

Proof: See Yang and Zhang (1999b).
Let $y \to +\infty$, then

$$\lim_{y\to+\infty} F(u,y) = \psi(u) \sim \frac{1-\frac{\lambda\mu}{p}}{\frac{\lambda}{p}\left(-\phi'(-R)-\frac{p}{\lambda}\right)} e^{-Ru}.$$

Again we obtain the same result as in Grandell (1991). Similar to Section 2, we can obtain the Laplace transform of $F_\delta(u,y)$.

Let

$$A_\delta(u,y) = \frac{F_\delta(u,y)-F_\delta(0,y)}{0-F_\delta(0,y)}.$$

Now we present the Lundberg type bound for $A_\delta(u,y)$, the result is stated in the following theorem.

Theorem 3.3. Let $\underline{u}_\delta = \frac{\frac{\lambda\mu}{F_\delta(0,y)}F_1(y)-p}{\delta}$ and $s = s_\delta(u,y)$ be the solution of equation $u\gamma_{\delta 1}(s,y) = -\gamma'_{\delta 1}(s,y)$, where $\gamma_{\delta 1}(s,y) = \int_0^\infty e^{-sv} dA_\delta(v,y)$. When the initial surplus u satisfies $u \geq \underline{u}_\delta$, then $1 - A_\delta(u,y)$ satisfies a Lundberg type inequality:

$$1 - A_\delta(u,y) \leq \frac{\frac{\lambda\mu}{F_\delta(0,y)}[\phi(s_\delta(u,y)) - \phi_y(s_\delta(u,y))] - \lambda\mu\phi(s_\delta(u,y))}{u\delta + p - \lambda\mu\phi(s_\delta(u,y))} \cdot e^{us_\delta(u,y)} \tag{3.5}$$

where ϕ and ϕ_y are same as before, $|s_\delta(u,y)|$ is called the adjustment function.

4 The Joint Distribution of Surplus before and after Ruin

In Sections 2 and 3 we discussed the distribution of surplus immediately before and after ruin, in this section we discuss the joint distribution of surplus before and after ruin:

$$\begin{aligned} H_\delta(u,x,y) &= P(T<\infty,\ U_\delta(T+) \geq -y \text{ and } U_\delta(T-) \leq x \mid U_\delta(0)=u) \\ &= P(T<\infty,\ |U_\delta(T+)| \leq y \text{ and } U_\delta(T-) \leq x \mid U_\delta(0)=u). \end{aligned} \tag{4.1}$$

This is the joint distribution of the surplus immediately before ruin and the deficit at ruin under interest force δ, where x, y are positive real numbers.

It is easy to see that

$$\lim_{x\to+\infty} H_\delta(u,x,y) = G_\delta(u,y),$$
$$\lim_{y\to+\infty} H_\delta(u,x,y) = F_\delta(u,x).$$

The following theorem gives an integral equation for $H_\delta(u,x,y)$:

Theorem 4.1

$$H_\delta(u,x,y) = \frac{p}{p+\delta u} H_\delta(0,x,y) + \frac{1}{p+\delta u}\int_0^u H_\delta(u-z,x,y)\big[\delta + \lambda(1-F(z))\big]dz$$
$$- \frac{\lambda}{p+\delta u}\Big[I_{(u\le x)}\int_0^u \big[F(v+y)-F(v)\big]dv + (1-I_{(u\le x)})$$
$$\int_0^x \big[F(v+y)-F(v)\big]dv\Big]. \tag{4.2}$$

Proof: See Yang and Zhang (1999c).

Let F_1 be the equilibrium distribution of F as defined in Section 2. A Cramér-Lundberg approximation type result is given in the following theorem.

Theorem 4.2. If $\delta = 0$ and the adjustment coefficient R exits, then we have for $u \le x$

$$\lim_{u\to+\infty} e^{Ru}H(u,x,y) = \frac{1 - e^{-Ry}\int_y^{+\infty}\frac{\lambda}{P}e^{Rz}(1-F(z))dz - \frac{\lambda\mu}{p}F_1(y)}{\frac{\lambda}{p}(-\phi'(-R) - \frac{p}{\lambda})} \tag{4.3}$$

and for $u \ge x$,

$$\lim_{u\to+\infty} e^{Ru}H(u,x,y) =$$
$$\frac{F^*(x) - e^{-Ry}(F^*(x+y) - F^*(y)) + \frac{\lambda\mu}{p}(F_1(x+y) - F_1(x) - F_1(y))}{\frac{\lambda}{p}(-\phi'(-R) - \frac{p}{\lambda})}. \tag{4.4}$$

Here $F^*(x)$ and ϕ are the same as that in Section 3.

Proof : See Yang and Zhang (1999c).

Similar to the previous sections, we also can solve the equation (4.2), and obtain the Laplace transform of $H_\delta(u,x,y)$. Lundberg type bounds were also obtained in Yang and Zhang (1999c).

5 The Probability of Ruin for Good

Dassios and Embrechts (1989) studied a piecewise-deterministic Markov process model, and pointed out that when the reserves of the insurance company becomes less than $-\frac{p}{\delta}$, then the interest payment will exceed the premium income and the surplus process will not able to come back to positive side again. Therefore it is meaningful to consider the following probability, we call it the probability of ruin for good (Dassios and Embrechts called it the absolute ruin). Define

$$\Psi_\delta(u) = P\left(\bigcup_{t\geq 0}\left(U_\delta(t) \leq -\frac{p}{\delta}\right) \Big| U_\delta(0) = u\right), \tag{5.1}$$

$$\overline{\Psi}_\delta(u) = 1 - \Psi_\delta(u) = P\left(U_\delta(t) > -\frac{p}{\delta} \quad \text{for all} \quad t \,\Big|\, U_\delta(0) = u\right). \tag{5.2}$$

Obviously

$$\Psi_\delta(u) \leq \psi_\delta(u)\ .$$

The probability of ruin for good measures the likelihood of occurrence of that surplus process will not come back to positive.

The following result provides an integral equation satisfied by the probability of ruin for good:

Theorem 5.1. For $-\frac{p}{\delta} < u < +\infty$,we have

$$\Psi_\delta(u) = \frac{1}{p+\delta u}\int_{-\frac{p}{\delta}}^{u} \Psi_\delta(z)[\delta + \lambda(1 - F(u-z))]\, dz - \frac{\lambda}{p+\delta u}\int_0^{u+\frac{p}{\delta}} (1 - F(z))\, dz, \tag{5.3}$$

$$\overline{\Psi}_\delta(u) = \frac{1}{p+\delta u}\int_{-\frac{p}{\delta}}^{u} \overline{\Psi}_\delta(v)\left[\delta + \lambda\big(1 - F(u-v)\big)\right] dv. \tag{5.4}$$

Moreover $\Psi_\delta(u)$ and $\overline{\Psi}_\delta(u)$ are continuous function for all real u.

Proof : If the initial surplus $U_\delta(0) = u > -\frac{p}{\delta}$, then it is obvious that the event $\left\{U_\delta(t) \leq -\frac{p}{\delta}\right\}$ can not occur before the first claim, so we condition on the first claim time T_1 and the first claim amount Y_1 , then

$$\begin{aligned}
\Psi_\delta(u) &= E\left[\Psi_\delta\left(u\, e^{\delta T_1} + p\cdot\frac{e^{\delta T_1}-1}{\delta} - Y_1\right)\right] \\
&= E\left(\Psi_\delta\left(u\, e^{\delta T_1} + p\cdot\frac{e^{\delta T_1}-1}{\delta} - Y_1\right) \Big| Y_1 < u\, e^{\delta T_1} + p\cdot\frac{e^{\delta T_1}-1}{\delta} + \frac{p}{\delta}\right) \\
&\quad P\left(Y_1 < u\, e^{\delta T_1} + \frac{p\, e^{\delta T_1}}{\delta}\right) + E\left(\Psi_\delta\left(u\, e^{\delta T_1} + p\cdot\frac{e^{\delta T_1}-1}{\delta} - Y_1\right) \Big| \right. \\
&\quad \left. Y_1 \geq u\, e^{\delta T_1} + p\cdot\frac{e^{\delta T_1}-1}{\delta} + \frac{p}{\delta}\right) P\left(Y_1 \geq u\, e^{\delta T_1} + \frac{p\, e^{\delta T_1}}{\delta}\right)
\end{aligned}$$

$$= \int_0^{+\infty} \lambda e^{-\lambda t} \int_0^{u e^{\delta t} + p\cdot\frac{e^{\delta t}-1}{\delta} + \frac{p}{\delta}} \Psi_\delta\left(e^{\delta t} u + p \cdot \frac{e^{\delta t} - 1}{\delta} - z\right) dF(z)\, dt$$
$$+ \int_0^{+\infty} \lambda e^{-\lambda t} \int_{u e^{\delta t} + p\cdot\frac{e^{\delta t}}{\delta}}^{+\infty} dF(z)\, dt\ .$$

By some similar calculation to Sundt and Teugels (1995), we have

$$\begin{aligned}
&(p+\delta u)\Psi_\delta(u) - \Big(p + \big(-\frac{p}{\delta}\big)\delta\Big)\Psi_\delta\big(-\frac{p}{\delta}\big) - \delta\int_{-\frac{p}{\delta}}^{u} \Psi_\delta(v)\, dv \\
&= \lambda \int_{-\frac{p}{\delta}}^{u} \Psi_\delta(v)\, dv - \lambda \int_{-\frac{p}{\delta}}^{u} \Big(1 - F\big(v + \frac{p}{\delta}\big)\Big)\, dv \\
&\quad +\lambda \int_{-\frac{p}{\delta}}^{u} \int_0^{v+\frac{p}{\delta}} \Psi_\delta(v-z) d(1-F(z))\, dv
\end{aligned} \tag{5.5}$$

and the last term of the right hand of (5.5) equals

$$\begin{aligned}
&\lambda \int_{-\frac{p}{\delta}}^{u} \int_0^{v+\frac{p}{\delta}} \Psi_\delta(v-z) d(1-F(z))\, dv \\
&= \lambda \int_{-\frac{p}{\delta}}^{u} \int_0^{v+\frac{p}{\delta}} (1-F(z)) \frac{d\,\Psi_\delta(v-z)}{dv}\, dz\, dv \\
&\quad +\lambda \int_{-\frac{p}{\delta}}^{u} (1-F(z))\Psi_\delta(v-z)\Big|_{z=0}^{v+\frac{p}{\delta}}\, dv \\
&= \lambda \int_{-\frac{p}{\delta}}^{u} \Big(1 - F\big(v+\frac{p}{\delta}\big)\Big)\Psi_\delta\big(-\frac{p}{\delta}\big)\, dv - \lambda \int_{-\frac{p}{\delta}}^{u} \Psi_\delta(v)\, dv \\
&\quad +\lambda \int_0^{u+\frac{p}{\delta}} (1-F(z)) \int_{z-\frac{p}{\delta}}^{u} \frac{d\,\Psi_\delta(v-z)}{dv}\, dv\, dz \\
&= \lambda \int_{-\frac{p}{\delta}}^{u} \Big(1 - F\big(v+\frac{p}{\delta}\big)\Big)\Psi_\delta\big(-\frac{p}{\delta}\big)\, dv - \lambda \int_{-\frac{p}{\delta}}^{u} \Psi_\delta(v)\, dv \\
&\quad +\lambda \int_0^{u+\frac{p}{\delta}} (1-F(z))\Psi_\delta(u-z)\, dz - \lambda \int_0^{u+\frac{p}{\delta}} (1-F(z))\Psi_\delta\big(-\frac{p}{\delta}\big)\, dz\ .
\end{aligned} \tag{5.6}$$

From the definition of $\Psi_\delta(u)$, we know that $\Psi_\delta\big(-\frac{p}{\delta}\big) = 1$. So plugging (5.6) into (5.5) yields

$$(p+\delta u)\Psi_\delta(u)-\delta\int_{-\frac{p}{\delta}}^{u}\Psi_\delta(v)\,dv=\lambda\int_0^{u+\frac{p}{\delta}}(1-F(z))\Psi_\delta(u-z)\,dz$$
$$-\lambda\int_0^{u+\frac{p}{\delta}}(1-F(z))\,dz$$

which can be rewritten to obtain (5.3).

In (5.3), let $u\to-\frac{p}{\delta}$, then

$$\lim_{u\to-\frac{p}{\delta}+}\Psi_\delta(u)=\frac{\lim_{u\to-\frac{p}{\delta}+}\left[\Psi_\delta(u)\left[\delta+\lambda(1-F(0))\right]-\lambda\left(1-F(u+\frac{p}{\delta})\right)\right]}{\lim_{u\to-\frac{p}{\delta}+}(p+\delta u)'}$$
$$=\frac{(\delta+\lambda)\lim_{u\to-\frac{p}{\delta}+}\Psi_\delta(u)-\lambda}{\delta},$$

so we have $\lim_{u\to-\frac{p}{\delta}+}\Psi_\delta(u)=1=\lim_{u\to-\frac{p}{\delta}-}\Psi_\delta(u)=\Psi_\delta\left(-\frac{p}{\delta}\right)$. Hence $\Psi_\delta(u)$ is a continuous function for all u with $\lim_{u\to+\infty}\Psi_\delta(u)=0$.

Let $\overline{\Psi}_\delta(u)=1-\Psi_\delta(u)$ be the survive probability, then $\overline{\Psi}_\delta(u)$ is a distribution function, and $\overline{\Psi}_\delta(u)$ satisfies (5.4). □

Let

$$\gamma_\delta(s)=\int_{-\frac{p}{\delta}}^{+\infty}e^{-sv}d\overline{\Psi}_\delta(v),$$

then by the integral equation (5.4), we have the following result:

$$\gamma_\delta(s)=e^{\int_0^s\frac{p-\lambda\mu\phi(v)}{\delta}dv}=e^{\frac{ps}{\delta}-\int_0^s\frac{\lambda\mu\phi(v)}{\delta}dv}. \tag{5.7}$$

Similar to Sundt and Teugels (1995), we can obtain the Lundberg type bound for the probability of ruin for good, we state the result in the following theorem.

Theorem 5.2. Denote $-\sigma$ the convergence abscissa of $\phi(s)$, supposed to be strictly negative, and assume moreover that $\phi(s)$ is convergence at the point $s=-\sigma$. Then for $u>\frac{\lambda\mu-p}{\delta}$, the Lundberg type inequality can be written as:

$$\Psi_\delta(u)\le e^{\int_0^\sigma\frac{\lambda\mu\phi(-z)}{\delta}dz}\cdot e^{s_\delta(u)(u+\frac{p}{\delta})} \tag{5.8}$$

Where $|s_\delta(u)|$ is the adjustment function.

Remark: For most of the results presented in this paper, we assumed that the adjustment coefficient exists. This means that we assume the moment generating function of the claim size distribution exists. It is well known that in actuarial science, questions involving extremal events (such as large insurance claims, reinsurance products) play a very important role. Embrechts, Klüppelberg and Mikosch (1997) provided a detailed study on the heavy-tailed distributions and their applications in risk theory. Klüppelberg and Stadtmuller (1998) studied the infinite time ruin probability in the presence of heavy-tails and interest rates. They proved that for a positive force of interest, the asymptotic ruin probability as the initial surplus tends to infinity is different from the non-interest model. Yang and Zhang (1999d) extended the paper of Kluppelberg and Stadtmuller (1998). We obtained some results on the distribution of surplus immediately before and after ruin, and on the probability of ruin for good in the presence of heavy-tails and interest rates.

6 Conclusion remarks

In this paper, we have briefly summarized the ruin problems of a continuous time compound Poisson surplus process with constant interest force. Using the renewal theory, integral equations satisfied by the distributions of surplus before and after ruin, the joint distribution of the surplus before and after ruin, and the probability of ruin for good are obtained. Lundberg approximations and Lundberg type bounds have been discussed. In particular, let the interest rate tend to zero, from our results, we obtained some new results on classical models. When the interest force is positive, the surplus process might not return to positive size if the claim is big enough. In classical models, we know that under usual assumptions, the surplus process will tend to infinity with probability one. In some senses, this property makes the investigation on ruin probability meaningless (since the surpluses will get back to positive for sure, ruin is just a temporal phenomenon). We analyzed the probability of ruin for good in our setup.

Acknowledgments

The authors would like to thank the referee for many helpful comments and suggestions. The work described in this paper was supported by a grant from the Research Grants Council of the Hong Kong Special Administrative Region, China (Project No. HKU 7168/98H).

References

1. J. Bühlmann, *Mathematical Methods in Risk Theory.* (Springer Verlag, Heidelberg, 1982).
2. H. Cramér, *On the Mathematical Theory of Risk.* (Skandia Jubilee Volume, Stockholm, 1930).
3. A. Dassios and P. Embrechts, *Martingales and insurance risk.* Commun. Statist. - Stochastic Models **5(2)**, 181-217 (1989).
4. C.D. Daykin, T. Pentikäinen and M. Pesonen, *Practical Risk Theory for Actuaries.* Chapman & Hall, London. 1994.
5. D.C.M. Dickson, *On the distribution of the surplus prior to ruin.* Insurance: Mathematics and Economics **11**, 191-207 (1992).
6. D.C.M. Dickson and A.D.E. dos Reis, *Ruin problems and dual events.* Insurance: Mathematics and Economics **14**, 51-60 (1994).
7. F. Dufresne and H.U. Gerber, *The surpluses immediately before and at ruin,and the amount of the claim causing ruin.* Insurance: Mathematics and Economics **7**, 193-199 (1988).
8. P. Embrechts, C. Klüppelberg and T. Mikosch, *Modelling Extremal Events for Insurance and Finance.* (Springer-Verlag 1997).
9. H. Gerber, *An Introduction to Mathematical Risk Theory.* (S.S.Huebner Foundation Monograph Series No.8. Distrbuted by R.Irwin, Homewood, IL, 1979).
10. H.U. Gerber, M.J. Goovaerts and R. Kass, *On the probability and severity of ruin.* ASTIN Bulletin **17**, 151-163 (1987).
11. H.U. Gerber and E.S.W. Shiu, *On the time value of ruin.* North American Actuarial Journal **2(1)**, 48-72 (1998).
12. H.U. Gerber and E.S.W. Shiu, *The joint distribution of the time of ruin, the surplus immediately before ruin,and the deficit at ruin.* Insurance; Mathematics and Economics **21**, 129-137 (1997).
13. J. Grandell, *Aspects of Risk Theory.* (Springer-Verlag 1991).
14. J. Grandell, *Mixed Poisson processes.* (Chapman and Hall, London. 1997).
15. S.A. Klugman, H.H. Panjer and G.E. Willmot, *Loss Models From Data to Decision.* (Wiley. 1998).
16. C. Klüppelberg and U. Stadtmuller, *Ruin probabilities in the presence of heavy-tails and interest rates.* Sand. Actuarial Journal **1**, 49-58 (1998).
17. F. Lundberg, *Approximerad Framstallning av Sannolikhetsfunktionen.* (Almqvist and Wiksell, Uppsala. 1903).
18. F. Lundberg, *Aterforsakring av Kollektivrisker.* (Almqvist and Wiksell, Uppsala.1903).

19. F. Lundberg, *Forsakringsteknisk Riskutjamning.* (F. Englunds boktryckeri A.B., Stockholm. 1926).
20. J. Paulsen and H.K. Gjessing, *Ruin theory with stochastic return on investments.* Advance in Applied Probability **29**, 965-985 (1997).
21. J. Paulsen, *Ruin theory with compounding assets-a survey.* Insurance: Mathematics and Economics **22**, 3-16 (1998).
22. T. Rolski, H. Schmidli, V. Schmidt and J. Teugels, *Stochastic Processes for Insurance and Finance.* Wiley & Sons, New York. 1999.
23. B. Sundt and J.L. Teugels, *Ruin estimates under interest force.* Insurance: Mathematics and Economics **16**, 7-22 (1995).
24. H. Yang, *Non-exponential bounds for ruin probability with interest effect included.* Scandinavian Actuarial Journal **1**, 66-79 (1999).
25. H. Yang and L. Zhang, *On the distribution of surplus immediately after ruin under interest force.* (Submitted, 1999a).
26. H. Yang and L. Zhang, *On the distribution of surplus immediately before ruin under interest force.* (Submitted, 1999b).
27. H. Yang and L. Zhang, *The joint distribution of surplus immediately before ruin and the deficit at ruin under interest force.* (Submitted, 1999c).
28. H. Yang and L. Zhang, *On some problems of ruin theory in the presence of heavy-tails and interest rates.* (Working paper, 1999d)

DETECTING STRUCTURAL CHANGES USING GENETIC PROGRAMMING WITH AN APPLICATION TO THE GREATER-CHINA STOCK MARKETS

X. B. ZHANG, Y. K. TSE

Department of Economics, National University of Singapore, 10 Kent Ridge Crescent, Singapore 119260

E-mail: ecszxb@nus.edu.sg and ecstseyk@nus.edu.sg

W. S. CHAN

Department of Statistics & Actuarial Science, The University of Hong Kong, Pokfulam Road, Hong Kong

E-mail: chanws@hku.hk

Structural changes usually refer to the changes in some parameters or in the structure of a chosen model that is postulated to describe the operation of a data generating process. However, the structure of the underlying data generating process may not necessarily be equivalent to a model. It may be a pattern or an operating mechanism identifiable by certain cognitive processes. Accordingly, structural changes are changes to the operating mechanism of the underlying system. This paper considers the application of genetic programming to the cognition of the operating mechanism of a dynamic system. Based on the knowledge accumulated in the cognition process, a diagnostic statistic is defined to detect structural changes in the system. This approach is model free since it is performed without reference to model specification. The effectiveness of the model-free approach is empirically illustrated through an application to four stock markets, namely the Greater-China markets.

1 Introduction

Consider a dynamic system composed of several time-dependent variables. The realization of the system can be regarded as a multivariate time series, and the observed multivariate time series are the outcomes of the activities of the individuals in the system. Generally speaking, the development of the dynamic system is governed by an operating mechanism which is a reflection of the dynamic relationship among the component variables. If the operating mechanism has been dramatically changed at a point or during a period, then a structural change occurs. In other words, structural changes are large changes to the operating mechanism of a dynamic system.

Model-specific structural changes of a system refer to the changes in some parameters or in the structure of a chosen model that is postulated to describe the operation of the system. However, models are often approximate descriptions of the underlying data generating process. A dynamic system is often adaptive and self-organized and is very difficult to model through parametric approaches. Thus, the definition of structural changes should not be restricted to the model-specific approach. Chen and Yeh (1997) presented a model-free definition of structural changes for a univariate time series. They pointed out that the idea of a "structure" may not necessarily be equivalent to a "model". It may be a pattern identifiable by certain cognitive processes. They concluded that structural changes simply refer to the loss of historical patterns and the appearance of a novel pattern. Even though this notion of structural changes may not be conventional in statistics and econometrics, it has a good intuitive meaning. Under this notion a structure need not have any precise representation nor any definite mathematical form. As a result, a reference model is not required if such a definition of structural change is accepted.

The operating mechanism of a univariate time series $\{x_t\}$ is reflected through the nonlinear autocorrelated relationship between x_t and its lagged terms. Chen and Yeh (1997) showed that this kind of operating mechanism can be recognized through recursive genetic programming. They presented a diagnostic statistic, which is based on the learning performance, to detect structural changes in $\{x_t\}$. The operating mechanism of a multivariate time series is often reflected through the dynamic relationship among the component time series, and the meaning of structural changes is different from that for univariate time series. Sun, Zhang and Zhang (1999) discussed the application of recursive genetic programming in the case of multivariate time series. This approach was applied to the detection of structural changes among several sectors in the Shanghai stock market. The algorithm is computationally intensive and is more difficult than that for a univariate time series.

Basically, the structure of a dynamic system may not have any precise form of representation or any definite mathematical form. When we consider the detection of structural changes in the operation of a dynamic system, we should not put any restriction on the types of the dynamic relationship among the component variables. Indeed, the dynamic relationship can be recognized through certain intelligent cognitive processes. This paper considers the application of genetic programming to the cognition of the operating mechanism of a dynamic system. On the basis of the cognition process a diagnostic statistic is defined to detect structural changes in the system. The effectiveness of this model-free approach is empirically illustrated with an application to the detection of structural changes among the stock indexes of the so-called Greater-China stock markets.

This paper is organized as follows. Section 2 discusses the notions of model-free structural changes of a dynamic system. In Section 3 we briefly describe the concepts of genetic programming and summarize the procedure for the detection of structural changes using genetic programming. Section 4 presents an application of the model-free approach to the detection of structural changes among the stock indexes of the Hong Kong, Shanghai, Shenzhen and Taiwan stock markets. Section 5 gives the conclusions of this paper.

2 Model-Free Structural Changes

Suppose the realization of a system can be represented by a multivariate time series which are the outcomes of the activities of the participants. The participants of the system are called *individuals*, and the collection of all individuals is called the *population*. Suppose also that the system has n_t individuals at time t and the m-dimensional multivariate time series is denoted as $X_t = (x_{1t}, x_{2t}, \cdots, x_{mt})'$ for $t = 1, 2, \cdots, T$. The activities of the individuals are described by the *activity functions* $f_i(X_t)$, for $i = 1, 2, \cdots, n_t$. Let the operating mechanism of the system be recognized as

$$\hat{g}\left(f_1(X_t), f_2(X_t), \cdots, f_{n_t}(X_t)\right), \tag{1}$$

which is based on the activity function of each individual.

The individuals are participants of the competition in the learning process of the system, and the success or failure of each individual is subject to its activity function $f_i(X_t)$, for $i = 1, 2, \cdots, n_t$. An evaluation function is also defined to estimate the competition ability of each individual. During the intelligent learning process, the individuals accumulate their own experience and those of others. Accordingly, they modify their activity functions. Thus, the activity of each individual is intelligent and adaptive. Some individuals will be driven out of the system due to failures, while some new participants will be included into the system as new entrants.

Structural changes in a system are the "surprises" or "shocks" which have significantly changed the operating mechanism of the system but cannot be dealt with by some intelligent cognition methods such as adaptive and self-organization training. Thus, the system $\{X_t\}$ is said to have undergone a structural change at time t if given a tolerance level $\delta > 0$,

$$d\left(X_t, \hat{g}(\cdot)\right) > \delta, \tag{2}$$

where $d(\cdot)$ is a function of distance measure.

Consider a short period around a time point k^* denoted as the k^*-th period. The cognition of the operating pattern of this period is based on the knowledge accumulated in previous periods. If there is no structural change during the k^*-th period, the operating pattern for this period will be similar to that of the previous periods. Thus, $d(\cdot)$ will be small for the k^*-th period.

However, if the recognized pattern of this period is quite different from previously observed patterns, $d(\cdot)$ will be large. It can then be concluded that there is a structural change during the k^*-th period. Similarly, the cognition of the operating pattern of the next period, i.e. the $(k^* + 1)$-th period, is based on the knowledge accumulated up to the k^*-th period. If there is no structural change in the $(k^* + 1)$-th period, its recognized pattern will be similar to its previously observed pattern again, so that $d(\cdot)$ will be small for the $(k^* + 1)$-th period.

The main difference between model-free structural changes and model-specific structural changes is the sensitivity to the perturbation of the existing pattern. It is clear that model-free structural changes are less sensitive to perturbations than the model-specific one, since the cognitive process of the existing pattern through a model-free approach must be adaptive. Any small perturbation to the existing pattern cannot be detected as a structural change due to the adaptive and self-organization of the underlying data generating process. From the view point of the model-free approach, the notion of structural changes is equivalent to "surprises", "shocks", or "breakdowns" in the operating mechanism of a dynamic system.

3 Detection of Model-Free Structural Changes

3.1 Genetic Programming

Holland (1992) showed how an evolutionary process can be used to solve problems by means of a highly parallel technique called *genetic algorithm.* The goal of parallel programming is to find a way to break a job into several units that can be executed concurrently. This approach permits the parallel execution of arithmetic operations and is able to handle a great deal of information. Genetic algorithm transforms a population of individual objects, each with an associated value of fitness, into a new generation of the population. The transformation is based on the Darwinian principle of survival and reproduction of the fittest. It is analogous to naturally occurring genetic operations such as crossover (sexual recombination) and mutation.

Genetic programming is an extension of the conventional genetic algorithm, in which the structures undergoing adaptation are hierarchical computer programs of dynamically varying sizes and shapes. In applying genetic programming to a problem, there are five major preparatory steps, which involve (i) determining the set of terminals, (ii) the set of primitive functions, (iii) the fitness measure, (iv) the parameters for controlling the run, (v) and the method for designating a result and the criterion for terminating a run. Each run of genetic programming requires the specification of a termination criterion for deciding when to terminate a run and a method of result designation. We usually designate the best-so-far individual as the result of a run. Once these steps for preparing to run the genetic programming have been established, a run can be made. In genetic programming, thousands of computer generated populations are bred genetically. This breeding is done using the Darwinian principle of survival and reproduction of the fitness along with a genetic crossover operation appropriate for mating computer generated populations. The population generating procedure that solves a given problem may emerge from this combination of Darwinian natural selection and genetic operation.

3.2 Recursive Genetic Programming

The recursive genetic programming (RGP) used in this paper is an extension of the basic genetic programming (BGP) proposed by Koza (1992). Chen and Yeh (1997) suggested that model-free structural changes in a univariate time series can be detected through RGP, where the *moving window technique* is used to obtain a sequence of sub-samples. For all the sub-samples, BGP algorithm is applied. Thus, the programming is called recursive genetic programming.

Suppose that $\{X_t : t = 1, 2, \cdots, T\}$ is the observed multivariate time series. Let the window size be n_1 and the moving step be n_2. The first sub-sample S_1 consists of the first n_1 observations of $\{X_t\}$, and the second sub-sample is the modification of S_1 by pushing it forward by n_2 steps. In general, S_j is the modification of S_{j-1} in a similar manner, that is,

$$S_j = \{X_t\}_{t=(j-1)n_2+1}^{n_1+(j-1)n_2}, \qquad j = 1, 2, \cdots, L, \tag{3}$$

where $L = [(T - n_1)/n_2] + 1$ with $[z]$ denoting the largest integer smaller than z.

Given the sequence of sub-samples $S = \{S_1, S_2, \cdots, S_L\}$, BGP is applied to S_1 to learn the operating pattern. The initial generation is chosen randomly and denoted as $GP_1^{(0)}$. When the training process is over for the first sub-sample, the last generation, namely the n-th generation $GP_1^{(n)}$, is obtained. The fitness of the GP-trees in the last generation $GP_1^{(n)}$ can be computed through a fitness function $fit(\cdot)$, which is usually residual-based. In this paper $fit(\cdot)$ is taken as the sum of the squared residuals.

The fitness of each GP-tree in the generation $GP_1^{(n)}$ in S_1 is ranked as $fit_1(\cdot) \leq fit_2(\cdot) \leq \cdots$. Then we choose the best q GP-trees with the smallest fitness and designate them as the representative GP-trees for $GP_1^{(n)}$. The representative GP-trees are denoted as Q_1, and the average fitness of Q_1 is defined as

$$\overline{fit}_1 = \frac{1}{q}\sum_{i=1}^{q} fit_i(\cdot). \tag{4}$$

The same process is applied to the second sub-sample S_2 with the initial generation being $GP_2^{(0)} = GP_1^{(n)}$. Suppose that the last generation $GP_2^{(n)}$ is obtained and the representative GP-trees, Q_2, is chosen based on the fitness function $fit(\cdot)$, then the average fitness of Q_2 is computed and denoted as $\overline{fit}_2$.

This training process continues with subsequent sub-samples. In the end we obtain a sequence of average-fitness functions, $\{\overline{fit}_k : k = 1, 2, \cdots, L\}$. We define a diagnostic statistic as

$$D_k = \frac{\overline{fit}_k - \overline{fit}_{k-1}}{\overline{fit}_{k-1}}, \qquad k = 1, 2, \cdots, L, \tag{5}$$

with initial value $\overline{fit}_0 = \overline{fit}_1$. D_k reflects the relative change in average fitness between two adjacent sub-samples.

The cognition of the operating mechanism or pattern in a sub-sample is based on the knowledge accumulated in former sub-samples since the initial generation is taken as the last generation in the former sub-sample. Suppose that there is a structural change in the k^*-th sub-sample, then the recognized operating pattern cannot give an appropriate description of the operating pattern based on former knowledge. Consequently, the average fitness $\overline{fit}_{k^*}$ will be much larger than $\overline{fit}_{k^*-1}$, so that the statistic D_{k^*} is much larger than zero. After the k^*-th sub-sample, the system will learn the operating pattern of the $(k^* + 1)$-th sub-sample based on the knowledge accumulated in the k^*-th sub-sample, and the average fitness will be similar to that of the k^*-th sub-sample. Then the statistic D_k will be close to zero again.

We propose to use D_k as an exploratory diagnostic for possible structural changes. A major advantage of this approach is that model assumptions are not required for the generation of the data. If desired, formal significance tests could be constructed with additional assumptions. This issue, however, will not be pursued in this paper.

3.3 RGP Algorithm for the Detection of Structural Changes

We now summarize the steps used to detect structural changes using the recursive genetic programming technique.

Step 1: Let $X_t = \{(x_{1t}, x_{2t}, \cdots, x_{mt})' : t = 1, 2, \cdots, T\}$ be a m-dimensional multivariate time series, and let the window size of each sub-sample be n_1 and the moving step be n_2.

Step 2: Define the function set for the configuration of the activity functions associated with the individuals in the system. The function set F may be defined as:

$$F = \{+, -, \times, \div, \sin, \cos, \log, \exp, \cdots\}. \tag{6}$$

The largest lagged order in the individual activity function is denoted as h. Let the *terminal set* for the leaves in the genetic process be

$$\kappa = \{x_1(t-i), x_2(t-i), \cdots, x_m(t-i) : i = 1, 2, \cdots, h\}. \tag{7}$$

The probabilities of the selection, crossover, mutation, and reproduction of each GP-tree are denoted as p_s, p_c, p_m, and p_r, respectively. Let the maximum number of generation be denoted as $MaxGen$ and the maximum length of a GP-tree be denoted as $MaxLen$. Let the number of the GP-trees in the last generation be r, and the size of the representative GP-trees for the last generation be q ($q \leq r$).

Step 3: Generate the initial generation for the first sub-sample. At the beginning we choose an arbitrary function $f(\cdot)$ from the function set F and denote it as the *root terminal.* Define the number of variables for the selected function $f(\cdot)$ as $N(f)$. For example, the "+" operation is a two-variable operation, while the "*log*" operation is a univariate operation. Then the selected terminal is connected with the $N(f)$ terminals in the next layer. Choose an element from the set $B = F \cup \kappa$ as the final terminal. If the selected element is a function from F, repeat the same procedure so that the GP-tree in this branch is kept growing. If the selected element is a terminal from κ, it is regarded as a final terminal, and the GP-tree at this branch terminates. This process continues until r GP-trees are generated. Then we calculate the fitness of each GP-tree in the initial GP-tree population.

Step 4: Let k be the label of the sub-sample. The initial value is $k = 1$.

Step 5: Let j be the serial number of the GP-trees in the initial generation. The initial value is $j = 1$.

Step 6: For the k-th sub-sample, we proceed with the selection, crossover, mutation and reproduction operations on the GP-trees in the $(j-1)$-th generation to obtain the j-th generation of GP-trees.

Step 7: For the k-th sub-sample, we calculate the fitness of each GP-tree. Suppose the i-th GP-tree in the j-th generation can be represented as

$$f_i^{(j)}\Big(x_1(t), x_2(t), \cdots, x_m(t);\quad x_1(t-1), x_2(t-1), \cdots, x_m(t-1);$$
$$\cdots;\quad x_1(t-h), x_2(t-h), \cdots, x_m(t-h)\Big). \tag{8}$$

The sum of the squared residuals for the i-th GP-tree is,

$$fit_i^{(k)} = \sum_{t=n_2(k-1)+1}^{n_2(k-1)+n_1} \left(x_m(t) - f_i^{(j)}(\cdot) \right)^2, \tag{9}$$

for $i = 1, 2, \cdots, r$. The inverse of $fit_i^{(k)}$ represents the fitness of the i-th GP-tree.

Step 8: If $j < MaxGen$, let $j = j + 1$ and go to Step 6; otherwise, go to the next step.

Step 9: For the last generation $GP_k^{(MaxGen)}$ in the k-th sub-sample, rank the fitness values associated with the GP-trees as

$$fit_{i_1}^{(MaxGen)}(\cdot) \leq fit_{i_2}^{(MaxGen)}(\cdot) \leq \cdots \leq fit_{i_r}^{(MaxGen)}(\cdot). \tag{10}$$

The GP-trees corresponding to the first q smallest fitness values are chosen as the representative GP-trees. The average fitness for the last generation is

$$\overline{fit}_k = \frac{1}{q} \sum_{p=1}^{q} fit_{i_p}^{(MaxGen)}(\cdot). \tag{11}$$

Step 10: Calculate the relative change in average fitness between the k-th sub-sample and the $(k-1)$-th sub-sample,

$$D_k = \frac{\overline{fit}_k - \overline{fit}_{k-1}}{\overline{fit}_{k-1}}. \tag{12}$$

Step 11: The last generation $GP_k^{(MaxGen)}$ in the k-th sub-sample is regarded as the initial generation for the $(k+1)$-th sub-sample.

Step 12: If $k < L$, let $k = k+1$ and go to Step 5; otherwise, the training process is terminated.

4 Application to the Greater-China Stock Markets

Chen and Yeh (1997) discussed the application of recursive genetic programming to the detection of structural changes in the univariate time series of S&P 500 and Nikkei 225. They found that the two time series experienced structural changes during the sample period. This model-free approach was further developed by Sun, Zhang and Zhang (1999) to the case of multivariate time series. Using genetic programming, they examined the structural changes in the dynamic relationship among the sub-indexes within the Shanghai stock market. This model-free approach works well in the detection of structural changes in the operating mechanism for both univariate and multivariate time series. In the multivariate case, the detection of structural change will be based on the relationship between the components of the time series.

In this paper we consider the dynamic structure of stock indexes of the four so-called Greater-China stock markets, namely the Hong Kong, Taiwan, Shanghai and Shenzhen stock markets. Stock prices are commonly used as a leading indicator of economic conditions. Thus, the extremely dynamic stock market activities in the Greater-China region are a reflection of her economic vitality. By the end of the first quarter in 1997, there have been 599 listed companies in the Stock Exchange of Hong Kong. The total market value was HKD 3,399.6 billion (USD 439 billion) and the average daily turnover was HKD 10.1 billion (USD 1.3 billion). Despite being a larger economy than Hong Kong, the Taiwan stock market is smaller. However, trading has been very active. By the end of the first quarter of 1997, there have been 387 listed companies with total market capitalization of NTD 8,845 billion (USD 323 billion) and an average daily turnover of NTD 85.8 billion (USD 3.1 billion). The two organized stock markets in China have experienced phenomenal growth in recent years. By the end of March 1997, the total market capitalization of the listed stocks in the Shanghai Stock Exchange was RMB 750 billion (USD 93.8 billion) with average daily turnover of RMB 3.69 billion (USD 0.46 billion). Two types of stocks are listed: A share and B share. A shares are available only to Chinese residents while foreign investors are allowed to trade the B shares in foreign currencies. By the end of March 1997, there were 43 B shares and 296 A shares listed in the exchange. The trading in the Shenzhen Stock Exchange has been equally active. By the end of March 1997, the total market capitalization was RMB 657.40 billion (USD 82.1 billion) with an average daily turnover of RMB 7.81 billion (USD 0.98 billion). Similar to the arrangement in the Shanghai Stock Exchange, A shares and B shares are traded. By the end of March 1997, 44 stocks were listed as B shares and 258 stocks were listed as A share.

The dynamic system of interest, which is also viewed as a multivariate time series, is composed of the stock indexes of the four markets. They are: (1) *The Hong Kong Hang Seng Index*; (2) *The Taiwan Stock Exchange Capitalization Weighted Stock Index*; (3) *The Shanghai B Share Index* and (4) *The Shenzhen B share Index*. Genetic programming is applied to the detection of structural changes in the operating mechanism of the Greater-China stock markets.

The sample ranging from June 1997 to May 1999 with 481 daily observations is obtained from the Datastream database. Figure 1 shows the time series plots of these four indexes. The indexes of the Hong Kong and Taiwan markets have been scaled down by a factor of 100.

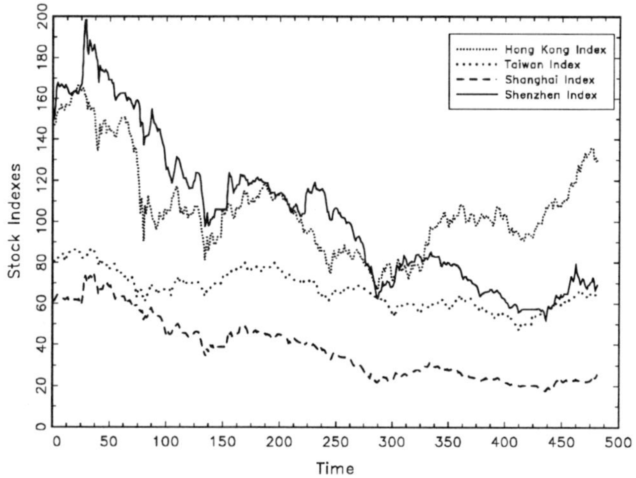

Figure 1: Stock indexes of the four Greater-China markets

Recursive genetic programming begins with the moving window technique, where the size of the windows is $n_1 = 20$ and the moving step is $n_2 = 3$. The parameters in the genetic programming are given in Table 1. As we discussed in Section 3, the dynamic relationship among the stock indexes may not have any definite mathematical form. It can be recognized through recursive genetic programming. Structural changes can be detected by locating large values of $\{D_k : k = 1, 2, \cdots, L\}$.

First, we consider the bivariate time series of the Shanghai and Shenzhen indexes. The diagnostic statistics D_k are computed and plotted in Figure 2. We can see that there are clusters of windows with relatively large D_k. The clusters are summarized in Table 2, in which the corresponding window numbers and the calendar periods are given. It is noted that the periods at the end of 1997 and the beginning of 1998, as well as July/August 1998 represent episodes of obvious structural change. The first episode corresponds to the drop in the Shenzhen market following the developments in the Asian crisis. In comparison, the Shanghai market was relatively unaffected by the crisis. The second episode occurred when the Shenzhen market suffered a sharp fall due to the failure of the Junan Securities. Many investors lost confidence in the Shenzhen market when Junan Securities was found involved in illegal trading and was heavily penalised. Again, the Shanghai market was not much affected by this incident.

Table 1: Parameters of RGP algorithm

population size (r)	100
function set	$\{+, -, \times, \div, \sin, \cos, \exp, r \times \log\}$
maximum lag order in GP-tree(h)	10
terminal set	$\{x_1(t-1), x_2(t-1), \cdots, x_{m-1}(t-1), \cdots\cdots,$ $x_1(t-h), x_2(t-h), \cdots, x_{m-1}(t-h)\}$
dimension of multivariate time series	4
prob. of selecting a univariate operator	0.1
sample size	481
window size	20
moving step	3
prob. of crossover	0.9
prob. of mutation	0.0
prob. of reproduction	0.1
size of the representative GP-trees	10
number of generation (MaxGen)	10
prob. of leaf selection	0.4
maximum depth in GP-trees	26

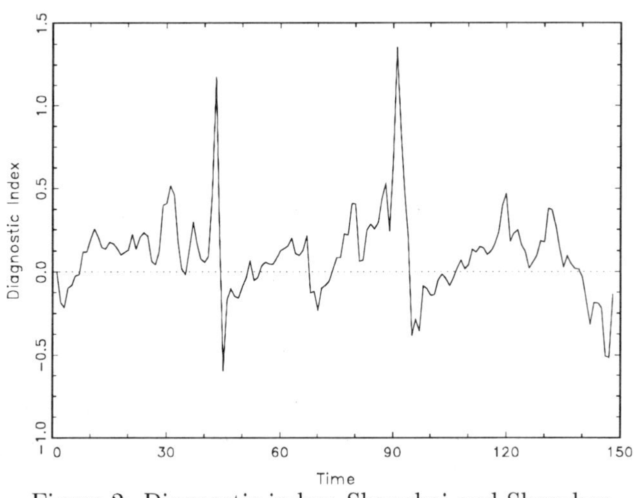

Figure 2: Diagnostic index: Shanghai and Shenzhen

Table 2: Clusters of Diagnostics: Shanghai versus Shenzhen

Cluster	Window number	Time period
1	28 ~ 32	30/10/97 ~ 12/12/97
2	42 ~ 43	29/12/97 ~ 02/02/98
3	88 ~ 93	09/07/98 ~ 27/08/98
4	119 ~ 120	17/11/98 ~ 17/12/98
5	131 ~ 132	06/01/99 ~ 04/02/99

Second, we consider the bivariate time series of the Hong Kong and Taiwan stock indexes. The diagnostic statistics D_k are computed and plotted in Figure 3. Similar to Table 2, we summarize the clusters of the diagnostics in Table 3. The most notable peak occurred at window 26, which corresponds to the second week in November 1997. During this period, the Hong Kong market went through a rough ride, plagued by persistent worries over the regional turmoil, the peg of the Hong Kong dollar and the rising interest rate. In contrast, the Taiwan market was relatively unaffected. Indeed, the Taiwan market remained very calm throughout the whole sample period. Thus, the sharp fall in the Hong Kong market caused a notable structural change across the two markets.

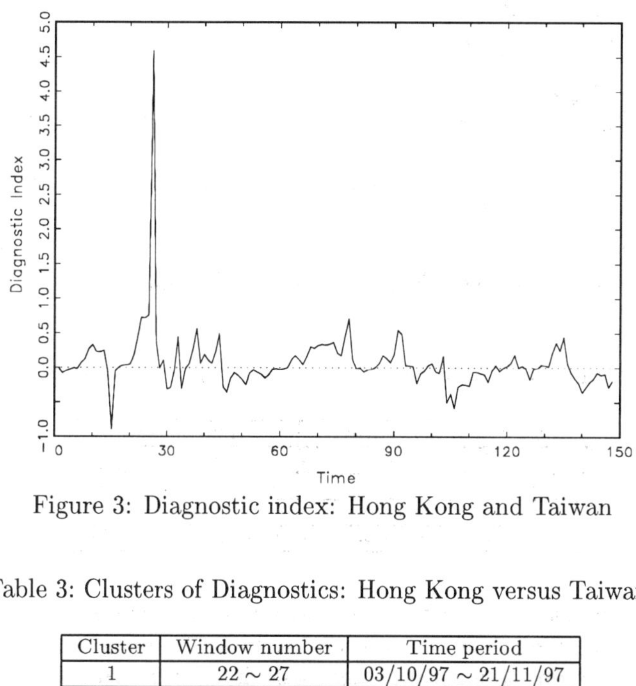

Figure 3: Diagnostic index: Hong Kong and Taiwan

Table 3: Clusters of Diagnostics: Hong Kong versus Taiwan

Cluster	Window number	Time period
1	22 ~ 27	03/10/97 ~ 21/11/97
2	33, 38 and 44	19/11/97 ~ 02/02/98
3	77 ~ 78	22/05/98 ~ 23/06/98
4	91 ~ 92	21/07/98 ~ 21/08/98
5	133 ~ 135	13/01/99 ~ 17/02/99

Third, we consider the trivariate time series of the Hong Kong, Shanghai and Shenzhen indexes. The computed diagnostic statistics D_k are shown in Figure 4 and the clusters of diagnostics are summarized in Table 4. Here we observe that the November 1997 turmoil in Hong Kong dominated the picture. The two peaks of diagnostics found in the Shanghai/Shenzhen markets are no longer apparent.

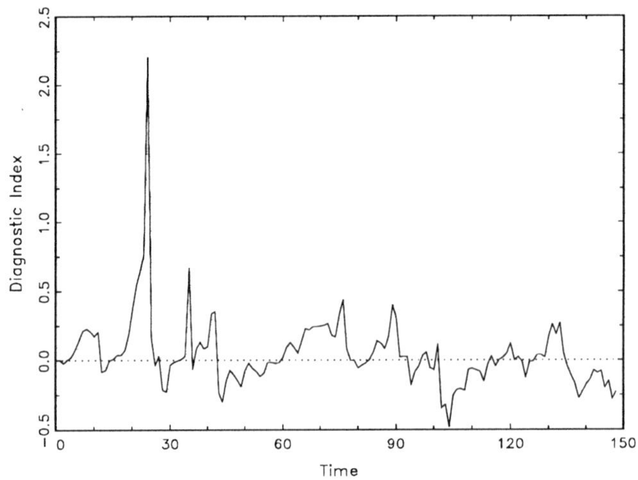

Figure 4: Diagnostic index: Hong Kong, Shanghai and Shenzhen

Table 4: Clusters of Diagnostics: Hong Kong, Shanghai and Shenzhen

Cluster	Window number	Time period
1	$21 \sim 24$	$30/09/97 \sim 10/11/97$
2	35	$27/11/97 \sim 24/12/97$
3	$75 \sim 76$	$14/05/98 \sim 15/06/98$
4	$89 \sim 90$	$15/06/98 \sim 15/07/98$

Finally, we consider the four-dimensional vector time series of the Hong Kong, Shanghai, Shenzhen and Taiwan indexes. The diagnostic statistics D_k are plotted in Figure 5, with the clusters of diagnostics summarized in Table 5. An interesting result is that the fall in November 1997 in the Hong Kong market is no longer a major episode of structural break. The main structural breaks occurred at windows 32 and 91, which correspond to the periods of November/December 1997 and July/August 1998, respectively. Thus, the fall-out in the market in Shenzhen caused by the illegal trading of Junan Securities emerged as the main structural change across the four markets over the sampling period.

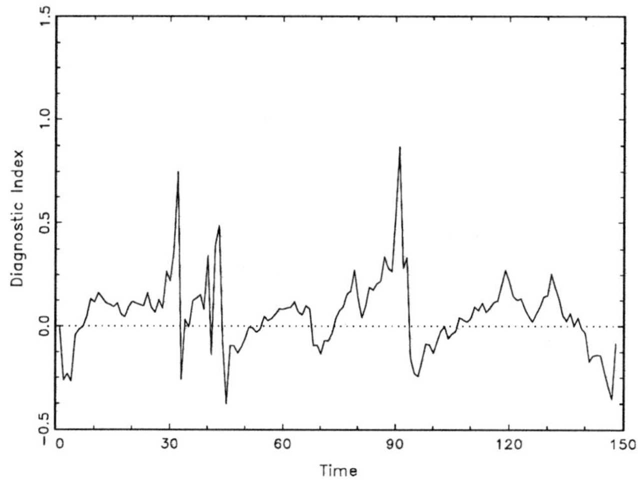

Figure 5: Diagnostic index: all four markets

Table 5: Clusters of Diagnostics: Four Markets

Cluster	Window number	Time period
1	31 ∼ 32	11/11/97 ∼ 11/12/97
2	42 ∼ 43	26/12/97 ∼ 27/01/98
3	90 ∼ 91	16/07/98 ∼ 17/08/98

5 Conclusion

In this paper we argue that model-free structural changes of a dynamic system or a multivariate time series can be interpreted as large changes in the dynamic relationship (or the operating mechanism) among the component variables. Due to its complexity, the dynamic relationship may not have any definite mathematical form or representation. However, it can be recognized through recursive genetic programming. Thus, the structural changes in the dynamic relationship can be detected based on the cognition process. The detection approach based on recursive genetic programming is model free. Thus, it is not necessary to specify a model in order to describe the operating mechanism of the data generating process. In this sense, the proposed approach overcomes the limitations of most model-specific approaches and can be performed empirically with no difficulties. Empirical examples show that the model-free approach works well in locating structural changes in multivariate time series.

References

1. Chen, S.-H. and Yeh, C.-H (1997), "Detecting structural changes with recursive genetic programming", presented at *The Far Eastern Meeting of the Econometric Society 1997*, Hong Kong.

2. Holland, J.H. (1992), *Adaptation in natural and artificial systems: an introductory analysis with applications to biology, control, and artificial intelligence*, Massachusetts, Cambridge: The MIT Press.

3. Koza, J.R. (1994), *Genetic programming II: automatic discovery of reusable programs*, Massachusetts, Cambridge: The MIT Press.

4. Sun, Q., X. Zhang and S. Zhang (1999), "A model-free method for structural change detection in Multivariate nonlinear time series", presented at *The Far Eastern Meeting of the Econometric Society 1999*, Singapore.